AF443243

*Biosensors and
Their Applications*

Biosensors and Their Applications

Edited by

Victor C. Yang
University of Michigan
Ann Arbor, Michigan

and

That T. Ngo
AMDL, Inc.
Tustin, California

KLUWER ACADEMIC / PLENUM PUBLISHERS
NEW YORK, BOSTON, DORDRECHT, LONDON, MOSCOW

ISBN 0-306-46087-4

©2000 Kluwer Academic/Plenum Publishers, New York
233 Spring Street, New York, New York 10013

http://www.wkap.nl/

10 9 8 7 6 5 4 3 2 1

A C.I.P. record for this book is available from the Library of Congress

Contributors

S. Alegret • Sensor and Biosensor Group, Department de Química, Universitat Autónoma de Barcelona, 08193 Bellaterra, Barcelona, Spain.

Jun-ichi Anzai • Faculty of Pharmaceutical Sciences, Tohoku University, Aramaki, Aoba-ku, Sendai 980-8578, Japan.

S. R. Beckett • School of Biomedical Sciences, Medical School, Queen's Medical Centre, Nottingham NG7 2UH, United Kingdom.

L. J. Blum • Laboratoire de Génie Enzymatique, UPRESA CNRS 5013, Université Claude Bernard Lyon 1, F-69622 Villeurbanne Cedex, France.

Ruben G. Carbonell • Department of Chemical Engineering, North Carolina State University, Raleigh, North Carolina 27695.

Geun Sig Cha • Department of Chemistry, Kwangwoon University Seoul 139-701, Korea.

Chiyui Chan • Biosensor and Bioelectronics Laboratory, Department of Chemistry, Hong Kong University of Science and Technology, Clearwater Bay, Kowloon, Hong Kong.

Deborah Charych • Lawrence Berkeley National Laboratory, Materials Sciences Division, Center for Advanced Materials, Berkeley, California 94720. Present address: Chiron Technologies, Life Sciences Center, Emeryville, California 94608

Qiang Chen • Cygnus Inc. Redwood City, California 94063-4719.

P. R. Coulet • Laboratoire de Génie Enzymatique, UPRESA CNRS 5013, Université Claude Bernard Lyon 1, F-69622 Villeurbanne Cedex, France.

C. Domínguez • Departmento de Microsistemas y Technologia de Silicio, Instituto de Microelectrónica de Barcelona, Centre Nacionale de Microelectronica, 08193, Bellaterra, Barcelona, Spain.

C. Duan • Department of Chemistry, University of Michigan, Ann Arbor, Michigan 48109.

M. W. Ducey • Department of Chemistry, University of Michigan, Ann Arbor, Michigan 48109.

Bin Fu • Diagnostic Division, Bayer Corporation, Tarrytown, New York 10591.

C. A. Galán-Vidal • Centro de Investigaciones Químicas, Universdad Autónoma del Estado de Hidalgo, 42076 Pachuca, Hidalgo, Mexico.

Ralf W. Glaser • Institut für Molekularbiologie, Friedrich Schiller Universität, D-07708 Jena, Germany.

G. G. Guilbault • Laboratory of Sensor Development, Department of Chemistry, University College Cork, Cork, Ireland.

Adam Heller • Department of Chemical Engineering, University of Texas at Austin, Austin, Texas 78712.

Tomonori Hoshi • Faculty of Pharmaceutical Sciences, Tohoku University, Aramaki, Aoba-ku, Sendai 980-8578, Japan.

Teruaki Katsube • Department of Information and Computer Science, Faculty of Engineering, Saitama University, Urawa, Saitama 338, Japan.

Eugenii Katz • Institute of Chemistry, The Hebrew University of Jerusalem, Jerusalem 91904, Israel.

Gregory L. Kenausis • Laboratory for Surface Science and Technology, ETH Zurich, Zurich CH-8092, Switzerland.

Gotthard Kunze • Institute of Plant Genetics and Crop Plant Research, D-06466 Gatersleben, Germany.

Alex W. K. Kwong • Biosensor and Bioelectronics Laboratory, Department of Chemistry, Hong Kong University of Science and Technology, Clearwater Bay, Kowloon, Hong Kong.

Maria L. Lung • Department of Biology, Hong Kong University of Science and Technology, Clearwater Bay, Kowloon, Hong Kong.

C. A. Marsden • School of Biomedical Sciences, Medical School, Queen's Medical Centre, Nottingham NG7 2UH, United Kingdom.

Mark E. Meyerhoff • Department of Chemistry, University of Michigan, Ann Arbor, Michigan 48109.

J. Muñoz • Departmento de Microsistemas y Technologia de Silicio, Instituto de Microelectrónica de Barcelona, Centre Nacional de Microelectronica, 08193 Bellaterra, Barcelona, Spain.

Yuji Murakami • School of Materials Science, Japan Advanced Institute of Science and Technology, Hokuriku, Tatsunokuchi, Ishikawa 923-12, Japan.

Hakhyun Nam • Department of Chemistry, Kwangwoon University, Seoul 139-701, Korea.

That T. Ngo • AMDL, Inc., Tustin, California 92680-7017.

C. K. O'Sullivan • Laboratory of Sensor Development, Department of Chemistry, University College Cork, Cork, Ireland.

Tetsuo Osa • Faculty of Pharmaceutical Sciences, Tohoku University, Aramaki, Aoba-ku, Sendai 980-8578, Japan.

Klaus Riedel • Dr. Bruno Lange GmbH Berlin, D-40549 Düsseldorf 11, Germany.

Reinhard Renneberg • Biosensor and Bioelectronics Laboratory, Department of Chemistry, Hong Kong University of Science and Technology, Clearwater Bay, Kowloon, Hong Kong.

Ajit Sadana • Chemical Engineering Department, University of Mississippi, University, Mississippi 38677-9740.

Joseph S. Schoeniger • Sandia National Laboratories, Livermore, California 94551-0969.

S. Shoji • Department of Electronics, Information and Communication Engineering, Waseda University, Shinjuku, Tokyo 169-8555, Japan.

Anup K. Singh • Sandia National Laboratories, Livermore, California 94551-0969.

A. M. Smith • Department of Chemistry, University of Michigan, Ann Arbor, Michigan 48109.

R. Smith • Department of Chemistry, University of Michigan, Ann Arbor, Michigan 48109.

Eiichi Tamiya • School of Materials Science, Japan Advanced Institute of Science and Technology, Hokuriku, Tatsunokuchi, Ishikawa 923-12, Japan.

Hidekazu Uchida • Department of Information and Computer Science, Faculty of Engineering, Saitama University, Urawa, Saitama 338, Japan.

A. Waterfall • School of Biomedical Sciences, Medical School, Queen's Medical Centre, Nottingham NG7 2UH, United Kingdom.

Bilha Willner • Institute of Chemistry, The Hebrew University of Jerusalem, Jerusalem 91904, Israel.

Itamar Willner • Institute of Chemistry, The Hebrew University of Jerusalem, Jerusalem 91904, Israel.

Rosie B. Wong • Agricultural Research Division, American Cyanamide Company, Princeton, New Jersey 08543-0400.

Victor C. Yang • College of Pharmacy, University of Michigan, Ann Arbor, Michigan 48109-1065.

Xian-En Zhang • Wuhan Institute of Virology, Chinese Academy of Sciences, Wuchang, Wuhan 430071, P. R. China.

Preface

A biosensor is an analytical device made up of a biological sensing element and a transducer. The sensing element detects the presence of an analyte via a specific interaction and generates a signal whose intensity is either directly or inversely proportional to the number of interactions between the analyte and the sensing element over a given period of time. The transducer in turn receives the signal coming from the sensing element and produces a digital electronic signal that is directly proportional to the intensity of the signal received. In order to achieve effective communication, the sensing element and the transducer must be intimately connected. Therefore methodologies that bring them into close contact play an important role in the successful construction of a biosensor.

Many types of sensing elements have already been incorporated into biosensors, and these are listed below in the order of increasing molecular weight or molecular complexity:

- Simple low-molecular-weight carbohydrates, peptides, fragments of nucleic acids, and coenzyme derivatives
- Synthetic polymers and polyelectrolytes
- Hybrids of biomolecule and synthetic polymers and biochromic polydiacetylene membranes
- Enzymes
- Modified enzymes, "wired" enzymes, and holoenzymes
- Antibodies
- Receptors
- Tissues
- Organelles
- Cells
- Microorganisms

We also have today a rapidly expanding range of choices for transducers, such as:

- Optical [absorption, fluorescence (polarization, energy transfer, time-resolved, phase-resolved, evanescent wave) and bio- and chemiluminescence]
- Amperometric
- Potentiometric

- Surface photovoltaic
- Piezoelectrical
- Surface plasmon resonance
- Conductometric
- Colorimetric

The commonly used methodologies that bring the biosensing element and the transducer together are as follows:

- Covalent attachment of the sensing elements to the surface of the transducer, e.g., covalent immobilization of thiol-bearing sensing molecules to the gold surface of the transducer.
- Adsorption of a polymeric network containing the sensing elements, e.g., adsorption of cross-linked enzymes or cross-linked enzymes and "inert" proteins (bovine serum albumin).
- Immobilization of the sensing elements on the surface of the transducer via a pair of linking molecules, e.g., the avidin–biotin system.
- Attachment of an apoenzyme to its prosthetic group anchored onto the surface of the transducer, e.g., immobilization of apo-glucose oxidase onto the FAD group anchored on the surface of a gold electrode.

The unique reaction kinetics taking place at the interface between the surface of the transducer and the solution also requires careful consideration. In most instances the reaction kinetics is diffusion-controlled. In some cases, where the signal produced by the sensing element is not strong enough, it must be amplified through the use of a device such as a liposome.

The various chapters in this volume review the aspects of biosensors that have not been covered or have been dealt with only superficially in earlier books on biosensors and describe recent novel advances to stimulate further research and development in this rapidly expanding field.

Victor C. Yang
That T. Ngo

Contents

4. Layered Functionalized Electrodes for Electrochemical Biosensor Applications

Itamar Willner, Eugenii Katz, and Bilha Willner

5. Biosensors Based on "Wired" Peroxidases

Qiang Chen, Adam Heller, and Gregory L. Kenausis

6. Nonseparation Electrochemical Enzyme Immunoassay Using Microporous Gold Electrodes

M. W. Ducey, A. M. Smith, R. Smith, C. Duan, and M. E. Meyerhoff

7. Liposomes as Signal-Enhancement Agents in Immunodiagnostic Applications

Anup K. Singh, Joseph S. Schoeniger, and Ruben G. Carbonell

8. Recent Development in Polymer Membrane-Based Potentiometric Polyion Sensors

Bin Fu, Mark E. Meyerhoff, and Victor C. Yang

Piezoelectric Immunosensors: Theory and Applications

C. K. O'Sullivan and G. G. Guilbault

10. Surface Photovoltage-Based Biosensor

Yuji Murakami, Eiichi Tamiya, Hidekazu Uchida, Teruaki Katsube

11. Surface Plasmon Resonance Biosensors

Ralf W. Glaser

12. Luminescent Biosensors

L. J. Blum and P. R. Coulet

13. Micromachining for Biosensors and Biosensing Systems

S. Shoji

14. Simultaneous Determination of Glucose and Analogous Disaccharides by Dual-Electrode Enzyme Sensor System

Xian-En Zhang

15. Application of Biosensors to the Measurement of Neurotransmitter Function

C. A. Marsden, S. R. Beckett, and A. Waterfall

16. Biosensors for Agrochemicals

Rosie B. Wong

17. Thick-Film Biosensors

C. A. Galán-Vidal, J. Muñoz, C. Domínguez, and S. Alegret

18. Alternative Polymer Matrices for Potentiometric Chemical Sensors

Hakhyun Nam and Geun Sig Cha

19. Rapid Measurement of Biodegradable Substances in Water Using Novel Microbial Sensors

Reinhard Renneberg, Alex W. K. Kwong, Chiyui Chan, Gotthard Kunze, Maria L. Lung, and Klaus Riedel

*Biosensors and
Their Applications*

1

Biochromic Polydiacetylene Synthetic Membranes

Deborah Charych

1.1. INTRODUCTION

Biosensor design research links well-known biological "lock-and-key" interactions with a variety of cleverly engineered signal transduction mechanisms.[1] Molecular recognition in biology assumes many forms, including enzyme–substrate, receptor–ligand, and antibody–antigen interactions, and all of these motifs can be incorporated into the design of biosensors. The way in which these "capture" molecules are coated onto or integrated with the transducer surface is itself the subject of numerous research efforts and is often referred to as interfacial design or surface modification.[2]

When considering interfacial design motifs, it is often fruitful to borrow upon Nature's most elegantly designed interface — the cell membrane. The biological cell membrane is a highly evolved self-assembled nanostructure that integrates molecular recognition and signal transduction functions. The surface of the membrane is heavily functionalized with recognition molecules, primarily in the form of glycosylated lipids and proteins (Fig. 1.1). These recognition sites often serve as "antenna" for specifically recognizing other molecules or other cell surfaces, resulting in a cascade of events such as the opening of ion channels, activation of cellular enzymes, or increasing/decreasing the rate of transport, secretion, or oxidative metabolism. While there are several examples of whole-cell biosensors in the literature,[3–5] we will focus our attention on simpler synthetic systems that mimic to one degree or another the self-organization and functionalization of the cell membrane. These membrane mimetic systems have been used in a number of biosensor design strategies, primarily as the molecular recognition function of the biosensor. For example, self-assembled membranelike films can be coated on a

---

Deborah Charych • Lawrence Berkeley National Laboratory, Materials Sciences Division, Center for Advanced Materials, Berkeley, California 94720. Present address: Chiron Technologies, Life Sciences Center, Emeryville, California 94688.

Biosensors and Their Applications, edited by Yang and Ngo, Kluwer Academic/Plenum Publishers, New York, 1999.

1

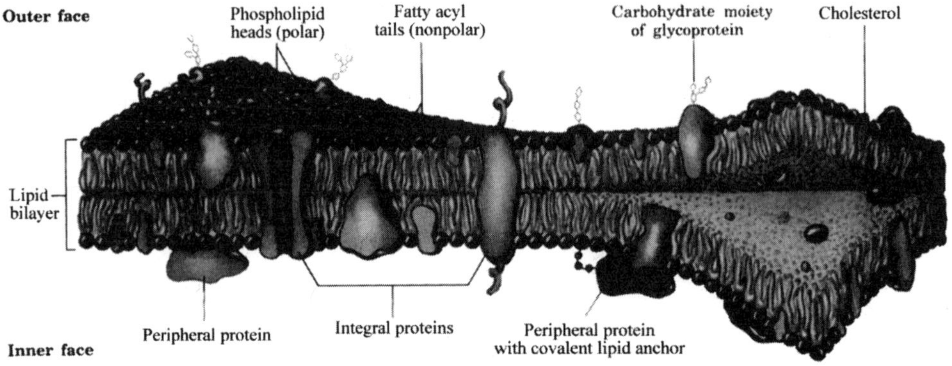

Figure 1.1   The fluid mosaic model for biomembrane structure. The carbohydrate groups attached to proteins and lipids are presented at the membrane surface and are often the site of specific molecular recognition interactions that are transduced into cellular messages. (From Albert L. Lehninger, David L. Nelson, and Michael M. Cox, *Principles of Biochemistry*, 2nd Ed. Worth Publishers, New York, 1993. Reprinted with permission.)

variety of device surfaces that provide signaling. These include measurement of conductivity in the case of electrode-modified surfaces or measurement of change in mass for quartz crystal microbalance (QCM) or surface acoustic wave (SAW)-modified surfaces.

In this chapter, we will briefly review these methods and also discuss an alternative transduction method that is based on lipophilic self-assembling conjugated polymers (CPs). The surface of the lipid polymer material is chemically modified by biospecific ligands and the π-conjugation of the polymer's backbone signals analyte binding by undergoing a color transition. At a basic level, synthetic ligand-modified CPs are analogous to the cell membrane in that molecular recognition is directly linked to signal transduction within a single supramolecular assembly.

1.2.   ASSEMBLING THE SYSTEM

Many approaches to biosensor design rely on systematically controlling the interfacial region of the biosensor device. This is readily achieved using processes widely referred to as molecular self-assembly (or supramolecular chemistry). The self-organization of molecules via noncovalent interactions occurs ubiquitously in nature; however, chemists have learned to manipulate these interactions to form a variety of interesting structures such as molecular "pencil cases," metal-binding "tongs," or linked molecular assemblies (catenanes).[6,7] In the context of this review, we refer to molecular self-assembly in a narrower sense. Here, the molecules comprising the system are surfactant-like and typically self-organize into micelles, multilayers, monolayers, or vesicles. The increased entropy of water drives the formation of these materials in an aqueous environment. Because of their resemblance to natural cell membranes, these structures are sometimes referred to as membrane mimetic (Fig. 1.2).[8,9] When appropriately functionalized with biospecific recognition molecules, these synthetic membranes form a model platform for biosensor interface design. The high surface-to-volume ratio enhances sensitivity while providing a more "natural" environment for many proteins, enzymes, or naturally derived lipophilic compounds.

Within the arena of surface-immobilized membrane mimetic materials, two broad categories may be considered: Langmuir–Blodgett films and self-assembled monolayers (or SAM films as they are often referred to in the literature).[9–12] The latter term often refers to alkylsilane or alkylthiol molecules that attach to surfaces simply by dipping the surface into a solution containing the SAM-forming molecules. A well-studied class of SAM films are the alkylthiols.[9,13–15] A solution of alkylthiols in polar solvent adsorb irreversibly to gold or platinum metal surfaces (ca. 150 kJ/mole). The resulting self-organized film is of monomolecular thickness with the hydrocarbon tails oriented nearly perpendicular to the surface in a close-packed, all-trans conformation (Fig. 1.2). Other common reagents for SAM film formation are the alkylsilanes. Alkyltrichloro- or alkyltrialkoxysilanes react with surface hydroxyl groups or surface water films to form a siloxane, M–O–Si, linkage. Silanization is a widely used method that alters the wettability and surface chemistry of the underlying material. It also provides a means to further modify the surface by way of a reactive SAM precursor film. For example, alkylamine silanes are often

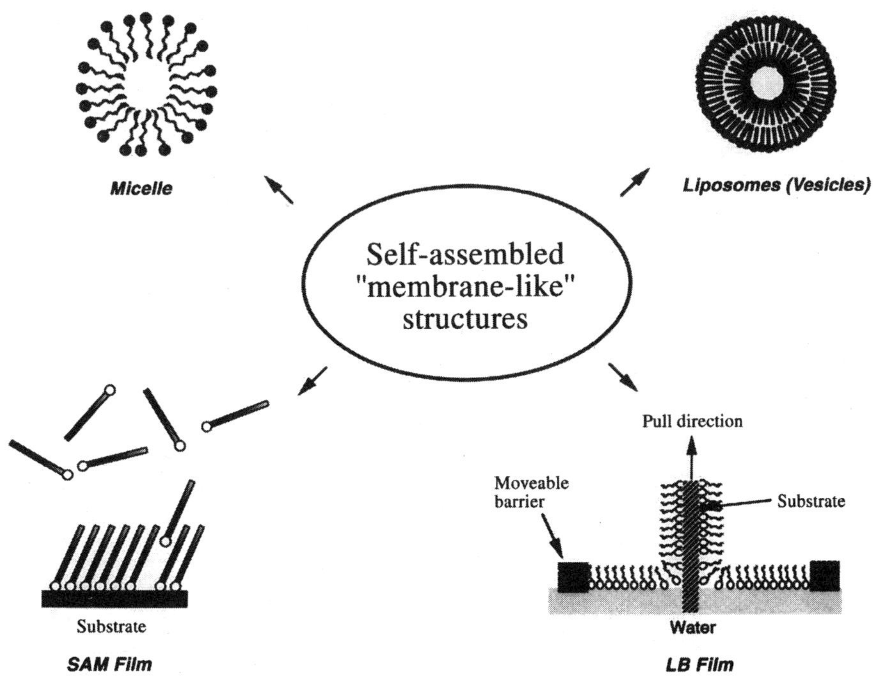

Figure 1.2. A sampling of the self-assembled nanostructures formed by long-chain amphiphilic compounds.

subjected to further reaction to allow attachment of a variety of molecules, including biological macromolecules (e.g., enzymes or antibodies). These biologically modified surfaces can then function as the molecular recognition component of a biosensor device.

Many useful amphiphilic molecules do not contain a surface-reactive anchoring group. These include naturally derived phospholipids, gangliosides, or fatty acid molecules. A variety of interesting amphiphilic compounds can be purchased commercially and are generally easier to synthesize than the reactive SAM-forming compounds described above. Coating a solid support with these materials is often accomplished using Langmuir–Blodgett techniques, although other methods have also been described.[16,17] Briefly, the amphiphilic compounds are dissolved in a suitable volatile organic solvent and aliquoted by syringe onto an aqueous surface contained within a Langmuir film balance apparatus (or Langmuir trough). The floating thin film (known as the Langmuir film) is compressed by a movable barrier. The change in water surface tension or lateral surface pressure (owing to 2-D densification of the molecules at the air–water interface) is recorded as a function of mean molecular area. At a chosen molecular area and surface pressure, the molecules can be transferred to a solid support by vertically passing the support through the air–water interface (see Fig. 1.2). The solid support may be "dipped" any number of times to produce transferred multilayers (known as LB films). A variation on the theme is the Langmuir–Schaefer method, wherein the solid support is passed through the interface horizontally.

1.3. MEMBRANELIKE STRUCTURES IN BIOSENSING

The membranelike materials discussed above have been used extensively as "coupling layers," primarily for the immobilization of enzymes to transducer surfaces (e.g., ISFET, electrode, QCMs, or SAW devices). For example, Willner et al.[18] utilized an amino-modified SAM film as a base layer for reaction to a bifunctional isothiocyanate coupling reagent. This surface, in turn, was conjugated to the enzyme glucose oxidase to build an electrochemical glucose sensor. Multilayers of the enzyme could be built up by further reaction of the enzyme amino groups with the coupling reagent. The multilayer configuration increases the sensitivity owing to the increased density of enzyme immobilized onto the electrode surface.

Electrochemical sensors for xanthine (an indicator of meat freshness) was devised from a biotinylated self-assembled membrane of biotinylated phospholipids.[19] Streptavidin-modified xanthine oxidase was coupled to the lipid layer. The assay is based on the electrochemical detection of enzymatically generated hydrogen peroxide. More recently, surface-modified QCMs have been used to detect viral antibody.[20] An antigenic peptide of picornavirus VP-1 capsid protein was modified with an alkylthiol "tail." [Picornavirus causes foot-and-mouth disease (FMDV), in cattle]. The peptide modified thiol lipid was coassembled with the underivatized alkylthiol. The continuous epitope was selectively recognized by monoclonal and polyclonal antibodies against FMDV and detected by the change in mass at the

QCM surface. In a related approach, Vogel et al.[21] derivatized the VP-1 peptide with three palmitoyl chains (lipopeptide) to mimic the structure of the coat protein of the virus. Surface modification with the peptide was accomplished by either Langmuir–Schaefer film transfer or by unfolding of lipopeptide vesicles. The controlled presentation of the peptide antigen allowed detection of the antibodies by surface plasmon resonance spectroscopy. This latter approach appeared to be more sensitive than the QCM transducer.

It is apparent that the flexible chemistry afforded by self-assembled thin films provides a useful platform for biosensors when coupled to suitable transduction devices. However, it should be noted that there are also many examples where straightforward adsorption of biomacromolecules can afford equal sensitivity and selectivity. Nonspecific protein or antibody adsorption onto a variety of surfaces (e.g., 96-well polystyrene microtiter plates) is common practice in immunology and in the development of enzyme-linked immunosorbent (ELISA) assays. These methods can generally be applied to QCM surfaces. For example, Guilbault et al.[22] nonspecifically coated the surface of a QCM with antibodies against Vibrio cholerae 0139. Binding of the antigen increases the mass on the sensor and is measured as a change in the resonant frequency of the crystal. Sensitivity limits of 10^5 cells/mL or 10^3 cells/crystal were obtained. However, at these levels, the degree of specificity between strains was reduced. A similar method was used for the detection of Listeria monocytogenes.[23,24]

The above examples illustrate the powerful coupling of biological molecular recognition to transduction device surfaces. The biological components may be directly immobilized to such surfaces or, when appropriate, incorporated into biomimetic membranes obtained from relatively simple "dipping procedures." In the next section, we focus on alternative signal transduction pathways and discuss how conjugated polymers may fit this role.

1.4. SENSORS AND BIOSENSORS BASED ON CONJUGATED POLYMERS

1.4.1. Basic Properties

A large volume of literature exists on the application of conjugated polymers to "molecule-based" devices.[25–28] This area of research encompasses many disciplines and will not be reviewed here except to outline the properties of CPs that make them attractive for biosensing purposes.

Nonconjugated polymers are comprised of a backbone consisting of only σ-bands or localized π-electronic levels. These materials are typically electrically insulating and transparent to visible light. The lowest-energy electronic transition can be excited only by UV light. On the other hand, π-conjugated polymers are intrinsically colored. The backbone atoms are made up of regular sequences in which multiple bonds (e.g., doubly or triply bonded carbon) are separated by no more than one single bond. The interesting electronic and optical properties of CPs arises from the delocalization of the π-electron wave functions over the polymer chain, leading to a relatively small electronic band gap E_g (approx. 1–2.5 eV) and low-energy electronic excitations. The last two decades has seen a surging interest in CP materials (sometimes referred to as "synthetic metals").[25–32] The increased interest stems, in part, from their potential use in the fabrication of light-emitting diodes, electrochromic and thermochromic devices, and light-weight batteries.

1.4.2. Conjugated Polymer-Based Sensors

Many CP materials exhibit unusual chromic properties such as thermochromism, ionochromism, or mechanochromism. With respect to sensors, CPs have been extensively investigated for the detection of metal ions, primarily alkali metal ions. Marcella et al.[29] synthesized polythiophenes modified with calix[4]arene ion receptors. The material exhibited an ionochromic response to Na^+ due to an increase in the effective conjugation length of the polymer backbone. Although a number of polymer responses were modified to ion binding (voltammetric, chromic, fluorescent, and resistive), the ionoresistive response was the most sensitive.

A few recent examples suggest the feasibility of using conjugation-length changes to detect other metal ions besides the alkali metals (e.g., transition-metal ions or main-group metal ions). In one interesting study, Wasielewski et al.[33] demonstrated that CPs based on bipyridyl-phenylenevinylene undergo chromatic transitions due to binding of transition-metal ions or main-group metal ions to the bipyridyl group. Binding results in increased conjugation length and therefore red-shifting of the visible absorption spectra. The increased conjugation arises from nonplanar to planar conformational transition of the polymer backbone. Synthetic methods for the preparation of head-to-tail polythiophenes functionalized with propionic acid groups have also led to interesting ionochromatic effects.[34] A 40-nm wavelength shift was observed as the size of the acid's counterion was increased from Li^+ (0.60 Å) to Cs^+ (1.69 Å).

1.4.3. Polydiacetylenes and Chromic Effects

Among the most extensively studied CPs are the polydiacetylenes (PDAs).[35] One reason for the intense interest is that PDAs can be prepared as single-crystal polymers — providing a means to study the fundamental properties of solid-state polymerization reactions. Once crystallized, the monomers can be polymerized via a 1,4-addition reaction by UV light (254 nm), ionizing radiation, or heat. The product is the conjugated, planar ene–yne backbone (Fig. 1.3). The nature of the solid-state or topochemical polymerization reaction was first described by Wegner[36–39] and requires that the monomeric units be well packed at specific distances and orientations with respect to one another. If this condition is not met, the material fails to polymerize. In addition to single crystals, asymmetrically substituted diacetylene monomers with a hydrophilic headgroup region and a hydrophobic side chain have been self-assembled into LB films,[40–48] cast films,[16,17] SAM films,[49–51] or bilayer vesicles.[52–57] For polymerization to proceed, the lipids must be in the solid-analogous state — below the main-phase transition temperature T_m of the material.

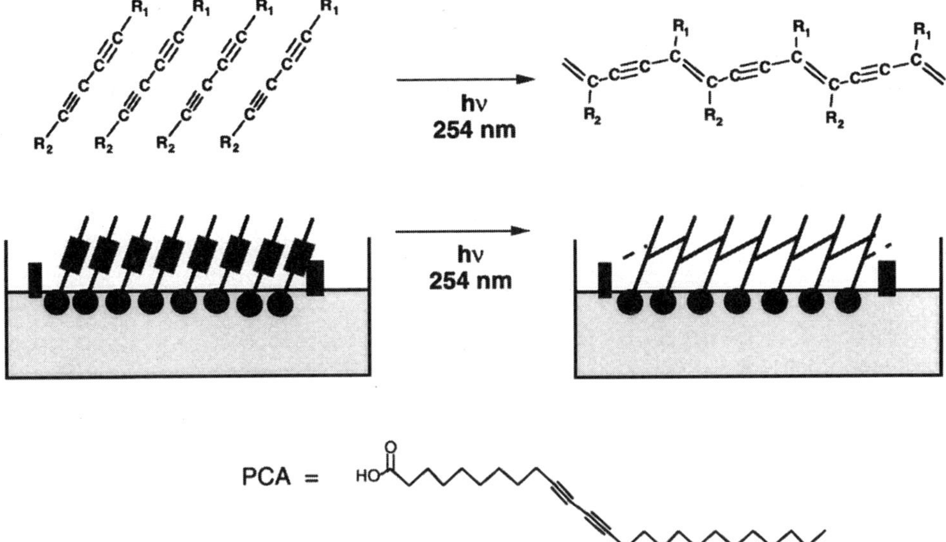

Figure 1.3. Polymerization reaction of diacetylene to form the conjugated ene–yne polydiacetylene. Amphiphilic derivatives of diacetylene are readily polymerized at the air–water interface using the Langmuir film balance apparatus. PCA = 10,12-pentacosadiynoic acid.

Other unique properties of the PDAs are their color and chromism. The energy of electronic excitations, and therefore the color of the material, can be dependent upon many factors, such as the original packing state of the monomers and the exposure of the polymeric material to environmental perturbations such as heat (thermochromism),[58–71] mechanical stress (mechanochromism),[72–77] or solvent (solvatochromism).[65,68,78,79] Typically, substituted PDAs undergo dramatic color changes from blue (absorption maximum ca. 630 nm) to red (absorption maximum ca. 540 nm). Of the various chromic effects, the thermochromic transition of polydiacetylene materials has been the most exhaustively studied, and much of the PDA literature is devoted to developing an understanding of the thermochromic blue-to-red color transition. Some basic threads emerge from this work: (1) the blue-to-red transition is associated with a conformational change in the PDA backbone from planar to nonplanar, and (2) side-chain conformation appears to play a critical role in the planar to nonplanar conformational change of the ene–yne backbone. However, teasing apart the exact molecular interactions between the side chains and elucidating their effect on the backbone remains an active area of current research.

In one of the first studies of PDA thermochromism, it was proposed that the acetylenic backbone converts to the butatrienic form[70]:

Subsequent NMR and laser Raman spectroscopic analyses of PDA single crystals have proven that a stable butatriene form is absent in the red-phase PDA.[63,64] Other reports have suggested that in the case of strongly hydrogen-bonded substituents, heating above the thermochromic transition serves to break the hydrogen bonding, changing the side chain conformation and straining the backbone from a planar to a nonplanar form.[69] Later studies of poly-(ETCD) [R = $(CH_2)_4$–O–C(O)–NH–C_2H_5] proved that the side chains remain hydrogen-bonded while undergoing a gauche-trans conformational transition.[64,80] The conformational transition of the side chains places a strain on the conjugated polymer backbone that produces the nonplanar form that absorbs shorter wavelengths.

With respect to LB films of PDA, many studies have focused on 10,12-penta-cosadiynoic acid (**1**, in Table 1.1). Results based on FTIR spectroscopy, DSC, and visible absorption spectroscopy have suggested that the thermochromic transition arises from an order–disorder transition of the alkyl side chains.[67] In our own laboratory, however, we have studied the molecular-level morphology of PDA thin films using atomic force microscopy in combination with FTIR spectroscopy. It appears from these data that the side chains actually become more ordered in the red phase, and assume a nearly perfect hexagonal packing (Fig. 1.4). It was proposed that the thermochromism arises from formation of an all-trans conformation of the alkyl side chain with maintenance of hydrogen bonding in the headgroup region — analogous to the results obtained for poly(ETCD) single crystals. The conformational transition of the alkyl side chain twists the planar ene–yne backbone into a nonplanar conformer.

1.4.4. PDA-Based Biosensors

Building on the body of research that describes chromic transitions of PDA materials, it appeared reasonable that this property might be useful in the design of simple, optically based biosensors. Given the strong relationship between side chain conformation and backbone conjugation, it should be possible to induce the color transition by biospecific interaction of the side chains with specific biological macromolecules. In order to explore this possibility, a wide range of single-chain and double-chain amphiphilic diacetylene monomers were synthesized and their ability to self-assemble and polymerize was investigated. Tables 1.1 and 1.2 list a sampling of the kinds of molecules studied. By judicious choice of the headgroup portion of the molecule, it is possible to design neutral, positively charged, or negatively charged interfaces. In certain cases, headgroups of more hydrophobic character are also possible. More complex functionalities such as amino acids or carbohyrates can be built into the system.

The LB film balance apparatus makes it possible to characterize the 2-D behavior of the synthetic lipids at the air–water interface. The apparatus measures the 2-D surface pressure of the film as a function of the mean molecular area. From these isotherms, one can determine the mean molecular area of the molecules at their most incompressible point (known as the limiting area). After a certain point, the film collapses into a multilayered structure. Tables 1.1 and 1.2 summarize the isotherm results and the ability of the lipid to polymerize in monolayers or multilayers. The limiting area of the acid derivative **1** is greater than its saturated fatty acid counterparts such as stearic acid (limiting area $18\,\text{Å}^2$/molecule). This increased area arises from the kink introduced in the chain by the diacetylene functionality. Most of the single-chain PDA derivatives with small headgroups have limiting areas in the range 22–$29\,\text{Å}^2$. The single-chain derivatives that contain an ethylene glycol spacer have much larger limiting areas than the derivatives without spacer groups, indicating increased interactions between the headgroups at very large molecular areas. The bisPDA derivatives occupied areas between 50 and $60\,\text{Å}^2$, as expected for these double-chain compounds.

The data also suggest that efficient headgroup packing is as important as the packing in the hydrocarbon chains, i.e., tight packing alone is insufficient for the polymerization reaction to occur, as evidenced by lack of reactivity of compound **4**. Strong hydrodren-bonding interactions in the headgroup region provide the best intermolecular packing arrangement for the 1,4-addition reaction. For example, replacement of the acid proton by the methyl ester results in a nonpolymerizable film. The bisPDA derivatives also indicate the importance of strong headgroup interactions to packing arrangements that are optimal for polymerization. The simple amino-modified bisPDA (**11**) does not polymerize, whereas the amino acid derivatives **12–14** show facile 1,4-addition.

Many of the compounds shown in Table 1.1 can be hydrated to form liposomes as well as thin films. The liposomes are made by probe sonication methods and are subsequently polymerized to form the polydiacetylene backbone. Liposomes offer some advantages over the LB films: (1) they can be made more simply and reproducibly, (2) one can utilize 96-well plate formats for optical studies and diagnostics, and (3) the structures more closely resemble the 3-D arrangement of cell membranes. Surface contamination also becomes less of an issue for liposomes as opposed to thin films. The liposome solutions are stable over a period of years and do not show any evidence of agglomeration or fusion. Table 1.3 outlines the ability of some of the PDA lipids to hydrate and form liposomes and also indicates the color of the material formed in water or PBS buffer. The colors range from blue to purple to even black solutions. The correlation between molecular structure and color of the liposome solution is not completely understood; however, it is reasonable to assume that the color formed is related to the initial packing state of the lipids. Lipid packing should have an overall influence on the effective conjugated length and planarity of the ene–yne backbone. For example, compound **4**, which lacks an extended hydrogen bonded network, appears to form only shorter oligomers in liposomes yielding an orange material (absorbs bluer wavelengths). Other factors such as the intensity and distance of the UV source also play a considerable role in the color development of the material. For example, it was found that compound **2** can polymerize to the blue form provided the sample is cooled during the polymerization reaction. Heating of the sample during irradiation induces partial thermochromic transition to the red form. Therefore, one has to be careful to distinguish between molecular and environmental effects when correlating molecular structure to color.

Table 1.1. The Area/Molecule (Å^2) and the Collapse Pressure (mN/M) of Synthetic Lipids Determined as Monolayers at the Air–Water Interface (an * indicates an overcompressible monolayer

	Compound*	Area, Å^2	Π (mN/M)	Polymerize
Acidic Headgroup	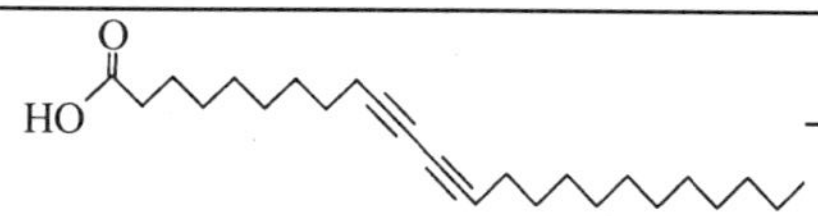 1	28	16	Y
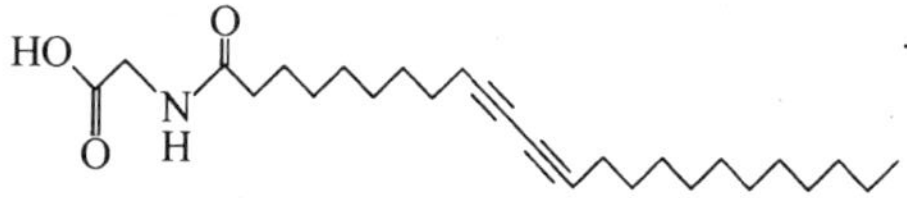	2	21	26–32*	Y
Neutral headgroup	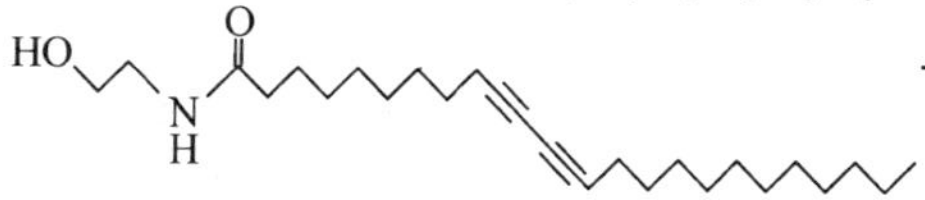 3	22	55	Y
	4	17	49	N
Basic headgroup	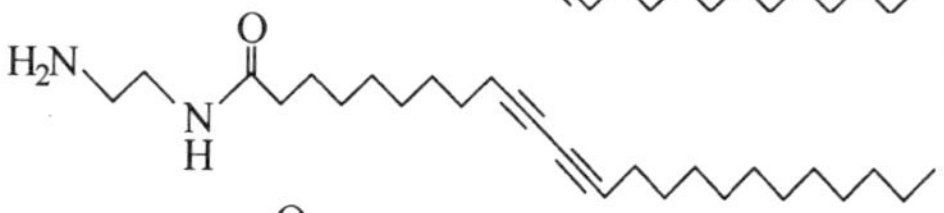 5	23	31–38*	Y
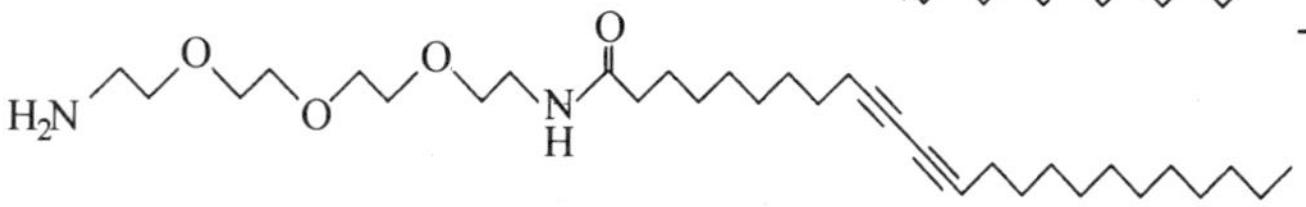	6	59	43–45*	Y
Hydrophobic Headgroup	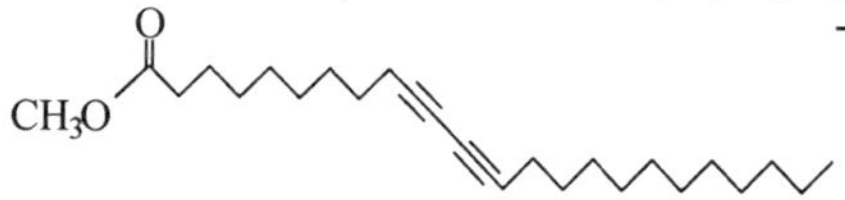 7	24	23–26*	N
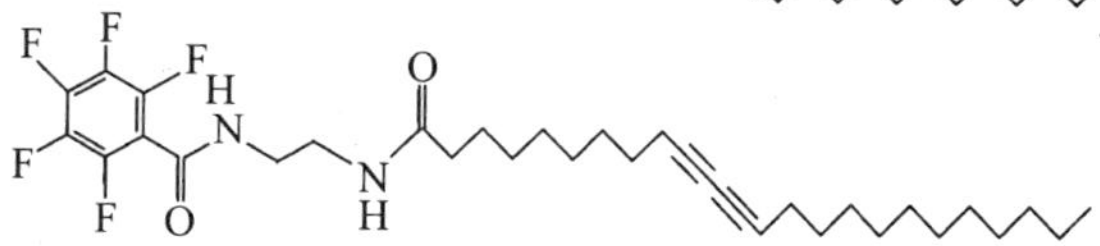	8	23	32–38*	Y

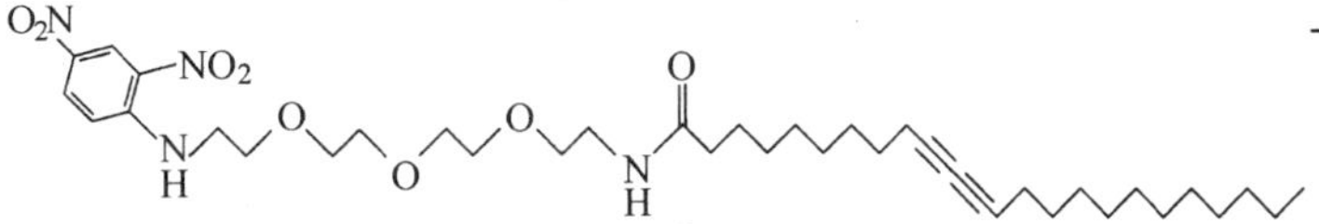 **9**	82	33	N
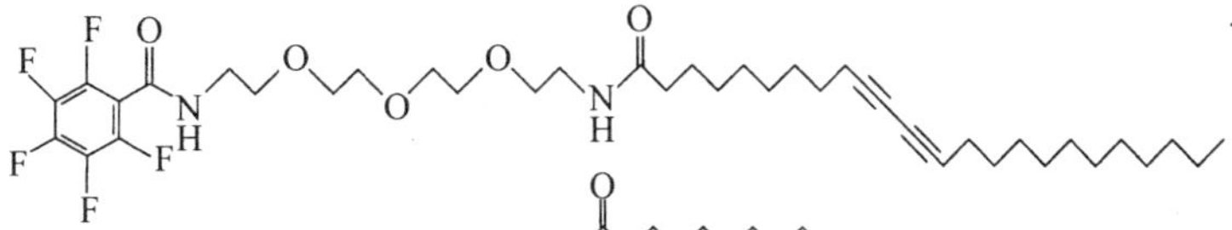 **10**	95–105	32–34*	Y
 **11**	62	40–41*	N
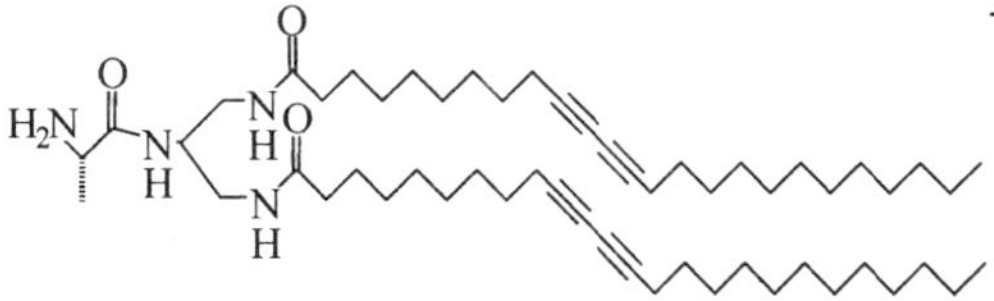 **12**	52	36*	Y
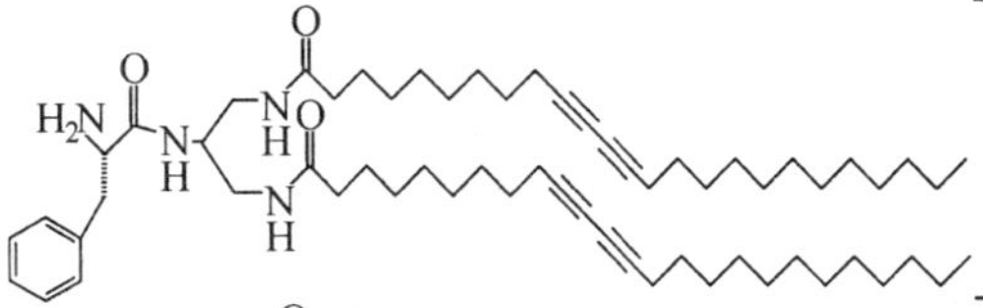 **13**	64	37	Y
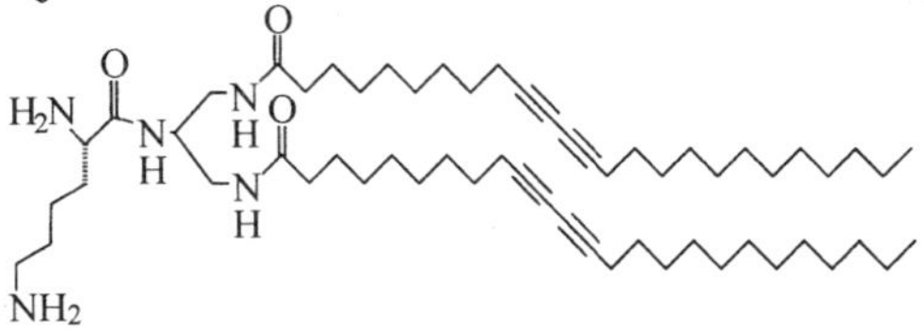 **14**	56	43*	Y

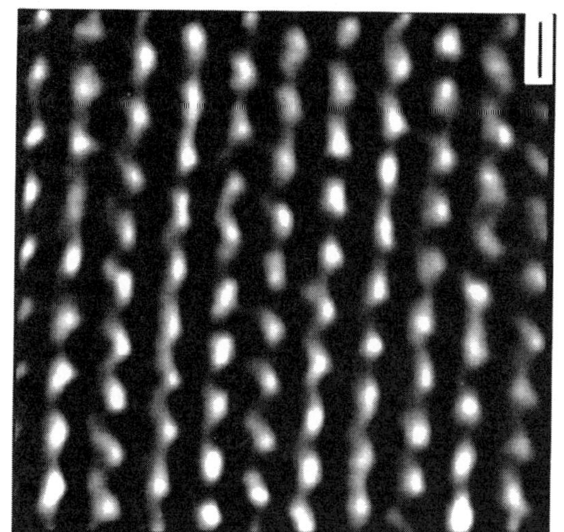

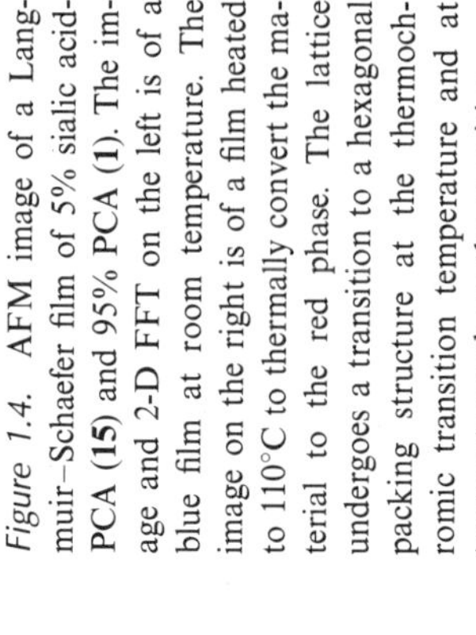

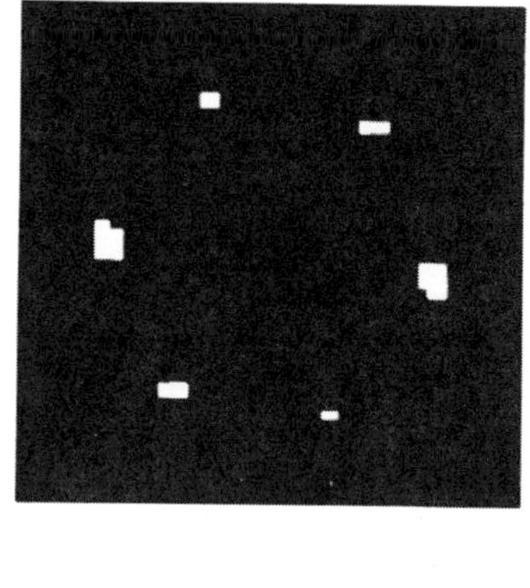

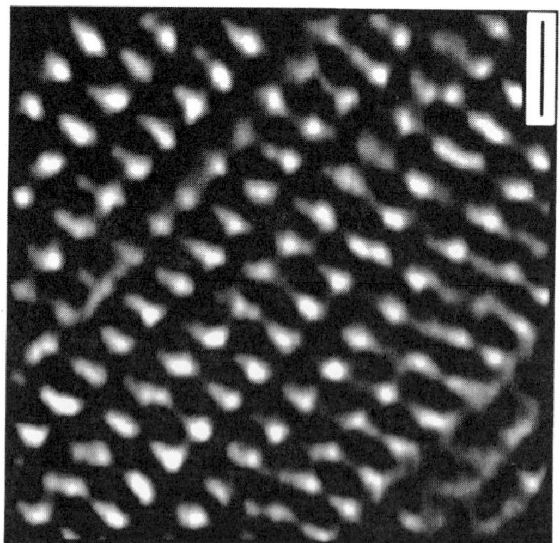

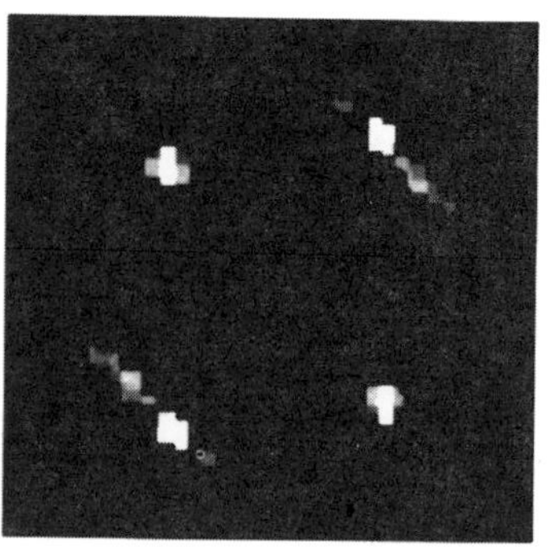

Figure 1.4. AFM image of a Langmuir–Schaefer film of 5% sialic acid-PCA (**15**) and 95% PCA (**1**). The image and 2-D FFT on the left is of a blue film at room temperature. The image on the right is of a film heated to 110°C to thermally convert the material to the red phase. The lattice undergoes a transition to a hexagonal packing structure at the thermochromic transition temperature and at temperatures above the transition.

Table 1.2. Liposome Formation in H_2O and PBS Buffer

Compound	Hydration[a]		Polymerization		Stability	
	H_2O	PBS	H_2O	PBS	H_2O	PBS
1	Y	Y	Blue	Orange	Y	Y
2	Y	Y	Red	Purple	Y	N
3	Y	N	Red	—	Y	—
4	Y	N	Orange	—	Y	—
5	Y	N	Red	—	N	—
6	Y	Y	Purple	Blue	Y	N
7	Y	N	Red	—	Y	—
8	Y	Y	N	N	—	—
9	Y	N	Black	—	Y	—
10	Y	N	Red	—	N	—

[a] Hydration refers to the formation of unpolymerized liposomes as determined by light scattering. Polymerization refers to the formation of polymerized liposomes as determined by visible spectroscopy. Stability was determined by light scattering, 24 h after polymerization.

Table 1.3.   Examples of Biospecific Ligand Lipids

15

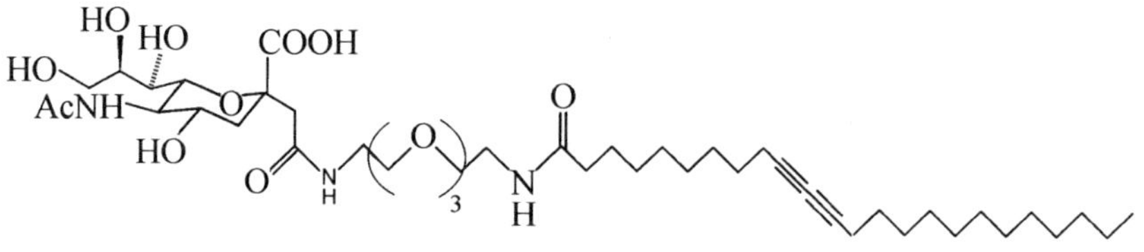

16

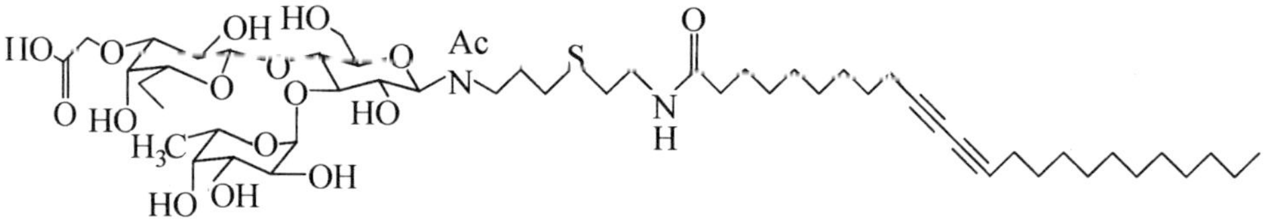

17

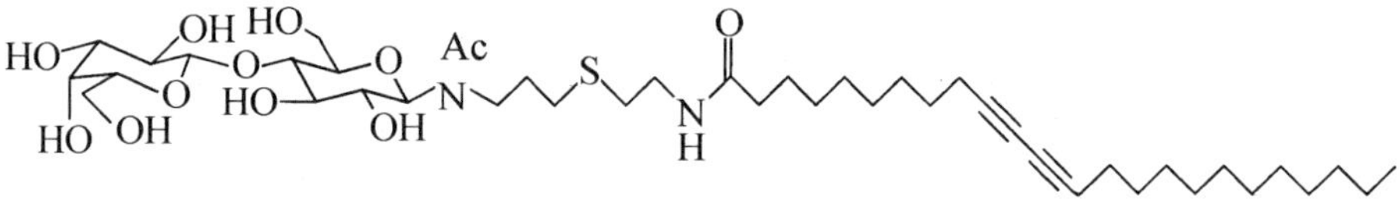

18

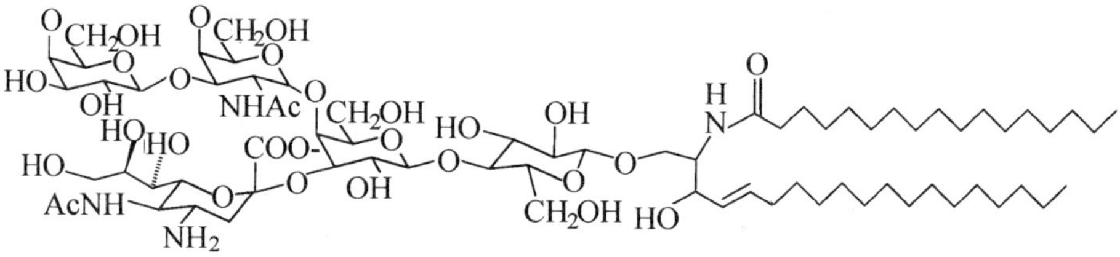

19

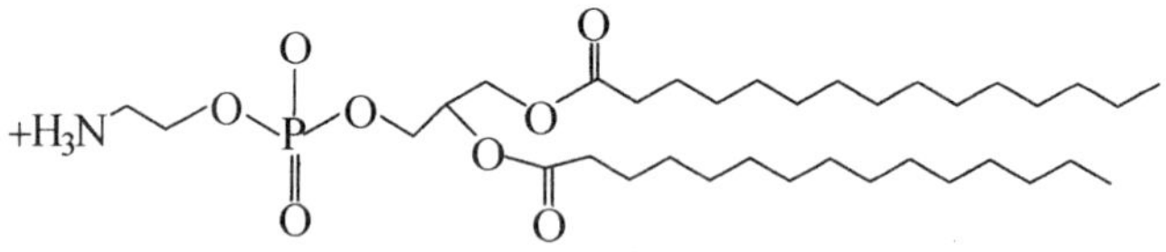

Biospecificity of the polymerized nanostructures is accomplished by coassembling any of the polymerizable lipids (referred to here as a "matrix lipid") of Tables 1.1 and 1.2 with a lipid that specifically binds the target of choice. Table 1.3 shows examples of some of the ligand lipids synthesized. The matrix lipids such as compounds **1** and **2** are the major components in mixed films and serve to dilute the ligand lipids so that binding is not sterically hindered. The carbohydrate **15** is an analogue of sialic acid, the specific ligand that binds to influenza virus hemagglutinin. Carbohydrate **16** is an analogue of sialyl lewis X and can be used to bind *p*-selectin, an important lectin involved in the inflammatory response. Liposomes that present this compound at their surface bind to *p*-selectin at nanomolar concentrations.[55] In comparison, the free (nonlipophilic) sialyl lewis X analog binds only at millimolar levels. Naturally derived ligands can readily be coassembled with the polymerizable matrix lipids. Examples of these include the ganglioside compound **18** and the phospholipid **19** (see below).

Ultimately, the goal of the biosensor design is to induce the PDA blue-to-red transition by specific binding of the biological target (biochromism). The resulting biosensor is an "integrated" molecular assembly, i.e., the molecular recognition component (e.g., carbohydrate or antigen) is directly linked to the signal transduction component. It is not required that secondary reagents (such as antibody-conjugated enzymes) be added to obtain a color response.

The earliest system studied using this approach was a thin film made up of ligand lipid **15** and matrix lipid **1**. The film was prepared by the Langmuir–Schaefer method as described above and typically resulted in trilayer formation (Figs. 1.4 and 1.5a). The initially blue film has a visible absorption maximum of 620 nm. When the film is incubated with influenza A virus, the binding of the viral hemagglutinin (HA) to the sialic acid residues on the film surface results in a blue-to-red color transition. The color change can be observed qualitatively with the naked eye or quantified by visible absorption spectroscopy (Fig. 1.6). After incubation with the virus, the red-phase peak at 550 nm increases with a concurrent decrease in the blue-phase peak at 620 nm, producing a red-colored film. Spectroscopically, the color change is similar to the thermochromic transition described above. In the thermochromic effect, it appears that conformational transition of the side chain produces strain on the polymer backbone and the concomitant transformation from the planar to the nonplanar state. It is reasonable to postulate that multipoint binding by the virus particle (one particle contains approximately 500 HA molecules on its surface) similarly distorts the side chain packing and therefore the backbone planarity. Interestingly, the fusogenic acitivity of the viral HA trimer may also play a role in the distortion of the PDA membrane. Each HA_2 chain is anchored near its carboxy terminus in the viral membrane and contains a hydrophobic region located at the amino terminus (HA_2 residues 1–23). This segment, called the fusion peptide, is thought to play an important role in promoting viral-cell fusion and is a natural precursor to cell infection.[81] Insertion of this domain and/or the viral membrane itself into the synthetic membrane of PDA may similarly disturb the PDA layers, resulting in the color change.

The specific nature of the interaction between the influenza virus and the sialoside film surface can be confirmed by competitive inhibition assays (Fig. 1.7).

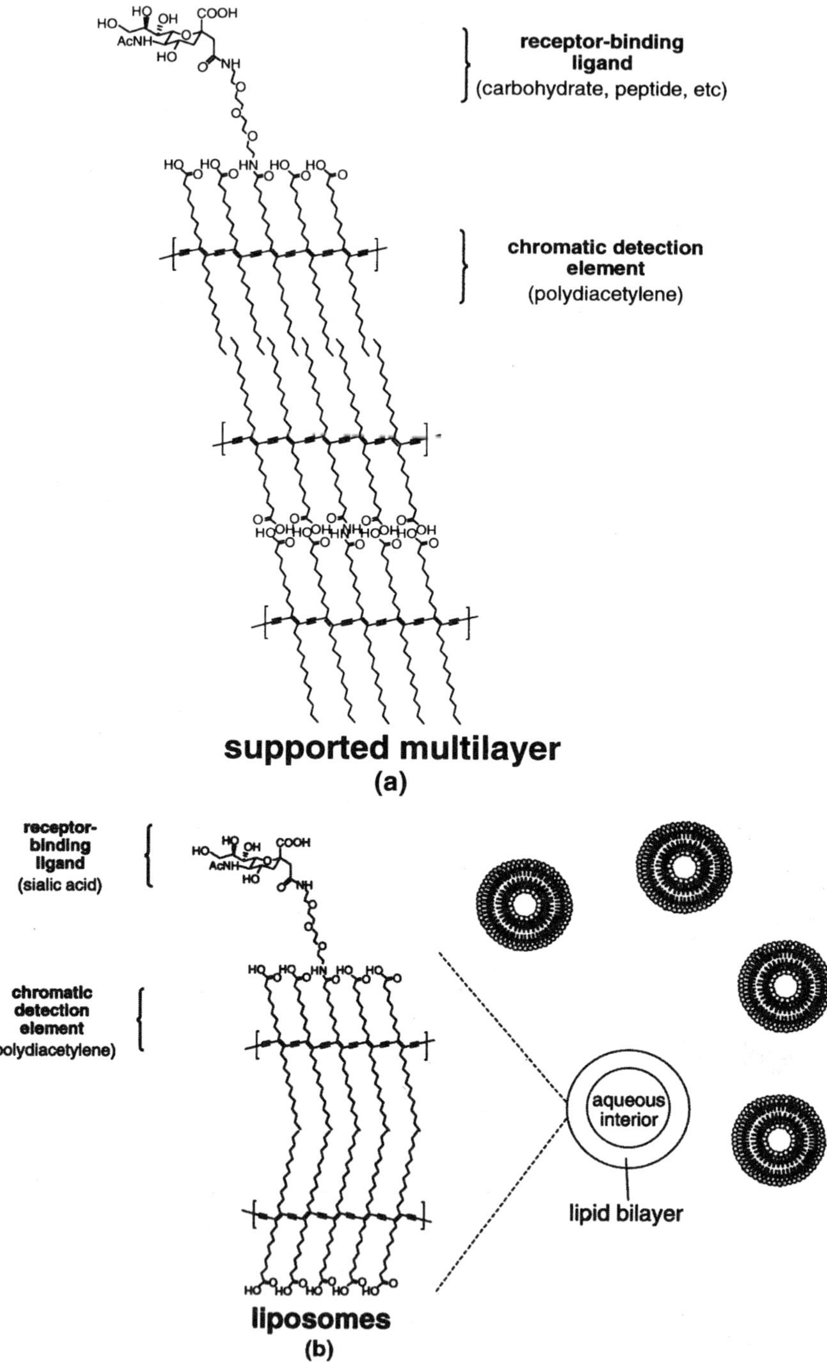

Figure 1.5. Two types of PDA molecular assemblies: (a) Langmuir–Schaefer film of matrix lipid and ligand lipid modified by sialic acid headgroups; the ligand lipid specifically interacts with influenza viral lectin, hemagglutinin. (b) Assembly of the same materials to form polymerized vesicles.

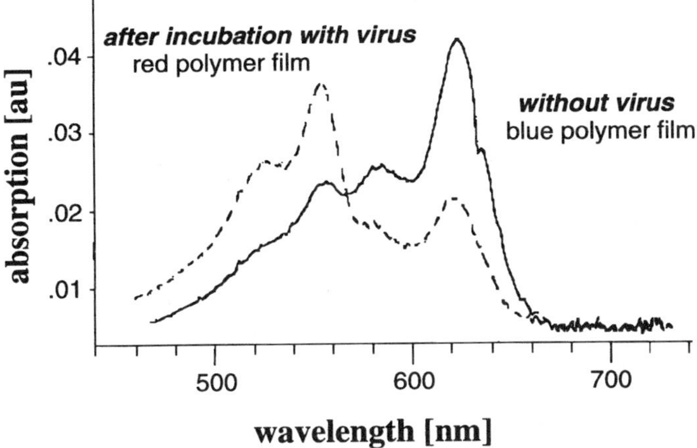

Figure 1.6. The visible absorption spectrum of the assembly prior to (solid line) and after (dashed line) viral incubation. The thin film was coated on a glass slide and inserted into a quartz cuvette containing PBS buffer (pH 7.4), and the absorption spectrum was obtained. Addition of influenza virus in PBS buffer (pH 7.4) resulted in a chromatic transition following a 30-min incubation period. Although the film color begins to change within seconds after exposure to the virus, 20–30 min was found to be the average length of time required to reach a plateau value in a nonstirred solution.

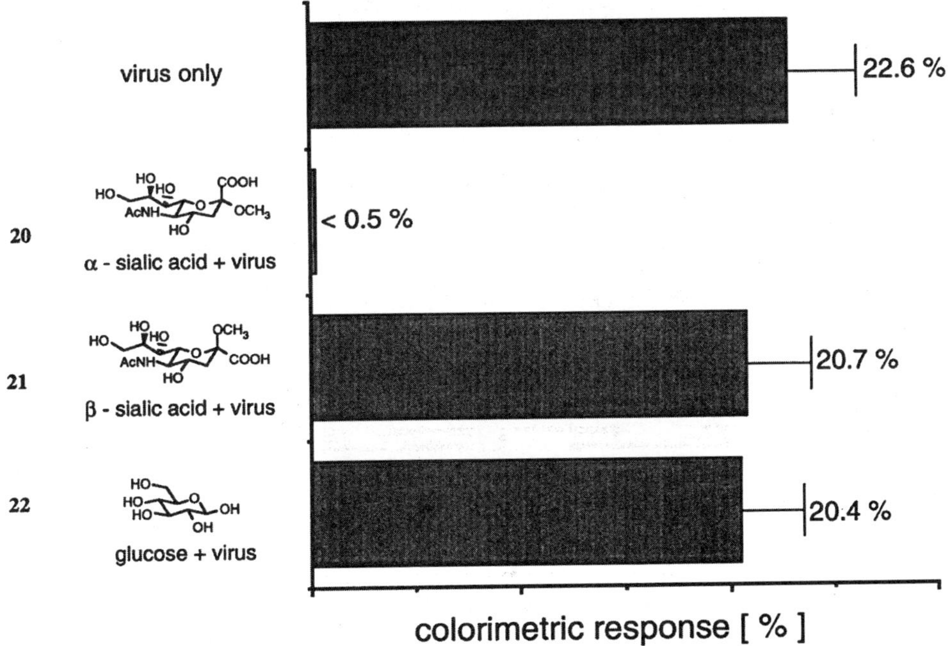

Figure 1.7. Colorimetric response of inhibited and noninhibited virus using a sialic modified PDA thin film coated on glass substrate. Compound **20** is a known inhibitor, whereas **21** and **22** are not.

The known inhibitor of influenza hemagglutination, α-o-methyl neuraminic acid (**20**) has a dissociation constant K_i of 2 mM as determined by a standard hemagglutination inhibition assay (HAI). Incubation of the sialic acid–PDA film assembly with influenza virus in the presence of this known binding inhibitor results in a film that remains blue and does not convert to the red phase. However, when the blue film is exposed to the same quantity of influenza virus in the presence of a noninhibitor such as the β-isomer of sialic acid (**21**) ($K_i > 50$ mM), or glucose (**22**), the color change is identical to a film exposed to the influenza virus alone.

In subsequent studies, it was found that the same ligand lipid and matrix lipid components could be assembled as liposome suspensions using probe sonication methods (see Fig. 1.5b). The liposomes are advantageous for 96-well microtiter plate studies and can often be prepared with high optical density. The optical absorption properties of the liposomes can be controlled to a certain extent by the polymerization time. Typically, blue liposomes turn pink, while purple liposomes turn orange upon addition of target analyte. Similar to the films, the color change can be inhibited by soluble compounds known to block the viral HA (e.g., **20** in Fig. 1.7). For both films and liposomes, the colorimetric response is not an all-or-nothing effect — it increases with increasing amounts of virus. No color change is obtained if the specific sialic acid ligand is removed from the PDA molecular assembly. In addition, no color change is observed if the PDA assembly is exposed to a variety of nonspecific proteins such as bovine serum albumin, blood plasma, human throat swabs, or human nasal washes. (Throat swabs and nasal washes are typically used in the clinical evaluation of influenza virus.)

The viral inhibition studies demonstrate that CP-based biosensors might be useful in high-throughput drug screening. Modern pharmaceutical drug discovery often involves screening hundreds of thousands of compounds for biological activity such as receptor-blocking or enzyme inhibition. Although state-of-the-art robotics provides the necessary hardware for carrying out high-throughput compound screening, assay development is often the rate-limiting step. Typical receptor-binding or enzyme-inhibition assays often require the use of radioactively tagged ligands or substrates. These materials incur significant costs associated with handling and storage of the radioactive compounds and may also constitute a safety hazard. The CP liposome assay offers an alternative approach to high-throughput compound screening because effective compounds are identified simply by observation of a blue color in the wells of a standard 96-well microtiter plate. Quantitation is obtained by measuring the visible absorption spectra of the wells using standard plate readers. Higher-density plate arrays can also be envisioned.

To futher demonstrate the utility of this approach to the development of potential high-throughput enzyme-inhibition assays, we recently studied the hydrolysis reaction of the enzyme phospholipase A_2 (PLA_2) on phospholipid substrates incorporated into PDA liposomes. PLA_2 cleaves membrane phospholipids by hydrolysis of the lipid C-2 ester bond (Fig. 1.8). The resulting fatty-acid derivatives are precursors for the formation of eicosanoids. Eicosanoids are involved in a variety of biological activities, such as smooth-muscle contraction and platelet aggregation. The eicosanoid synthesis pathway is a primary target for a large number of therapeutic drugs because of its role in pain, fever, and inflammation. For example,

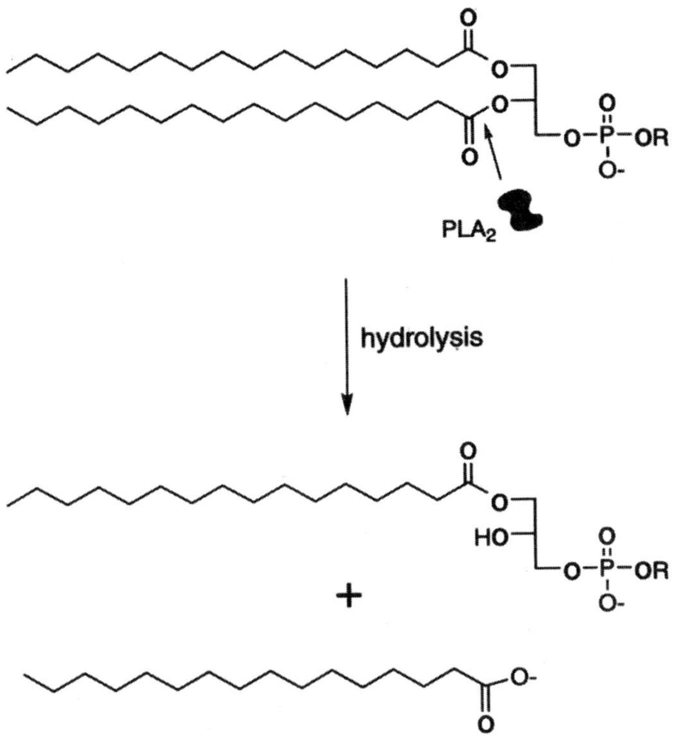

Figure 1.8. Schematic showing PLA_2 action on phospholipid molecules. The reaction products are fatty acid and lysolecithin.

corticosteroid hormones such as cortisone inhibit the activity of phospholipase in the first step of the eicosanoid synthesis pathway and is widely used to treat noninfectious inflammatory diseases, such as some forms of arthritis. Toward a first demonstration of this principle, the naturally derived phospholipid, dimyristoyl phosphatidylcholine (DMPC) was coassembled with matrix lipid **1** and formed into liposomes. Up to 40% DMPC could be incorporated into the vesicles while maintaining efficient polymerization of the PDA matrix. Upon addition of the enzyme, a rapid color change to red is observed in the wells of the microtiter plate. Inhibition of the enzyme with Zn^{2+} results in complete inhibition of the blue-to-red color change and the wells remain blue (Fig. 1.9). The differences between the inhibited and noninhibited wells is clearly visible to the naked eye (differences of only 10% in colorimetric response are often detected visually). A variety of enzyme–substrate systems can be envisioned and current efforts are focused on expanding the scope of enzyme-inhibitor screening.

Recently, it was shown that modifications of CP-based biosensors allow detection of toxin molecules if the PDA membrane contains gangliosides. Gangliosides are a complex subclass of sphingolipids and are abundant in animal nerve and brain cells. One representative member of this family is the G_{M1}

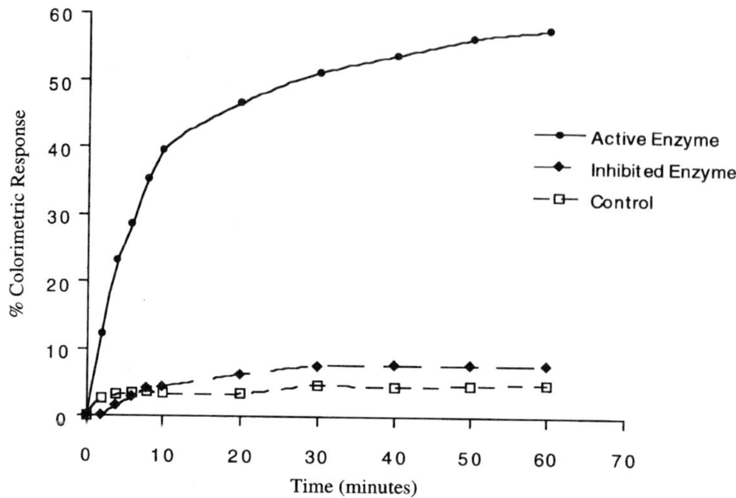

Figure 1.9. Colorimetric reaction of DPPC-modified PDA liposomes incubated with PLA_2. Inhibition of the enzyme with Zn^{2+} significantly suppresses the blue-to-red transition. The colorimetric response is the percentage change of the material from the blue to the red form. Control is DPPC-modified liposomes incubated in buffer only.

ganglioside (**18** in Table 1.4), found on the surface of intestinal cells. This ligand is the primary target of the bacterial toxin that causes the debilitating diarrhea of cholera. Cholera toxin only enters cells that have G_{M1} on their surface (e.g., intestinal epithelial cells) and increases intracellular cyclic AMP. We[82] found that cholera toxin can be detected by PDA-based thin-film biosensors if the G_{M1} ligand is mixed with matrix lipid **1** and a "promotor" lipid such as compound **15**. In the presence of these three components, the film rapidly changes from blue to red upon exposure to the toxin. If the promotor lipid is removed, the color response is significantly reduced. The most sensitive layers are composed of a mole ratio of 90% **1**, 5% **18**, and 5% **15**. The promotor lipid may serve as a destabilization factor, reducing the activation barrier of the blue-to-red transition. Another possibility is the synergistic effect of carbohydrate clustering, similar to that proposed for lectin binding.

Another avenue to enhance the blue-to-red response by toxin binding is to alter the positioning of the diacetylene group closer to the interfacial binding region. Liposomes composed of 5,7-docosadiynoic acid were mixed with the G_{M1} ganglioside to produce the carbohydrate-modified interface. Incubation of this material with cholera toxin resulted in a rapid colorimetric reaction.[57] Titration of the color change with increasing amounts of cholera toxin results in a sigmoidal binding curve that saturates at a concentration that corresponds to the density of G_{M1} ligand sites (Fig. 1.10). The sigmoid behavior of the curve suggests cooperative effects of the color change upon binding of toxin.

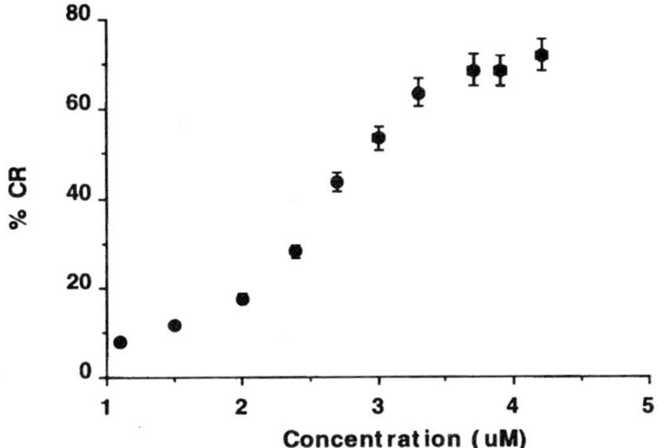

Figure 1.10. Colorimetric response of polymerized liposomes of 5% **18** and 95% 5,7-docosadiynoic acid after successive additions of cholera toxin. The liposomes were incubated with toxin for 2 min after each addition and the spectrum recorded.

1.5. CONCLUSION

In this chapter, we have discussed the use of self-assembling nanostructures as components for biosensor design. Very often, SAM or LB films are used as coupling layers to immobilize biological reagents such as enzymes or antigenic peptides. The biologically modified layer coats the surface of a variety of transduction devices such as QCMs or electrodes.

Transduction can also be directly incorporated into the self-assembling material itself. Conjugated polymers such as polydiacetylene can be modified with biospecific ligand molecules to bind and detect the presence of virus, enzyme, or toxin. Signaling occurs by a change in conformation of the polymer backbone from planar to nonplanar, producing a distinct color change in the material. Biosensors of this type integrate molecular recognition and signal transduction into a single supramolecular assembly and may offer a novel alternative to biosensor design.

REFERENCES

1. Taylor R, Schultz, J. *Handbook of Chemical and Biological Sensors.* Philadelphia: IOP Publishing Ltd, 1996.
2. Mallouk T, Harrison DJ *Interfacial Design and Chemical Sensing.* Washington, D.C.: American Chemical Society 1993, vol. 561.
3. Peter J, Hutter W, Stollnberger W, et al. Detection of chlorinated and brominated hydrocarbons by an ion sensitive whole cell biosensor. *Biosens Bioelectron* 1996;11:1215–1219.
4. Katrlik J, Svorc J, Rosenberg M, Miertus, S. Whole cell amperometric biosensor based on aspergillus niger for determination of glucose with enhanced upper linearity limit. *Analy Chim Acta* 1996;331:225–232.
5. Reshetilov A, Iliasov P, Donova M, et al. Evaluation of a gluconobacter oxydans whole cell biosensors for amperometric detection of xylose. *Biosens Bioelectron* 1997;12:241–247.
6. Atwood J. *Comprehensive Supramolecular Chemistry.* New York: Pergamon, 1996.

7. Ball P. *Designing the Molecular World: Chemistry at the Frontier.* Princeton: Princeton University Press, 1994.
8. Yen TF, Gilbert RD, Fendler JH. *Advances in the Applications of Membrane-Mimetic Chemistry.* New York: Plenum Press, 1994.
9. Ulman A. *An Introduction to Ultrathin Organic Films from Langmuir–Blodgett to Self-Assembly.* San Diego: Academic Press, 1991.
10. Roberts G. *Langmuir–Blodgett Films.* New York: Plenum Press, 1990.
11. Barlow WA. *Langmuir–Blodgett Films.* New York: Elsevier, 1980.
12. Gaines GL. *Insoluble Monolayers at Liquid-Gas Interfaces.* New York: Interscience Publishers, 1966.
13. Whitesides G, Laibinis P. *Langmuir* 1990;6:87.
14. Dubois L, Nuzzo R. *Ann. Rev. Phys. Chem.* 1992;43:437.
15. Bain C, Troughton E, Yao Y, et al. *J. Am. Chem. Soc.* 1989;111:321.
16. Kuo T, O'Brien DF. Free standing polydiacetylene films cast from bilayer membranes. *J Am Chem Soc* 1988;110:7571–7572.
17. Kuo T, O'Brien DF. Cast polydiacetylene films from diacetylenic glutamate lipids. *Langmuir* 1991;7:584–589.
18. Willner I, Riklin A, Shoham, D, et al. Development of novel biosensor enzyme electrodes: glucose oxidase multilayer arrays immobilized onto self-assembled monolayers on electrodes. *Adv Mat* 1993;5:912.
19. Rehak M, Snejdarkova M, Otto, M. Application of biotin-streptavidin technology in developing a xanthine biosensor based on a self-assembled phospholipid membrane. *Biosens Bioelectron* 1994;9:337–341.
20. Rickert J, Weiss T, Kraas W, et al. A new affinity biosensor: self-assembled thiols as selective monolayer coatings of quartz crystal microbalances. *Biosens Bioelectron* 1996;11:591–598.
21. Boncheva M, Duschl C, Beck W, et al. Formation and characterization of lipopeptide layers at interfaces for the molecular recognition of antibodies. *Langmuir* 1996;12:5636–5642.
22. Carter R, Mekalanos J, Jacobs M, et al. Quartz crystalmicrobalance detection of *Vibrio cholerae* 0139 serotype. *J Immunol Meth* 1995;187:121–125.
23. Jacobs M, Carter R, Lubrano C, et al. A piezoelectric biosensor for *Listeria* monocytogenes. *Am Lab* 1995;27:26–28.
24. Minunni M, Mascini M, Carter R, et al. A quartz crystal microbalance displacement assay for *Listeria* monocytogen. *Anal Chim Acta* 1996;325:169–174.
25. Salaneck W, Stafstrom S, Bredas JL. *Conjugated Polymer Surfaces and Interfaces.* New York: Cambridge University Press, 1996.
26. Inganas O, Gustafsson G. Thermochromism in poly(3-alkylthiophenes) and their polymer blends. *Syn Met* 1990;37:195–205.
27. Liang W, Karasz FE. Solid-state thermochromic transition in poly(2-methoxyphenylene vinylene). *Polymer* 1993;34:2702–2706.
28. Rubner MF, Skotheim TA. *Controlled Molecular Assemblies of Electrically Conductive Polymers.* Amsterdam: Academic Press, 1991, pp 363–403.
29. Marsella MJ, Newland RJ, Carroll PJ, et al. Ionoresistivity as a highly sensitive sensory probe: investigations of polythiophenes functionalized with calix[4]arene-based ion receptors. *J Am Chem Soc* 1995;117:9842–9848.
30. Leclerc M, Roux C, Bergeron J-Y. Structural effects on the thermochromic properties of poly-thiophene derivatives. *Syn. Met* 1993;55–57:287–292.
31. Sandman DJ. Semiconducting polymers and their solid-state properties: are these materials semicon-ductors or large conjugated molecules? *TRIP* 1994;2:44–55.
32. Roux C, Faid K, Leclerc M. Polythiophene derivatives: smart materials. *Polym News* 1994;19:6–10.
33. Wang B, Wasielwski M. Design and synthesis of metal ion-recognition-induced conjugated polymers: an approach to metal ion sensory materials. *J Am Chem Soc* 1997;119:12–21.
34. McCullough R, Ewband P, Lowe R. Self-assembly and dissassembly of regioregular, water soluble polythiophenes: chemoselective ionchromatic sensing in water. *J Am Chem Soc* 1997;119:633–634.
35. Bloor D, Chance RR. *Polydiacetylenes.* London: Martin Nijhoff, 1985.
36. Wegner G. *Z Naturforsch B* 1969;24:824.
37. Wegner G. *Polym Lett* 1971;9:133.

38. Wegner G. *Makrmol Chem* 1971;145:85.

39. Wegner G *Discuss Faraday Soc* 1980;68:494.

40. Day D, Ringsdorf H. Polymerization of diacetylene carbonic acid monolayers at the gas–water interface. *J Polym Sci: Polym Lett Ed* 1978;16:205–210.

41. Tieke B, Lieser G, Weiss K. *Thin Sol Films* 1983;99:95.

42. Tieke B, Lieser G. Polymerization of diacetylenes in mixed multilayers. *J Colloid Inter Sci* 1981;83:230.

43. Tieke B, Enkelmann V, Kapp H, et al. Topochemical reactions in Langmuir–Blodgett multilayers. *J Macromol Chem* 1981;A15:1045–1058.

44. Tieke B, Wegner G, Naegele D et al. Polymerization of Tricosa-10,12-diynoic acid in multilayers. *Angew Chem Int Ed Engl* 1976;15:764.

45. Yuan C W, Lu CL, Wang L. Formation of Langmuir–Blodgett films composed of polydiacetylene and copper tetracynoquinodimethane. *Polymer* 1992;33:3525.

46. Yanusova L, Klechkovsskaya V, Sveshnikova L, et al. Investigation of the structure of polymer diacetylene Langmuir–Blodgett films. *Liq Cryst* 1993;14:1615–1620.

47. Robinson KM, Hernandez C, Adin J, et al. In situ X-ray diffraction of diacetylene monolayers. *Thin Solid Films* 1992;210/211:73–75.

48. Olmsted J, Strand M. Fluorescence of polymerized diacetylene bilayer films. *J. Phys. Chem* 1983;87:4790–4792.

49. Batchelder DN, Evans SD, Freeman TL, et al. Self-assembled monolayers containing polydiacetylenes. *J Am Chem Soc* 1994;116:1050–1053.

50. Kim T, Chan KC, Crooks RM. Polymeric self-assembled monolayers: 4. Chemical, electrochemical, and thermal stability of w-functionalized, self-assembled diacetylenic and polydiacetylenic monolayers. *J Am Chem Soc* 1997;119:189–193.

51. Saremi F, Maassen W, Tieke B. Self-assembled alternating multilayers built up from diacetylene bolaamphiphiles and poly(allylamine hydrochloride): polymerization properties, structure, and morphology. *Langmuir* 1995;11:1068–1071.

52. Hub H, Hupfer B, Koch H, et al. *Angew Chem Int Ed Engl* 1980;19:938–946.

53. O'Brien D, Whitesides T, Klingbiel R. *J Polym Sci: Polym Lett Ed* 1981;19:95–101.

54. Reichert A, Nagy JO, Spevak W, et al. Polydiacetylene liposomes functionalized with sialic acid bind and colorimetrically detect influenza virus. *J Am Chem Soc* 1995;117:829–830.

55. Spevak W, Foxall C, Charych DH, et al. Carbohydrates in an acidic multivalent assembly: nanomolar p-selectin inhibitors. *J Med Chem* 1996;38:1018–1020.

56. Spevak W, Nagy JO, Charych DH, et al. Polymerized liposomes containing C-glycosides of sialic acid: potent inhibitors of influenza virus *in vitro* infectivity. *J Am Chem Soc* 1993;115:1146–1147.

57. Pan J, Charych D. Molecular recognition and colorimetric detection of cholera toxin by poly(diacetylene) liposomes incorporating G(m1) ganglioside. *Langmuir* 1997;13:1365.

58. Kuriyama K, Kikuchi H, Kajiyama T. Chromatic phase and molecular packings of polydiacetylene Langmuir–Blodgett films. *Chem Lett* 1995, p 1071–1072.

59. Deckert AA, Horne JC, Valentine B, et al. Effects of molecular area on the polymerization and thermochromism of Langmuir–Blodgett films of Cd^{2+} salts of 5,7-diacetylenes studied by UV–visible spectroscopy. *Langmuir* 1995;11:643–649.

60. Beckham HW, Rubner MF. On the origin of thermochromism in cross-polymerized diacetylene-functionalized polyamides. *Macromolecules* 1993;26:5198–5201.

61. Hankin SHW, Downey MJ, Sandman DJ On the structural origins of solid-state urethane polydiacetylene thermochromism: thermal and structural properties of IPUDO monomer and polymer. *Polymer* 1992;33:5098–5101.

62. Mino N, Tamura H, Ogawa K. Analysis of color transitions and changes on Langmuir–Blodgett films of a polydiacetylene derivative. *Langmuir* 1991;7:2336–2341.

63. Wenzel M, Atkinson GH Chromatic properties of polydiacetylene films. *J Am Chem Soc* 1989;111:6123–6127.

64. Tanaka H, Gomez MA, Tonelli AE. Thermochromic phase transition of a polydiacetylene, Poly(EETCD), studied by high-resolution solid-state ^{13}C-NMR. *Macromolecules* 1989;22:1208–1215.

65. Batchelder DN. Colour and chromism of conjugated polymers. *Contemp Phys* 1988;29:3–31.

66. Tanaka H, Thakur M, Gomez MA, et al. Study of the thermochromic phase transition of polydiacetylene by solid-state ^{13}C-NMR. *Macromolecules* 1987;20:3094–3097.

67. Singh A, Thompson RB, Schnur JM. Reversible thermochromism in photopolymerized phosphatidyl-choline vesicles. *J Am Chem Soc* 1986;108:2785–2787.
68. Chance RR Chromism in polydiacetylene solutions and crystals. *Macromolecules* 1980;13:386–399.
69. Chance RR, Patel GN, Witt JD. Thermal effects on the optical properties of single crystals and solution-case films of urethane substituted polydiacetylenes. *J Chem Phys* 1979;71:206.
70. Chance RR, Baughman RH, Muller H, et al. Thermochromism in a polydiacetylene crystal. *J Chem Phys* 1977;67:3616.
71. Exarhos GJ, Risen WM, Baughman RH. Resonance raman study of the thermochromic phase transition of a polydiacetylene. *J Am Chem Soc* 1976;98:481–488.
72. Rubner MF. Novel optical properties of polyurethane-diacetylene segmented copolymers. *Macro-molecules* 1986;19:2129–2138.
73. Galiotis C, Young RJ, Batchelder DN. The solid-state polymerization and physical properties of bis(ethyl urethane) of 2,4-hexadiyne-1,-diol: II. Resonance Raman spectroscopy. *J Polym Sci: Polym Phys Ed* 1983;21:2483.
74. Batchelder DN, Bloor D. Strain dependence of the vibrational modes of a diacetylene crystal. *J Polym Sci: Polym Phys Ed* 1979;17:569–581.
75. Mitra VK, Risen WM, Baughman RH. A laser Raman study of the stress dependence of vibrational frequencies of a monocrystalline polydiacetylene. *J Chem Phys* 1977;66:2731.
76. Tomioka Y, Tanaka N, Imazeki S. Effects of side-group interactions on pressure-induced chromism of polydiacetylene monolayer at a gas-water interface. *Thin Solid Films* 1989;179:27–31.
77. Hammond PT, Nallicheri RA, Rubner MF. An examination of the strain-induced orientation of hard segment domains in 4,4′-methylenebix(phenyl isocyanate)-based polyurethane-diacetylene segmented copolymers. *Mat Sci Eng* 1990;A126:281–287.
78. Nava AD, Thaku M, Tonelli AE. ^{13}C-NMR structural studies of a soluble polydiacetylene poly(4BCMU). *Macromolecules* 1990;23:3055–3063.
79. Patel GN, Khanna YP. One-dimensional order/disorder in a polydiacetylene. *J Polym Sci: Polym Phys Ed* 1982;20:1029–1035.
80. Rubner MF, Sandman DJ, Velazquez C. On the structural origin of the thermochromic behavior of urethane-substituted poly(diacetylenes). *Macromolecules* 1987;20:1296.
81. Hughson FM. Structural characterization of viral fusion proteins. *Curr Biol* 1995;5:265–274.
82. Charych D, Cheng Q, Reichert A, et al. A 'litmus test' for molecular recognition using artificial membranes. *Chem Biol* 1996;3:113–120.

Analysis of the Kinetics of Antigen–Antibody Interactions and Fractal Dimension in Biosensors

Ajit Sadana

2.1. INTRODUCTION

Sensitive detection systems (or sensors) are required to distinguish a wide range of substances. Sensors find application in the areas of biotechnology, physics, chemistry, medicine, aviation, and environmental control. Biosensor systems have different operating parameters, including stability, specificity, response time, selectivity, and regenerability. A better understanding of the mode of operation of these parameters would lead to increasing biosensor performance efficiency. Separation or the detection of reactants can be performed by the solid-phase immunoassay technique (antibody–antigen reaction) or by immobilizing enzymes on appropriate surfaces (enzyme–substrate reaction). There is a need to characterize the reactions occurring at the biosensor surface by paying attention to both the reaction and the sensing element (surface).

External diffusional limitations play a role in immunoassays.[1,2] There are limitations to classical modeling techniques that do not account for heterogeneity at the reaction surface. It is to be reasonably expected that the immobilization procedure utilized to attach the antigen or the antibody to the biosensor surface would yield an "irregular surface." Kopelman[3] indicates that surface diffusion-controlled reactions that occur on clusters or islands are expected to exhibit anomalous reaction orders and time-dependent rate (e.g., binding) coefficients. Fractals are disordered systems, and according to Pfeifer and Obert[4] the disorder is described by nonintegral dimensions. These authors further indicate that as long as

Ajit Sadana • Chemical Engineering Department, University of Mississippi, University, Mississippi 38677-9740.

Biosensors and Their Applications, edited by Yang and Ngo, Kluwer Academic/Plenum Publishers, New York, 1999.

surface irregularities show a scale invariance that is dilatationally symmetric they can be characterized by a single number—the fractal dimension. Surfaces of progressively higher irregularity are characterized by progressively higher fractal dimensions.[5]

Antibodies are heterogeneous and their immobilization on a fiber-optic surface, e.g., would result in some degree of heterogeneity. This is a good example of a "disordered system" for which a fractal analysis is appropriate. Fractal analyses using a single fractal dimension have provided physical clarification of the diffusion-controlled reactions at the surface.[6–8] It was noted in the above analysis that for some biosensor applications there was room for improvement in the "single-fractal" dimension analysis as far as fitting the data was concerned. More refined fits may be obtained if a multifractal instead of a single-fractal analysis is used to model the data.

A dual-fractal analysis is presented for the kinetics of binding of the antigen in solution to antibody immobilized on the biosensor surface. Different types of data are presented. Sometimes a single-fractal analysis is sufficient. Sometimes, a dual-fractal analysis provides a significant improvement in the fit compared to a single-fractal analysis. Also, in some cases there are not enough data points available to test for a reasonable fit based on statistical considerations. These data are not used in the present analysis.

2.2. THEORY

An analysis of the binding kinetics of antigen in solution to antibody immobilized on the biosensor surface is available.[9,10] Here we present a method of estimating actual fractal dimension values for antibody–antigen binding systems utilized in fiber-optic biosensors.

2.2.1. Single-Fractal Analysis

Havlin[11] indicates that the diffusion of a particle [antibody (Ab)] from a homogeneous solution to a solid surface [antigen (Ag)-coated surface], where it reacts to form a product [antibody–antigen complex (Ab.Ag)] is given by

$$(\text{Ab.Ag}) \sim \begin{cases} t^{(3-D_f)/2} & t < t_c \\ t^{1/2} & t > t_c \end{cases} \tag{1}$$

Here D_f is the fractal dimension of the surface. Equation (1) indicates that the concentration of the product Ab.Ag(t) in the reaction Ab + Ag → Ab.Ag on a solid fractal surface scales at short and intermediate time scales as Ab.Ag(t) as Ab.Ag $\sim t^p$, with the coefficient $p = (3 - D_f)/2$ at short time scales, and $p = \frac{1}{2}$ at intermediate time scales.

2.2.2. Dual-Fractal Analysis

The single-fractal analysis presented above is extended to include two fractal dimensions. In this case, the product (antibody–antigen complex, Ab.Ag) concentration on the biosensor surface is given by

$$(Ab.Ag) \sim \begin{cases} t^{(3-D_{f_1})/2} & t < t_1 \\ t^{(3-D_{f_2})/2} & t_1 < t < t_2 = t_c \\ t^{1/2} & t > t_c \end{cases} \tag{2}$$

2.3. RESULTS

Piehler et al.[12] indicate that the binding of small molecules (<1000 g/mole) to receptors is of significant interest. They have utilized reflectometric interference spectroscopy (RIFS) to detect the binding of biotin (244 g/mole) to thin films coated with streptavidin. The RIFS technique allows the accurate measurement of changes in the optical thickness of the transparent films with high resolution. Figure 2.1 shows the curves obtained using Eqs. (1) and (2) for the binding of polystreptavidin to an interference film pretreated with biotin-bovine serum albumin (BSA). Table 2.1 shows (1) the values of the binding rate coefficient k and the fractal dimension D_f using a single-fractal analysis [Eq. (1)], and (2) the values of the binding rate coefficients k_1 and k_2 and the fractal dimensions D_{f_1} and D_{f_2} using a dual-fractal analysis [Eq. (2)]. The binding rate coefficients k, k_1, and k_2 are a composite of the rate constants of the different reactions occurring on the biosensor surface. The values of the binding rate coefficient k presented in Table 2.1 were obtained from a regression analysis using Sigmaplot[13] to model the experimental data using Eq. (1), wherein $(Ab.Ag) = kt^p$. The value of p was also obtained from this regression analysis. The values were not shown to save space. The fractal dimension, D_f, value was obtained from $p = (3 - D_f)/2$ [see Eq. (1)]. Similarly, for a dual-fractal analysis the values of k_1, k_2, p_1, and p_2 were obtained by a regression analysis. The D_{f_1} and D_{f_2} values were obtained from p_1 and p_2, respectively. Figure 2.1 clearly shows that the dual-fractal analysis provides a better fit than the single-fractal analysis. Note that $k_1 < k$ and $D_{f_1} < D_f$. Also, $k_2 > k$ and $D_{f_2} > D_f$. As time increases, both the fractal dimensions ($D_{f_2} > D_{f_1}$) and the binding rate coefficient ($k_2 > k_1$) increase. A 32.4% increase in the fractal dimension leads to an increase in the binding rate coefficient by a factor of 4.3.

Domenici et al.[14] have recently analyzed the equilibrium behavior of a competitive system using a total internal reflection fluorescence (TIRF) immunosensor. One of the major advantages of the competitive equilibrium model is that it permits an analysis of nonspecific binding, which is a major hindrance in improving sensor performance. Figure 2.2 shows the curves obtained using Eqs. (1) and (2) for the binding of 100 nM FITC-labeled albumin in solution to IgG antialbumin immobilized on a fiber-optic TIRF sensor. Once again, the dual-fractal analysis provides a better fit than that obtained by a single-fractal analysis. Table 2.1 shows the values of the binding rate coefficients (k, k_1, and k_2) and the fractal dimensions (D_f, D_{f_1}, and D_{f_2}) using a single- and a dual-fractal analysis. Once again, note that $k_1 < k$, $D_{f_1} < D_f$, $D_{f_2} > D_f$, and $k_2 > k$. An increase in the fractal dimension by a factor of 24.2 from a value of 0.1020 to 2.4248 (D_{f_1} to D_{f_2}) leads to an increase in the binding

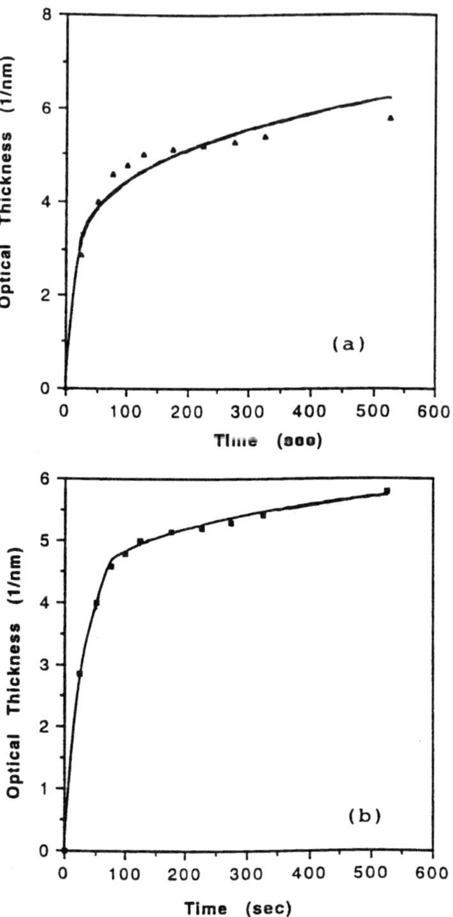

Figure 2.1. Theoretical curves using Eqs. (1) and (2) for the binding of polystreptavidin to an interference film pretreated with bovine serum albumin (BSA)[12]: (a) single-fractal analysis; (b) dual-fractal analysis.

rate coefficient by a factor of 279. The binding rate coefficient changes from a value of 0.0287 to 8.011 (k_1 to k_2).

Liliom et al.[15] have recently characterized tubulin–alkaloid interactions using ELISA (enzyme-linked immunosorbent assay). Antimitotic drugs, such as vinblastine (VBL) and vincristine (VCR) obtained from *Vinca* alkaloid (VA) are used in cancer therapy. These drugs inhibit the self-assembly of tubulin in microtubules.[16] Liliom et al.[15] wanted to develop a simple and sensitive procedure to quantitatively characterize tubulin–drug interactions under conditions of no tubulin interactions. These authors were able to analyze the binding relationships of binding domains and antibodies on the tubulin with monoclonal and polyclonal antibodies. This provided them with insights into the drug–tubulin interactions.

Table 2.1. Fractal Dimensions and Binding Rate Coefficients for Different Analyte–Receptor Reactions

Analyte concentration	k	D_f	k_1	k_2	D_{f_1}	D_{f_2}	Reference
Low-molecular-weight analyte	1.738 ± 0.1525	2.5926 ± 0.0606	0.6942 ± 0.021	2.996 ± 0.0351	2.1170 ± 0.0758	2.7932 ± 0.0016	Piehler et al.[12]
100 nM FITC-labeled albumin	0.4808 ± 0.1661	1.4562 ± 0.1354	0.0287 ± 0.0033	8.011 ± 0.1948	0.1020 ± 0.1454	2.4248 ± 0.0240	Domenici et al.[13]
Tubulin, 0 μM KAR-2	0.3100 ± 0.0615	1.7062 ± 0.0964	0.4554 ± 0.0123	0.3568 ± 0.0122	1.068 ± 0.036	2.2864 ± 0.0480	Liliom et al.[14]
Tubulin, 0.05 μM KAR-2	0.1698 ± 0.007	1.5777 ± 0.0215	0.2326 ± 0.0024	0.1698 ± 0.0034	1.2184 ± 0.0278	1.6037 ± 0.0177	Liliom et al.[14]

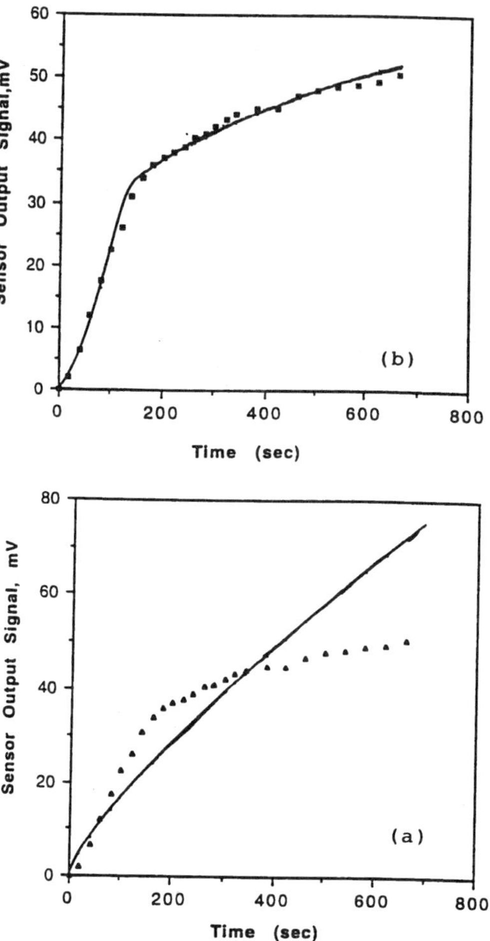

Figure 2.2. Theoretical curves using Eqs. (1) and (2) for the binding of 100 nM FITC-labeled albumin in solution to IgG antialbumin immobilized on a fiber-optic TIRF sensor[14]: (a) single-fractal analysis; (b) dual-fractal analysis.

KAR-2, a new semisynthetic derivative of bisindol alkaloids and an antimitotic agent, is a good inhibitor for tubulin polymerization and immunochemical formation.[15] Figure 2.3 shows the curves obtained using Eqs. (1) and (2) for the influence of KAR-2 on tubulin polymerization induced by 20 μM taxol. Table 2.1 shows the values of the binding rate coefficients (k, k_1, and k_2) and the fractal dimensions (D_f, D_{f_1}, and D_{f_2}) using Eqs. (1) and (2) for the effect of 0.05 μM of KAR-2 on tubulin polymerization induced by 20 μM taxol. It is of interest to note that when the drug KAR-2 is used, a single-fractal analysis provides a reasonable fit. Though data for only one concentration are presented, this is also true for three other concentrations. However, when KAR-2 is not used, a dual-fractal dimension analysis provides a better fit than that obtained by a single-fractal analysis.

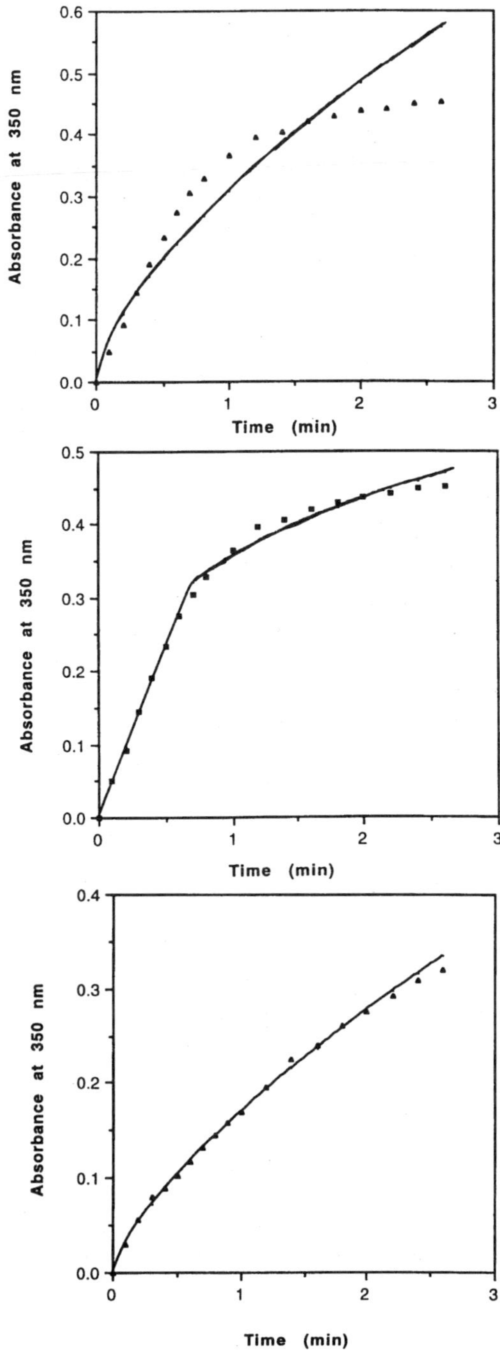

Figure 2.3. Theoretical curves utilizing ELISA to characterize tubulin–alkaloid interactions[15]: (a) no KAR-2 used [single-fractal analysis (top); dual-fractal analysis (bottom)]; (b) 0.05 μM KAR-2 used (single-fractal analysis is sufficient).

Table 2.1 shows the values of the fractal dimension and the binding rate coefficient when a single-fractal analysis is sufficient to model the data when KAR-2 is used. Table 2.1 shows the values of the binding rate coefficients (k, k_1, and k_2), and the fractal dimensions (D_f, D_{f_1}, and D_{f_2}) using a single- and a dual-fractal analysis. In this case, note that $k_1 < k$ and $k_2 < k$. However, $D_{f_2} > D_f$. An increase in the fractal dimension by a factor of 2.14 from a value of 1.068 to 2.2864 (D_{f_1} to D_{f_2}) leads to a decrease in the binding rate coefficient of about 22%. It is of interest to note that when KAR-2 is used, the binding can be adequately described by a single-fractal analysis. However, when KAR-2 is not used, the binding must be described by a dual-fractal analysis. As expected, the introduction of the drug significantly influences the polymerization of tubulin to microtubulins, and this is reflected in the fractal dimensions.

2.4. CONCLUSIONS

The fractal analysis of the binding of antigen (or antibody) in solution to the antibody (or antigen) immobilized on the biosensor surface provides a quantitative indication of not only the state of disorder and the binding rate coefficient on the surface, but also of the change in the state of disorder and the accompanying change in the binding rate coefficient. The analysis presented is, in general, also applicable to analyte–receptor systems. In some cases, the dual-fractal analysis provides an improved fit when compared to a single-fractal analysis. In general, as the experimental run proceeds, the state of disorder or the degree of inhomogeneity on the surface increases. This leads to an increase in the fractal dimension (D_{f_1} to D_{f_2}) and the binding rate coefficient (k_1 to k_2). There is also direct[8] and indirect evidence[17] to suggest that an increasing "roughness" on the surface would tend to increase the adsorption or binding rate coefficient.

The binding of KAR-2 and its influence on tubulin polymerization is of particular interest. Note that when KAR-2 is used the binding can be adequately described by a single-fractal analysis. This provides fresh insights into the nature of drug–tubulin interactions. Further analysis of other antigen–antibody (or, in general, analyte–receptor) interactions on biosensor surfaces (and in other reaction systems, such as cell-surface interactions) is required to further delineate the changes in the fractal dimension and in the binding rate coefficients in the same direction.

REFERENCES

1. Bluestein BI, Craig M, Slovacek R, et al. Evanescent wave immunosensors for clinical diagnostics. In: Wise D and Wingard Jr. LB, eds. *Biosensors with Fiberoptics.* Totowa, NJ: Humana Press, 1991, pp 181–223.
2. Eddowes MJ. Direct immunochemical sensing:basic chemical principles and fundamental limitations. *Biosensors* 1987/1988;3:1–15.
3. Kopelman R. Fractal reaction kinetics. *Science* 1988;241:1620–1626.
4. Pfeifer P, Obert M. Fractals:basic concepts and terminology. In: Avnir D, ed. *The Fractal Approach to Heterogeneous Chemistry: Surfaces, Colloids, Polymers.* New York: John Wiley & Sons, 1989, pp 112–123.

5. Seri-Levy A, Avnir D. Fractal analysis of surface geometry effects on catalytic reactions. *Surf Sci* 1991;248:258–270.
6. Sadana A, Beelaram A. A fractal analysis of antigen–antibody binding kinetics:biosensor applications. *Biotechnol Progr* 1994;10:291–298.
7. Sadana A, Beelaram A. Antigen–antibody diffusion-limited binding kinetics of biosensors: a fractal analysis. *Biosens Bioelectron* 1995;10:310–316.
8. Sadana A. Antigen–antibody binding kinetics for biosensors: the fractal dimension and the binding rate coefficient. *Biotechnol Progr* 1995;11:50–57.
9. Sadana A, Sii D. The binding of antigen by immobilized antibody: influence of a variable rate coefficient on external diffusional limitations. *J Colloid Interf Sci* 1992;151(1):166–177.
10. Sadana A, Sii D. Binding kinetics of antigen by immobilized antibody: influence of reaction order and external diffusional limitations. *Biosens Bioelectron* 1992;7:559–568.
11. Havlin S. Molecular diffusion and reaction. In: Avnir D, ed. *The Fractal Approach to Heterogeneous Chemistry: Surfaces, Colloids, Polymers.* New York: John Wiley & Sons, 1989, pp 251–269.
12. Piehler J, Brecht A, Gauglitz G. Affinity detection of low molecular weight analytes. *Anal Chem* 1996;68:139–143.
13. Sigmaplot (1993). Scientific Graphing Software, User's Manual. Jandel Scientific, San Rafael, CA.
14. Domenici C, Schirone A, Celebre M, et al. Development of a TIRF immunosensor: modeling the equilibrium behavior of a competitive system. *Biosens Bioelectron* 1995;10:371–378.
15. Liliom K, Lehotzky A, Molnar A, et al. Characterization of tubulin–alkaloid interactions by enzyme-linked immunosorbent assay. *Anal Biochem* 1995;228:18–26.
16. Correia JJ. *Pharmacol Ther* 1991;52:127–147.
17. Douglas JF. How does surface roughness affect polymer-surface interactions? *Macromolecules* 1989;22:3707–3716.

3

Avidin–Biotin Mediated Biosensors

Jun-ichi Anzai, Tomonori Hoshi, and Tetsuo Osa

3.1. INTRODUCTION

Much attention has been paid to the development of biosensors because of their usefulness, such as for an easy and fast measurement of analytes, as compared with rather the tedious pretreatment often required in conventional techniques for biochemical analyses. An electrochemical biosensor consists of an enzyme as the molecular recognition element and an electrode as the transducer that converts the chemical signal produced by the enzymatic reaction into an electric form. In the fabrication of enzyme sensors, the immobilization of the enzyme on the electrode surface is a crucially important process, which determines the performance characteristics of the sensor, because it is usually difficult to introduce an enzyme onto the surface of a metal electrode on which no reactive group is available. Many different procedures have been reported concerning the immobilization of enzymes on the surface of solid materials, including physical adsorption, covalent bonding, cross-linking with inert polymers or proteins, and simple entrapment within a polymeric matrix or a dialysis membrane.

However, there are problems to be addressed with each method, such as loss of the catalytic activity, change in the reaction parameters, and decreased stability. These problems are related to conformational changes within the enzyme, inadequate geometry of the enzyme and the electrode, and/or weak adhesion of the enzyme to its support. In addition, biosensor devices should be as small as possible with a view toward future use as implantable devices. In this context, it is anticipated that the output signal of a miniaturized sensor is rather weak or unstable because the amplitude of the output signal of an enzyme sensor depends basically on the total activity of the enzyme immobilized on the electrode as well as on the surface area of the electrode itself. The available surface area of a miniature sensor is so limited that

Jun-ichi Anzai, Tomonori Hoshi, and Tetsuo Osa • Faculty of Pharmaceutical Sciences, Tohoku University, Aramaki, Aoba-ku, Sendai 980-8578, Japan.

Biosensors and Their Applications, edited by Yang and Ngo, Kluwer Academic/Plenum Publishers, New York, 1999.

it is impractical to immobilize a sufficient amount of enzyme on the electrode using the conventional techniques mentioned above. In order to enhance the loading (or the total activity) of the enzyme of miniaturized electrodes, it is essential to immobilize the enzyme densely by regulating its molecular orientation and packing on the surface. For this purpose, we have developed a novel technique based on an avidin–biotin technology, and in this chapter, we describe the principle of the avidin–biotin system and its use for the fabrication of enzyme sensors in relation to the preparation of enzyme thin films in which enzymes are assembled into a layer-by-layer structure composed of monomolecular layers.

3.2. AVIDIN–BIOTIN SYSTEM

Avidin is a highly stable glycoprotein (molecular mass ca. 68,000) found in egg white, and is usually isolated as a tetramer of identical 128-residue polypeptide chains.[1] It contains carbohydrate chains composed of mannose, N-acetylglucosamine, and galactose, which are covalently connected to asparagine residue in the polypeptides and comprise ca. 10% of the total molecular weight of avidin. Owing to the carbohydrate chains, avidin exhibits a basic nature, with an iosoelectric point of ca. 10.5. Its molecular dimensions are ca. $6.0 \times 5.5 \times 4.0$ nm. Avidin's most characteristic feature is that each of its subunits contains a binding site for biotin and forms a highly stable complex with biotin and its derivatives, the binding constant being reported to be ca. $10^{15}\,M^{-1}$ (Fig. 3.1). Avidin assumes a tetramer with twofold symmetry and has four identical binding sites arranged in two pairs on opposite 6.0×5.5 nm faces, and can therefore maximally bind four biotin molecules. X-ray crystallographic studies have revealed that the exceptionally strong affinity of avidin for biotin arises from hydrophobic interactions of biotin and aromatic amino acids (phenylalanine and tryptophan) arranged in the binding pocket of avidin and a multiple hydrogen bonding between heteroatoms in the ureido ring of biotin and asparagine, serine, tyrosine, and threonine residues in avidin.[2] It is known that the carbohydrate chains in avidin do not play any significant role in the binding of biotin, and thus a deglycosylated form of avidin is sometimes used as an alternative

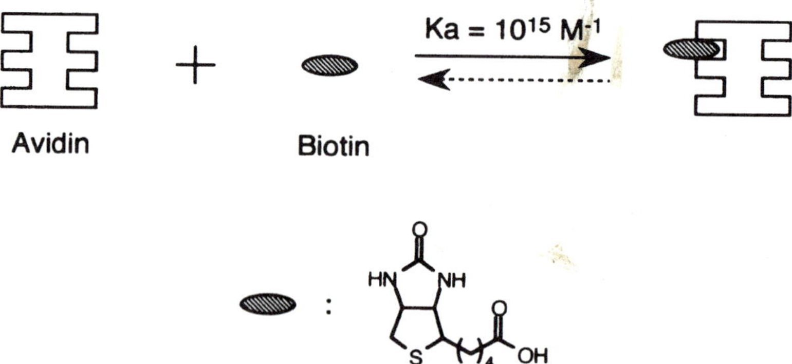

Figure 3.1.   A specific binding between avidin and biotin.

to avidin. Kinetic studies of avidin–biotin binding suggested that a diffusion-limited association of the complex and its very slow dissociation resulted in the high affinity.[3] The avidin–biotin complex is not dissociated in 9 M urea or 6 M guanidinium hydrochloride, even in the pH range from 2 to 13. This is the strongest known noncovalent bonding in nature. A homologous protein of avidin, streptavidin, is known to have almost the same strong affinity to biotin.[1] Streptavidin is also a tetramer protein and contains a biotin binding site in each subunit, the molecular dimensions being 5.6 × 4.2 × 4.2 nm. It does not contain a carbohydrate chain and, as a result, its isoelectric point (pI) is ca. 5–6, as compared with the basic nature of avidin (pI = 10.5).

Because of the highly specific and strong binding ability of avidin and streptavidin for biotin, the avidin–biotin system has been widely used in a variety of biochemical fields, including affinity chromatography, immunochemical strains, and binding assay. For these purposes, many kinds of biotin- or avidin-labeled reagents, including enzymes, lectins, binding proteins, DNA and RNA, and fluorophores are now commercially available. In addition, we can easily prepare the biotin-labeled reagents using reactive biotins such as an active ester of biotin, maleimide of biotin, or biotin hydrazide (Fig. 3.2).

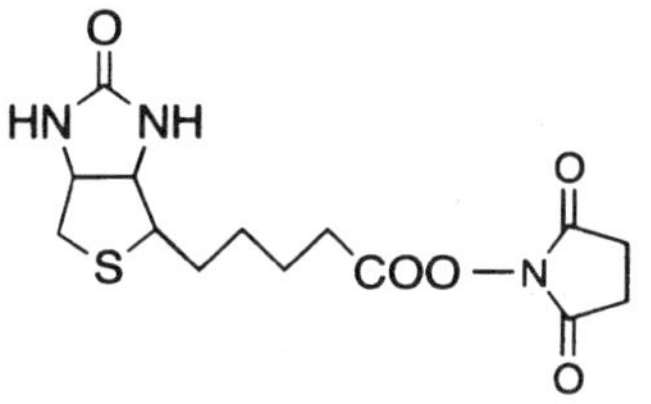

Biotin active ester

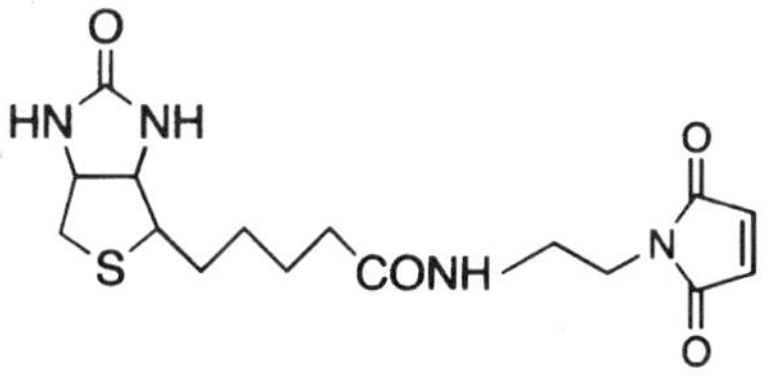

Biotin maleimide

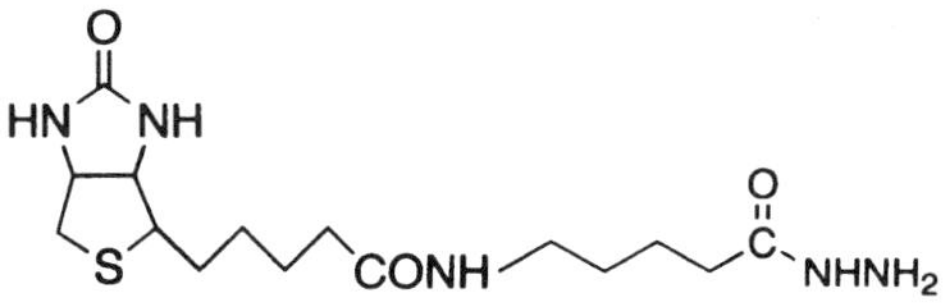

Biotin hydrazide

Figure 3.2. Biotin analogues.

3.3. IMMOBILIZATION OF ENZYMES THROUGH AVIDIN–BIOTIN COMPLEXATION

Several groups independently have used the avidin–biotin system for the preparation of biosensor devices. There are two different ways to immobilize biotin-labeled enzymes on the electrode surface through avidin–biotin complexation, and the two procedures are shown schematically in Fig. 3.3. The first method uses a biotin-modified electrode on which biotin-labeled enzymes are immobilized through avidin as a binder. Using this strategy, Walt et al. immobilized biotin-modified-enzymes (urease, esterase, and penicillinase) on the surface of biotin-modified optical fiber using avidin as a binder.[4] They demonstrated the general use of this procedure for immobilizing any kind of enzyme. Kuhr et al. anchored biotin residues on the surface of a carbon electrode and immobilized avidin and then biotin-labeled glutamate dehydrogenase to prepare carbon-fiber microsensors to study neurotransmitter dynamics.[5] Another way to modify the electrode surface with biotin based on a biotin-bearing phospholipid bilayer was developed by Snejdarkova et al.[6] The phospholipid-coated electrode, whose surface contains an appropriate number of biotin residues, was used to prepare a glucose sensor by coupling it with biotin-labeled glucose oxidase (GOx).

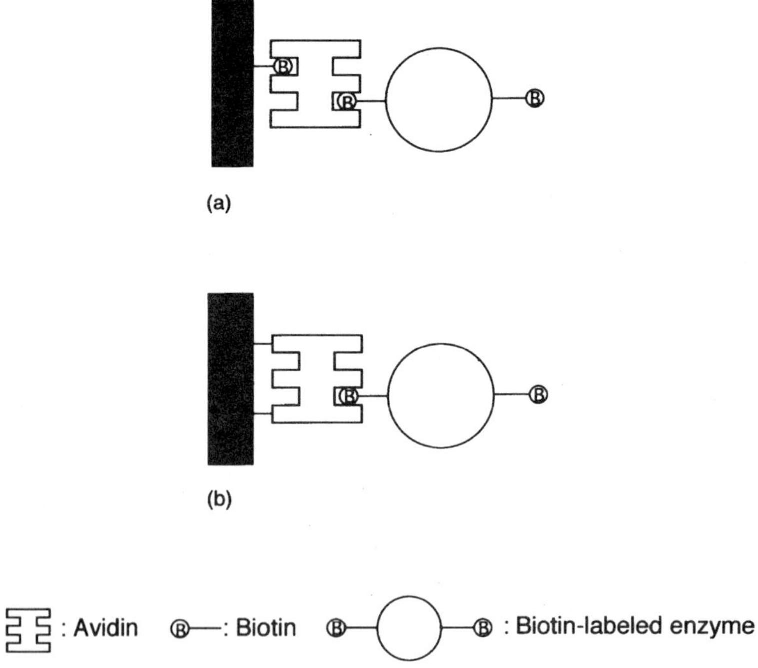

Figure 3.3. Two different routes for immobilizing biotin-labeled enzyme on the electrode: avidin is adsorbed to the biotin-modified surface (a) and directly to the electrode through physical adsorption (b).

An alternative way to immobilize enzymes involves the direct modification of the electrode surface with avidin. If avidin could be immobilized directly without loss of the binding activity to biotin, biotin-labeled enzymes could be loaded onto the electrode surface more easily. Gunaratna and Wilson prepared an enzyme column in which choline esterase and choline oxidase were immobilized through avidin–biotin complexation for the determination of acetylcholine.[7] They attached avidin covalently to the support and then coupled it with biotin-labeled enzymes. Avidin is known to be strongly adsorbed on the hydrophobic surfaces through hydrophobic interactions. Therefore, we tried to immobilize avidin by simple adsorption on the surface of a Langmuir–Blodgett (LB) monolayer film composed of stearic acid.[8] The surface of a metal electrode was first coated with the LB monolayer and the resulting hydrophobic electrode was immersed in an avidin solution to form an avidin layer. Then, the avidin-modified electrode was treated with biotin-labeled GOx to cover the surface of the electrode with a GOx monolayer. Alternatively, to immobilize a large amount of enzyme on the electrode, the LB-film-coated electrode was immersed in a diluted solution of the multiple complex of avidin and biotin-labeled GOx.

One can expect the first procedure to afford a monomolecular layer of enzyme on the electrode surface, while in the second the giant aggregates of proteins composed of avidin and enzyme molecules are adsorbed randomly. From the viewpoint of enzyme sensors, the sensors in the latter case would show a bigger response than in the former because the total activity of the enzyme (or total amount of enzyme) directly determines the rate of the catalytic reaction. As expected, the output current for the latter sensor was higher than that of the former. These results show that GOx is still active, even in the multiple complex with avidin. A colorimetric estimation of the catalytic activity of GOx at both electrodes revealed the rate of H_2O_2 generation to be 0.8×10^{-11} and 3.0×10^{-11} mole/cm^2 s, respectively, for the monolayer and multilayer GOx sensors. The use of native GOx in place of biotin-labeled GOx in the preparation of glucose sensors resulted in a negligibly small response to glucose, confirming that GOx is immobilized through avidin–biotin complexation.

In order to enhance the functionality of the LB film, we modified it with a redox-active ferrocene moiety and immobilized the GOx monolayer on it through avidin–biotin complexation.[9] The glucose sensor thus prepared showed an amperometric response to glucose at $+0.5$ V vs. Ag/AgCl, which is an inadequate potential for the electrochemical oxidation of H_2O_2 on the electrode used (InSnO electrode). Indeed, the glucose sensor based on the stearic acid LB film did not show any current response to glucose at this electrode potential. The electron relay was considered to be accelerated in the LB film/GOx assembly on the electrode as illustrated in Fig. 3.4.

It is known that adsorption kinetics and/or thermodynamics of proteins depend on the electric or electrochemical properties of solid supports on which the proteins are adsorbed, so we studied the effects of electrode potential on the adsorption behavior of avidin on the electrode surface. The potential of a Pt electrode was varied systematically in the range from -0.5 to $+2.0$ V in an avidin solution (pH 7.4). Although the data were somewhat scattered, the general trend that was observed indicated that the adsorption of avidin is suppressed by the application of

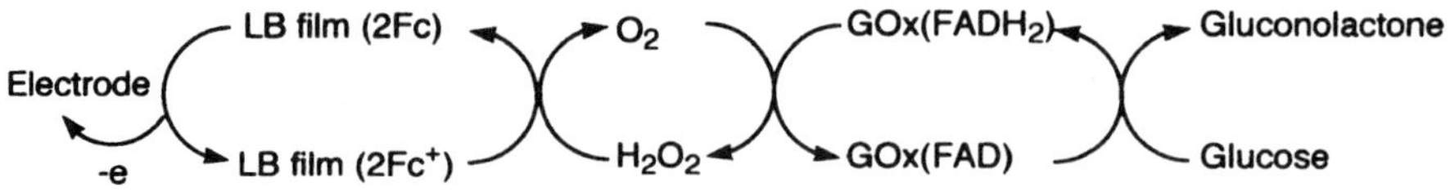

Electrode — LB film (2Fc) — O_2 — GOx(FADH₂) — Gluconolactone

-e — LB film (2Fc⁺) — H_2O_2 — GOx(FAD) — Glucose

Fc: Ferrocene

Figure 3.4. An electron relay from glucose to electrode through GOx and the ferrocene-modified LB film.

a positive potential ($+1.0 - +2.0$ V). This may be due to the fact that avidin is a basic protein and has a net positive charge in a neutral pH solution. In the potential range tested, no significant acceleration in the adsorption was induced. On the other hand, the adsorption of avidin was significantly accelerated under the influence of an alternating potential of triangular waveform from -0.5 to $+2.1$ V (200 V/s).[10] The adsorption of avidin was found to be accelerated more than 10-fold under the influence of the alternating potential, as monitored by a quartz crystal microbalance (QCM). The QCM data also demonstrated that the loading of avidin (or the thickness of the avidin layer) can be regulated by changing the adsorption time. The electrodeposited avidin film retained its binding activity to biotin, and enzyme sensors were successfully prepared using biotin-labeled enzymes. A disadvantage of this technique for preparing the avidin layer relates to the difficulty of controlling the orientation of avidin molecules in the surface layer.

3.4. LAYER-BY-LAYER STRUCTURE OF ENZYME MULTILAYERS

The avidin molecule (tetramer) is in a cubelike shape, about $6.0 \times 5.5 \times 4.0$ nm in size. The binding sites to biotin are arranged in two pairs on opposite faces of the avidin molecule. The two binding sites on the same face (6.0×5.5 nm) are separated by 2.0–2.5 nm, according to x-ray crystallography and electron microscope studies of bis-biotin-linked avidin polymer.[11] The shape and size of avidin, together with the favorable arrangement of binding sites make it a promising possibility for use as a building block for the construction of two- or three-dimensional protein architectures. If we use enzymes tagged with more than two biotin residues, an enzyme multilayer as illustrated in Fig. 3.5 would be constructed because avidin contains four biotin-binding sites per molecule. Fortunately, as the biotin-binding sites are located in two pairs on opposite faces of the avidin molecule, the use of e.g., 5-nm-thick biotin-labeled enzyme should result in the formation of a multiple protein layer composed of approximately 10-nm double layers of avidin plus enzyme. In other words, the thickness of the multilayer can be regulated solely by changing the number of electrodepositions.

Based on this strategy, we prepared a variety of enzyme multilayer-modified biosensors.[12–16] Prior to the fabrication of the biosensors, we determined whether the enzyme multilayers form as shown in Fig. 3.5 by spectrophotometric measure-

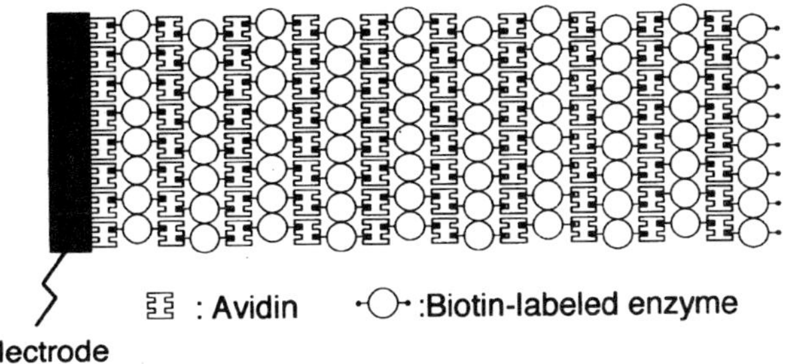

Figure 3.5. Idealized structure of an enzyme multilayer based on an avidin–biotin system.

ment of the multilayer cast on the surface of a quartz slide. For this purpose, we assembled fluorescein-5-isothiocyanate (FITC)-conjugated avidin and biotin-labeled enzyme alternately on the slide. Immediately prior to the formation of the protein multilayer, the slide was modified with dichlorodimethylsilane to make the surface hydrophobic. The silylated quartz slide was then immersed in FITC–avidin and biotin-labeled enzyme solutions alternately and repeatedly to coat both sides of the slide with protein multilayers. An increase in absorbance at 495 nm, originating from the FITC moiety of FITC–avidin, was monitored after each protein-layer deposition. A typical result for an ascorbate oxidase multilayer is shown in Fig. 3.6. The absorbance increased in proportion to the number of layers deposited, which suggests the formation of a multilayer structure on the quartz slide. This result indicates that FITC–avidin is immobilized in each layer as a roughly monomolecular layer considering the molecular dimensions of avidin and molar extinction coefficient of $176,000 \, M^{-1} \, cm^{-1}$ for FITC–avidin at 495 nm. Almost the same result was obtained for the other enzymes tested.

The enzyme multilayers were deposited on the surface of a Pt electrode to prepare the enzyme—glucose, lactate, alcohol, and acetylcholine—sensors. As an

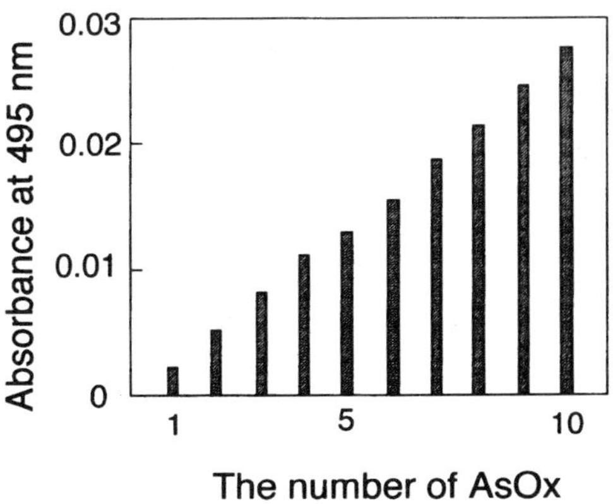

Figure 3.6. A layer-by-layer deposition of FITC–avidin and AsOx on the quartz slide as monitored by UV absorption spectra at 495 nm.

example, we will describe the performance of the alcohol sensors, which are based on alcohol oxidase (AOx). A 10-layer AOx-modified sensor was prepared and its amperometric response to ethanol was tested. Upon addition of ethanol to the buffer solution, the output current increased rapidly.[16] The amperometric response should arise from the electrochemical oxidation of H_2O_2, which is produced through the standard AOx-catalyzed enzymatic reaction:

$$CH_3CH_2OH + O_2 \xrightarrow{\text{AOx}} CH_3CHO + H_2O_2 \tag{1}$$

These results clearly show that the enzyme AOx is active in the multilayer assembly on the electrode surface. The rapid response of the sensor implies that the AOx multilayer is relatively porous, enabling ethanol and H_2O_2 to diffuse freely. The thinness of the AOx multilayer may be another reason for the rapid response of the sensor. For example, the thickness of the monolayer (one avidin plus one AOx layer) is roughly estimated at 15 nm, based on the molecular sizes of avidin and AOx. It was of interest to verify how the output current depends on the number of AOx layers on the electrode, so we measured the response current of the sensors to 10 mM ethanol after each deposition of AOx from 1 to 10 layers on the electrode. As expected, the response current was magnified linearly with the increase of AOx layers (Fig. 3.7). The increase in the output current per AOx layer was ca. 0.6 μA. The linear relationship between the output current and the number of the AOx layers suggests that the same amount of AOx is immobilized on each deposition, because the magnitude of the output current of the sensor should depend on the total activity of the enzyme on the electrode. It was found that the stability of the AOx multilayer-modified alcohol sensors was acceptable. In the test in which the sensor was operated once a day and stored in a refrigerator at 4°C, the response current to 10 mM ethanol retained ca. 95% activity of its original value after 4 weeks and thereafter the response gradually decreased down to 30% activity of the original value after 10

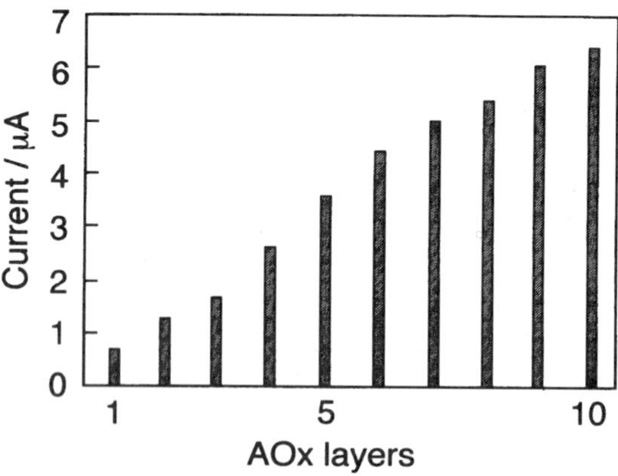

Figure 3.7. The relationship between the output current of alcohol sensors to 10 mM ethanol and the number of AOx layers deposited on the electrode.

Table 3.1. Comparison of Streptavidin- and Avidin-Based Lactate Sensors

Type and number of LOx layers	$i/\mu A^a$	K_m^{app}/mM
LOx/streptavidin		
1	4.0	0.30
5	7.3	0.36
10	11	0.43
LOx/avidin		
1	3.2	0.32
5	8.6	0.40
10	15	0.54

[a] The output current of the sensors of 3 mM lactate.

weeks. In the successive use for determining 10 mM ethanol, virtually no deterioration was observed in the response after 50 measurements.

In a similar procedure, we prepared glucose,[12] lactate,[13] and choline sensors.[17] In all cases, the output current was a linear function of the number of enzyme layers deposited on the electrode. In the course of preparing various types of enzyme multilayers, we found that streptavidin can also be used as a component of the enzyme multilayer in place of avidin. Typical data for a comparison of the properties of various sensors based on lactate oxidase (LOx) and both avidin and streptavidin are collected in Table 3.1.

In the enzyme multilayers, it is easy to assemble two or more different types of enzymes into the same multilayer simultaneously because the multilayer can be prepared layer by layer using the desired enzymes. We prepared GOx/ascorbate oxidase (AsOx) multilayers on the Pt electrode to eliminate the ascorbate interference of the glucose sensors.[14,15] A H_2O_2 detection-type glucose sensor often suffers from interference arising from the direct oxidation of redox-active substances in blood samples. We may be able to eliminate the interference originating from ascorbic acid using the GOx/AsOx multilayer-modified glucose sensors because ascorbic acid can be oxidized into the redox-inactive form, dehydroascorbic acid, by AsOx in the GOx/AsOx multilayer, according to the following reaction:

$$\text{Ascorbic acid} + \tfrac{1}{2}O_2 \xrightarrow{\text{AsOx}} \text{Dehydroascorbic acid} + H_2O \qquad (2)$$

Figure 3.8 illustrates the electrochemical responses of the GOx and GOx/AsOx multilayer-modified sensors to a physiological level of glucose and ascorbic acid. The GOx multilayer sensor gave an 11-μA response to 5 mM glucose, and a further interfering current of 2.2 μA was induced by the addition of 0.1 mM ascorbic acid to the sample solution. On the other hand, no interference was observed with the GOx/AsOx multilayer sensor upon addition of 0.1 mM ascorbic acid. It is likely that ascorbic acid was oxidized to dehydroascorbic acid by AsOx in the multilayer before being oxidized directly on the Pt surface.

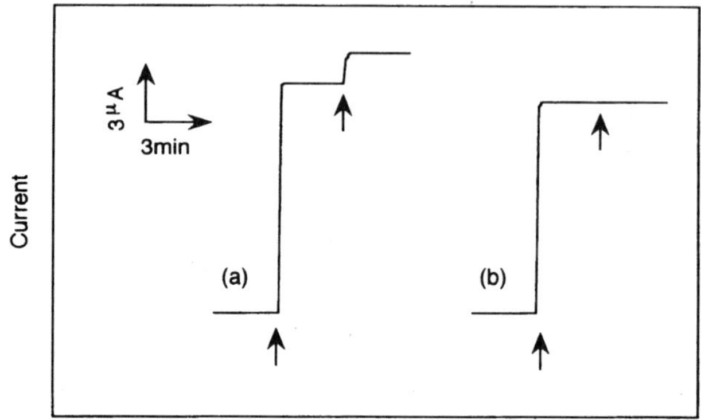

Figure 3.8.   Electrochemical response of the glucose sensors to glucose and ascorbate: 5 mM glucose and 0.1 mM ascorbic acid were measured successively with GOx-modified sensor (a) and GOx/AsOx-modified sensor (b).

Another example of the bienzyme multilayer-modified sensor includes acetylcholine sensors prepared using choline oxidase (ChOx) and choline esterase (ChE) assembled simultaneously on the Pt electrode.[17] With the ChOx/ChE multilayer-modified sensors, acetylcholine can be measured through the consecutive reactions catalyzed by ChE and ChOx:

$$\text{Acetylcholine} + H_2O \xrightarrow{\text{ChE}} \text{Choline} + \text{Acetate} \tag{3}$$

$$\text{Choline} + 2O_2 + H_2O \xrightarrow{\text{ChOx}} \text{Betaine} + 2H_2O_2 \tag{4}$$

Figure 3.9 illustrates a typical response of the ChE/ChOx multilayer-modified acetylcholine sensor as well as that of ChOx-modified sensor. The ChOx (10 layers)-modified sensor exhibited no response to 1×10^{-6}–3×10^{-5} M acetylcholine, though an appropriate response was observed to choline. This is reasonable because ChOx cannot hydrolyze acetylcholine to produce choline and no H_2O_2 is generated on the electrode surface. On the other hand, the ChOx (10 layers)/ChE (1 layer)-modified sensor showed an amperometric response to acetylcholine, confirming that the successive reactions [Eqs. (3) and (4)] occurred to produce H_2O_2 in the enzyme multilayer.

3.5. CONCLUSIONS

Much attention has long been devoted to the surface modification of metal and carbon electrodes with enzymes and other biological materials for the development of high-performance biosensors. These efforts resulted in the formation of well-defined monomolecular layer-modified biosensors. Unfortunately, in an enzyme sensor case, the conventional way of preparing electrodes sometimes suffers from low

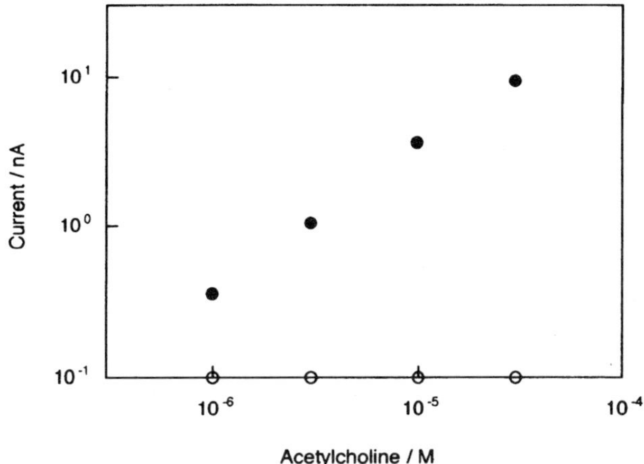

Figure 3.9. Typical calibration graphs of the acetylcholine sensors modified with ChOx (○) and ChOx/ChE (●).

output owing to an insufficient quantity bound enzyme: the size of the signal depends on the total activity of the enzymes immobilized on the electrode surface. In this context, the avidin–biotin system is useful for regulating the quantity of enzymes bound on the electrode. It is an outstanding feature of this technique that a desired type and quantity of enzyme can be assembled into an enzyme multilayer on the electrode almost at will. Thus we can systematically design the layer-by-layer structure of the enzyme multilayer, using two or more different types of enzymes for enhancing the function of the biosensor. Moreover, it is expected that this technique will open up a new possibility for the construction of a protein superlattice which is capable of highly sophisticated functions — more than the sum of those of the individual proteins.

REFERENCES

1. Green NM. Avidin and streptavidin. *Meth Enzymol* 1990;184:51–67.
2. Pugliese L, Coda A, Malcovati M, et al. Three-dimensional structure of the tetragonal crystal form of egg-white avidin in its function complex with biotin at 2.7 Å resolution. *J Med Biol* 1993;231:698–710.
3. Chilcoti A, Stayton PS. Molecular origin of the slow streptavidin–biotin dissociation kinetics. *J Am Chem Soc* 1995;117:10622–10628.
4. Luo S, Walt DR. Avidin-biotin coupling as a general method for preparing enzyme-based fiber-optic sensors. *Anal Chem* 1989;61:1069–1072.
5. Pantano P, Kuhr WG. Dehydrogenase-modified carbon-fiber microelectrodes for the measurement of neurotransmitter dynamics: 2. Covalent modification utilizing avidin-biotin technology. *Anal Chem* 1993;65:623–630.
6. Snejdarkova M, Rehak M, Otto M. Design of a microsensor based on streptavidin-glucose oxidase complex coupling with self-assembled biotinylated phospholipid membrane on solid support. *Anal Chem* 1993;65:665–668.

7. Gunaratna PC, Wilson GS. Optimization of multienzyme flow reactors for determination of acetylcholine. *Anal Chem* 1990;62:402–407.

8. Anzai J, Lee S, Hoshi T, et al. Use of the avidin-biotin system for immobilization of an enzyme on the electrode surface. *Sens. Act B* 1993;13/14:73–75.

9. Lee S, Anzai J, Osa T. Enzyme-modified Langmuir–Blodgett membranes in glucose electrodes based on avidin-biotin interaction. *Sens Act B* 1993;12:153–158.

10. Hoshi T, Anzai J, Osa T. Electrochemical deposition of avidin on the surface of a platinum electrode for enzyme sensor applications. *Anal Chim Acta* 1994;289:321–327.

11. Green NM, Konieczny L, Toms EJ, et al. The use of bifunctional compounds to determine the arrangement of subunits in avidin. *Biophys J* 1971;125:781–791.

12. Hoshi T, Anzai J, Osa T. Controlled deposition of glucose oxidase on platinum electrode based on an avidin/biotin system for the regulation of output current of glucose sensors. *Anal Chem* 1995;67:770–774.

13. Anzai J, Takeshita H, Hoshi T, et al. Regulation of output current of L-lactate sensors based on alternate deposition of avidin and biotinylated lactate oxidase on electrode surface through avidin/biotin complexation. *Chem Pharm Bull* 1995;43:520–522.

14. Anzai J, Takeshita H, Hoshi T, et al. Elimination of ascorbate interference of glucose sensors by use of enzyme multilayers composed of avidin and biotin-labeled glucose oxidase and ascorbate oxidase. *Denki Kagaku* 1995;63:1141–1142.

15. Anzai J, Du X, Hoshi T, Suzuki Y, et al. Enzyme sensors modified with avidin/biotin system-based protein multilayers. *Anal Sci Tech* 1995;8:591–596.

16. Du X, Anzai J, Osa T, et al. Amperometric alcohol sensors based on protein multilayers composed of avidin and biotin-labeled alcohol oxidase. *Electroanalysis* 1996;8:813–816.

17. Chen Q, Kobayashi Y, Takeshita H, et al. Avidin-biotin system-based enzyme multilayer membranes for biosensor applications: optimization of loading of choline esterase and choline oxidase in the bienzyme membrane for acetylcholine biosensors. *Electroanalysis* 1998; 10:94–97.

4

Layered Functionalized Electrodes for Electrochemical Biosensor Applications

Itamar Willner, Eugenii Katz, and Bilha Willner

4.1. INTRODUCTION

The application of redox enzymes as bioactive matrices for biosensor design is of substantial basic and practical importance.[1-4] Two basic configurations for the use of enzymes in biosensor devices are outlined in Fig. 4.1. In Fig. 4.1A, the biocatalyst generates a redox-active product that undergoes a redox transformation at the electrode interface and yields a current or a potential response. For example, biocatalyzed oxidation of substrates such as glucose or amino acids by molecular oxygen in the presence of glucose oxidase (GOx)[5-8] or L-amino acid oxidase (AAOx),[9] respectively, generates H_2O_2 as an electroactive product. The amperometric response due to the reduction of H_2O_2 is proportional to the substrate concentration.[10] Potentiometric detection of enzymatically produced H_2O_2 was also used.[11,12] Alternatively, the depletion of oxygen monitored at an oxygen-sensitive electrode represents a potential transduction of the substrate concentration.[13] This approach was used to develop the first generation of electrochemical biosensors and the methodology has been extensively reviewed.[14] The second approach to designing amperometric biosensors is shown in Fig. 4.1B. Electrocatalyzed oxidation (or reduction) of the enzyme redox center stimulates the oxidation (or reduction) of the substrate, and the resulting current is proportional to the substrate concentration. The development of such enzyme/electrode systems suffers from the basic limitation that redox enzymes usually lack direct electrical contact with electrode surfaces.[15]

Electron transfer theory[16] indicates that the electron-transfer rate constant (k_{et}) between a donor and acceptor couple is given by [Eq. (1)]:

$$k_{et} \propto \exp - [\beta(d - d_0)] \exp[-(\Delta G^\circ + \lambda)^2/4R\lambda T] \tag{1}$$

Itamar Willner, Eugenii Katz, and Bilha Willner • Institute of Chemistry, The Hebrew University of Jerusalem, Jerusalem 91904, Israel.

Biosensors and Their Applications, edited by Yang and Ngo, Kluwer Academic/Plenum Publishers, New York, 1999.

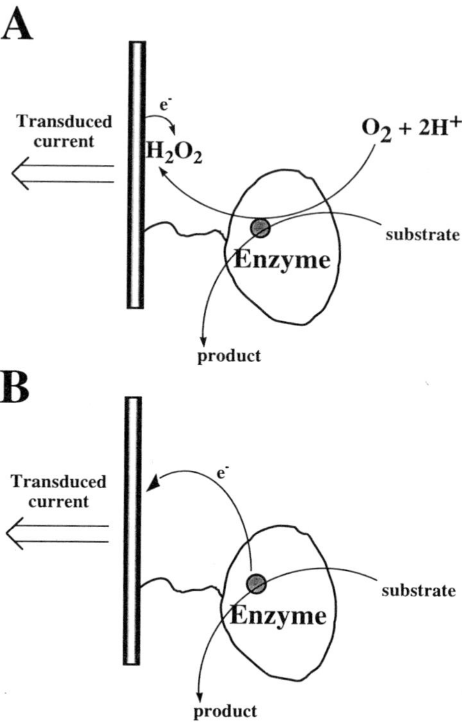

Figure 4.1. Transduction of a chemical signal (substrate concentration) into electrical current by an electrochemical biosensor: (A) intermediate formation of an electrochemically detectable product; (B) direct electrochemical communication with the enzyme.

where $\Delta G°$ and λ correspond to the free energy and reorganization energy accompanying electron transfer and d_0 and d are the van der Waals distance and actual distance separating the donor–acceptor centers, respectively. The electrode and the protein-embedded redox center can be viewed as a donor–acceptor pair. The thick protein layer surrounding the active site introduces a spatial separation of the donor–acceptor pair and a kinetic barrier for electron transfer. This leads to the electrical insulation of the active centers of most redox proteins. In nature, redox enzymes participate in electron-transfer reactions by several routes:

1. Low-molecular-weight substrates, e.g., oxygen, diffuse into the protein and reach short distances with respect to the active site that enables their oxidation.[17]
2. Low-molecular-weight cofactors such as 1,4-nicotinamide adenine dinucleotide (NAD⁺) or 1,4-nicotinamide adenine dinucleotide phosphate (NADP⁺), and the respective 1,4-dihydronicotinamide derivatives act as diffusional two-electron mediators that electrically communicate the biocatalytic redox centers with their environment.[18,19]

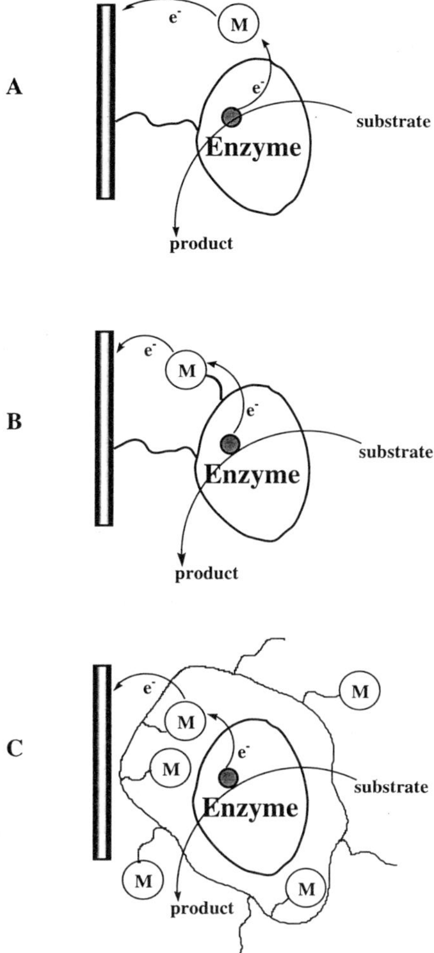

Figure 4.2. Mediated electrical communication between redox centers of enzymes and electrodes: (A) diffusional electron-transfer mediator; (B) electron-transfer mediator covalently linked to functional groups of the enzyme molecules; (C) electron-transfer mediator associated with a polymer-entrapped enzyme molecules.

3. Interprotein complexes lead to short spatial distances between the redox sites associated with two different proteins.[20]

Different methods were developed to electrically contact redox enzymes with electrodes in order to overcome the intrinsic barriers to direct electron transfer (Fig. 4.2). Diffusional electron mediators[18] such as N,N'-bipyridinium salts,[21] quinones,[22] and ferrocene derivatives,[23] were employed for electrical communication between the redox biocatalyst and electrode supports (Fig. 4.2A). In these systems, the electron mediator acts as a diffusional electron relay that accepts or donates the electron from or to the active center.[18] The electrochemical contact between the enzyme and the electrode support in the presence of the diffusional electron transfer mediators was assayed by amperometric or potentiometric methods.[24] The architecture of integrated bioelectrocatalytic electrodes lacking diffusional components has important implications in bioelectronic and biosensor technology, as it enables the use of the electrodes as invasive sensors for continuous *in situ* analysis of samples.[25,26] Several methods to electrically contact enzymes with electrodes by a nondiffusional route have been reported. These include the chemical modification of enzymes with redox-active electron mediators (Fig. 4.2B)[27–31] and the immobilization of the enzymes in redox-active polymers (Fig. 4.2C).[32–35] Tethering of redox relay groups to the protein, e.g., ferrocene[27–29] or N,N'-bipyridinium units,[30,31] provides a pathway for electron hopping (or tunneling) through the protein. Immobilization of the enzymes in redox copolymers exhibiting structural flexibility, such as Os(II/III)-polypyridine/poly-pyridinium[32,33] or N,N'-dialkyl-4, 4'-bipyridinium polythiophene,[34,35] allows dynamic interactions between the electron-relay sites and the protein active center that yield appropriate distances for mutual electron transfer. A combination of both approaches, i.e., incorporation of a redox-relay-modified enzyme into conductive polymers, was also applied to enhance the electrical contact between the biocatalyst active centers and the electrode.[36] The enzyme-electrode architecture for biosensor applications has to solve several problems:

1. The assembly of coupled enzyme electrodes for continuous use; i.e., the development of methods for the immobilization of the proteins on conductive support is essential.
2. There must be electrical contact in the enzyme-electrode assembly.
3. The enzyme electrode should exhibit the appropriate sensitivity and preferably a built-in tunability of the sensitivity in the method of preparation.
4. The enzyme electrode should display specificity and selectivity for the analyte substrate.
5. Tailoring of integrated, electrically contacted $NAD(P)^+$-dependent enzyme electrodes is essential for bioelectrocatalytic electrodes that require the participation of an $NAD(P)^+$-dependent enzyme.
6. Invasive, implanted, use of electrochemical biosensors requires the preparation of miniaturized enzyme electrodes. The transduced output from these microelectrodes should have a sufficient signal-to-noise ratio to enable precise analyses.

This chapter will address the assembly of electrically contacted layered enzyme electrodes for biosensor applications. The methods used to assemble these electrodes will be discussed with specific emphasis on the architecture of thiolated monolayers consisting of enzymes on gold electrodes. In the second part of the chapter, the architecture of electrochemical immunosensors and DNA sensors that use electroactive substrate redox enzymes as labels for the electrochemical probing of the analyte-sensing interface interactions will be described.

4.2. MONOLAYER ENZYME ELECTRODES

Covalent attachment of enzyme monolayers to conductive supports has been the subject of extensive research in the last decades. The attachment mode is usually controlled by the nature of the conductive surface — metal, metal oxide, or carbon — and the availability of surface functions for the immobilization of the biomaterial. Figure 4.3a illustrates surface hydroxyl functions associated with, e.g., SnO_2, In_2O_3, or oxidized carbon. These are functionalized by cyanuric chloride[37,38] to yield an active interface for immobilization of the enzymes. For example, glucose oxidase[9,39] and L-amino acid oxidase[9,11] were linked to a pyrographite electrode by this method. Functionalization of hydroxylated surfaces with amino, thiol, and other alkoxysilane derivatives yields modified surfaces for covalent linkage of the enzymes (Fig. 4.3b).[40–43] Glucose oxidase[8] and horseradish peroxidase[44] were coupled to Pt-oxide and SnO_2 electrodes, respectively, using this approach. Carboxylic functions are available at oxidized carbon surfaces[45] and these provide sites for the direct coupling of enzymes.[46] Activation of the carboxylic acid functions by the generation of acyl halides or anhydrides yields reactive interfaces[5,47,48] for the covalent linkage of enzymes (Fig. 4.3c). For example, glucose oxidase,[5–7,49–53] xanthene oxidase,[24] horseradish peroxidase,[54] lactoperoxidase,[54] and lactate dehydrogenase[55] were coupled to carbon electrodes by these anchoring groups. Synthetically useful amino functions can be introduced on a graphite electrode by treating the surface with a NH_3-plasma (Fig. 4.3c).[56–59] These surface amino groups can then be coupled to carboxylic functions of enzymes.[60]

Functionalized olefins or acetylenes react with metal surfaces (Pt)[61–64] or carbon surfaces[59] to yield functionalized interfaces for the subsequent covalent coupling of enzymes. Immobilization of olefins on Pt requires the reduction of the metallic surface, where the pretreatment of the surface with the formation of radical sites is important for the binding of the olefins to carbon (Fig. 4.3d). For example, Pt electrodes were activated with acrylic acid, and glucose oxidase was covalently linked to the resulting monolayer.[60] The strong $\pi-\pi$ interactions between the basal surface of graphite electrodes and polycyclic aromatic systems,[65–67] allows the assembly of enzyme monolayers by the modification of the graphite electrode with a functionalized polycyclic aromatic compound. For example, pyrene carboxylic acid was linked to graphite via $\pi-\pi$ interactions and the modified interface was used for the linkage of proteins (Fig. 4.3e).[68]

The self-assembly of thiols or disulfides as monolayers on various conductive supports, such as Pt,[69–71] Ag,[72–74] Cu,[75-77] and particularly Au,[78–80] or semiconductive materials, such as GaAs[81] and InP[82] has been studied extensively in recent years.[83,84] Formation of Au–thiolate monolayers on the electrode upon the interaction of thiols with Au surfaces was attributed to the oxidation of the Au surface by the thiol [Eq. (2)][83]:

$$Au_n + R\!-\!SH \rightarrow Au_{n-1}Au^+\!-\!{}^-SR + \tfrac{1}{2}H_2 \qquad (2)$$

A suggested alternative mechanism for the formation of the thiolate monolayer on

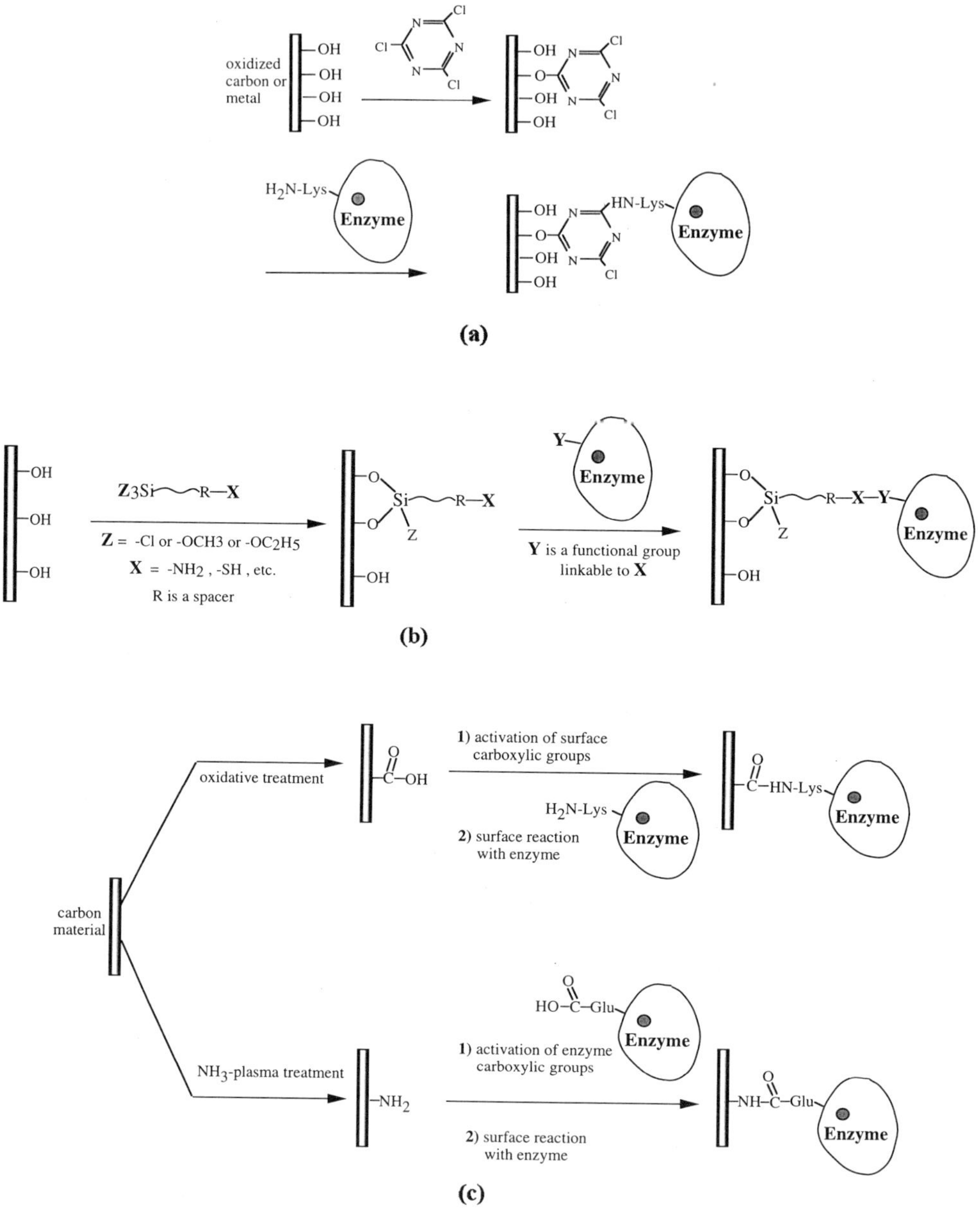

Figure 4.3. Different kinds of chemical attachment of enzyme molecules to functionalized electrode surfaces: (a) using cyanurylchloride; (b) using silanized surfaces; (c) using oxidized or aminated carbon surfaces; (d) using chemisorption of functionalized olefins; (e) using polyaromatic anchor groups adsorbed at a basal graphite surface; (f) using functionalized thiolated monolayers chemisorbed at a Au surface.

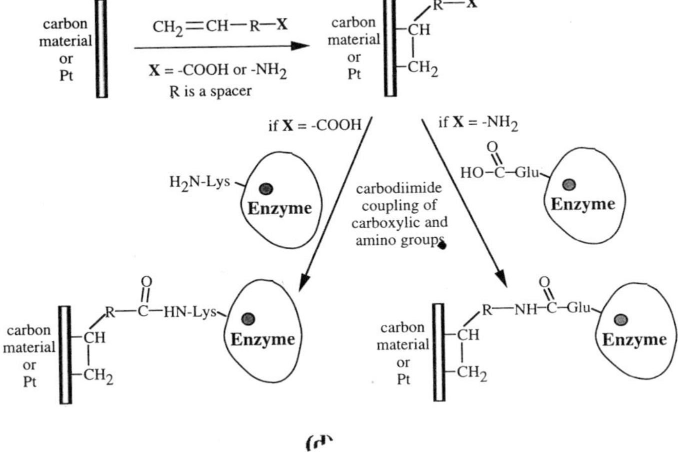

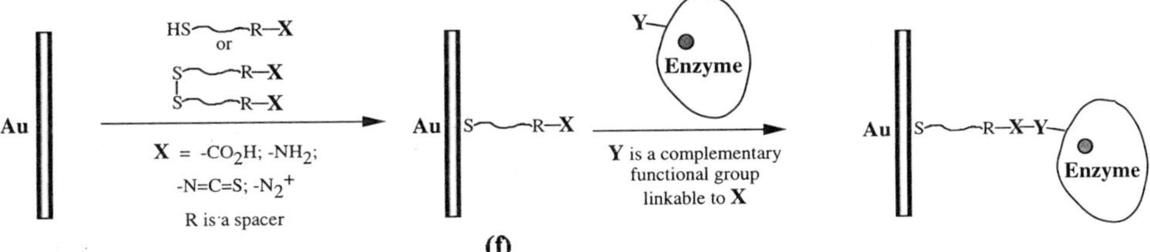

Figure 4.3.   Continued.

the surface involves the reduction of a surface oxide layer by the thiol [Eq. (3)][83]:

$$Au_{n-1}Au\text{—}OH + R\text{—}SH \rightarrow Au_{n-1}Au^{+}\text{—}^{-}SR + H_2O \qquad (3)$$

A similar thiolate monolayer is formed by the interaction of the Au electrode with a disulfide. This is attributed to the reduction of the S–S bond by the Au surface [Eq. (4)][83]:

$$Au_n + RS\text{—}SR \rightarrow Au_{n-2}(Au^{+}\text{—}^{-}SR)_2 \qquad (4)$$

Carboxylic acid residues,[85] active esters,[86] aldehyde,[88] amino,[71] isothiocyanate,[88] and diazonium ion[89] functions linked to thiolated organic substrates, were used to functionalize Au surfaces. The terminal functional groups of thiolated monolayers were employed for the covalent coupling of enzymes via complementary functional groups associated with the protein (usually lysine, glutamic/aspartic, or tyrosine residues) (Fig. 4.3f). For example, glutathione reductase (GR),[30,31] glucose oxidase (GOx),[90–92] glucose dehydrogenase,[93] or diaphorase[94] was covalently linked to monolayer-functionalized Au electrodes.

Enzyme electrodes can also be tailored by noncovalent interactions. Hydrophilic,[95] hydrophobic,[96–99] and electrostatic[100] interactions were used to immobilize proteins on functionalized surfaces. The organization of enzyme electrodes by complementary affinity interactions provides a general route for the assembly of stable enzyme electrodes. For example, an avidin-functionalized Pt electrode was used to bind biotin-labeled lactate oxidase (LOx), choline oxidase (ChO),[101] or alcohol oxidase (AOx)[102] (Fig. 4.4A). Similarly, horseradish peroxidase (HRP) was immobilized onto a carbon surface via the functionalization of the electrode with biotin and the conjugation of biotin-labeled HRP to the electrode by avidin (Fig. 4.4B).[103] The resulting enzyme electrode was used for the biocatalyzed oxidation of catechol. Antigen–antibody interactions enable an alternative approach to the tailoring of noncovalent enzyme electrodes.[104–106] For example, a glassy carbon electrode was functionalized with rabbit-IgG antibodies and a conjugate of glucose oxidase, and goat antirabbit IgG antibodies were linked to the electrode interface (Fig. 4.4C).[104] The resulting enzyme electrode was used for the bioelectrocatalyzed oxidation of glucose in the presence of ferrocene methanol as the diffusional electron mediator.

4.3. ELECTRICAL CONTACT OF MONOLAYER AND MULTILAYER ENZYME ELECTRODES

The development of amperometric biosensor devices requires electrical contact between the enzyme layer and the electrode surface.[15] Chemical modification of the enzyme with electron relay groups provides a method to "electrically wire" the insulating protein matrices.[27–31] Covalent attachment of electron mediators to the protein shortens the electron transfer distances and electron tunneling (or "hopping") through the relay units,[29] electrically communicates the enzyme redox site with its environment, i.e., electrode surface. The enzyme glutathione reductase (GR)

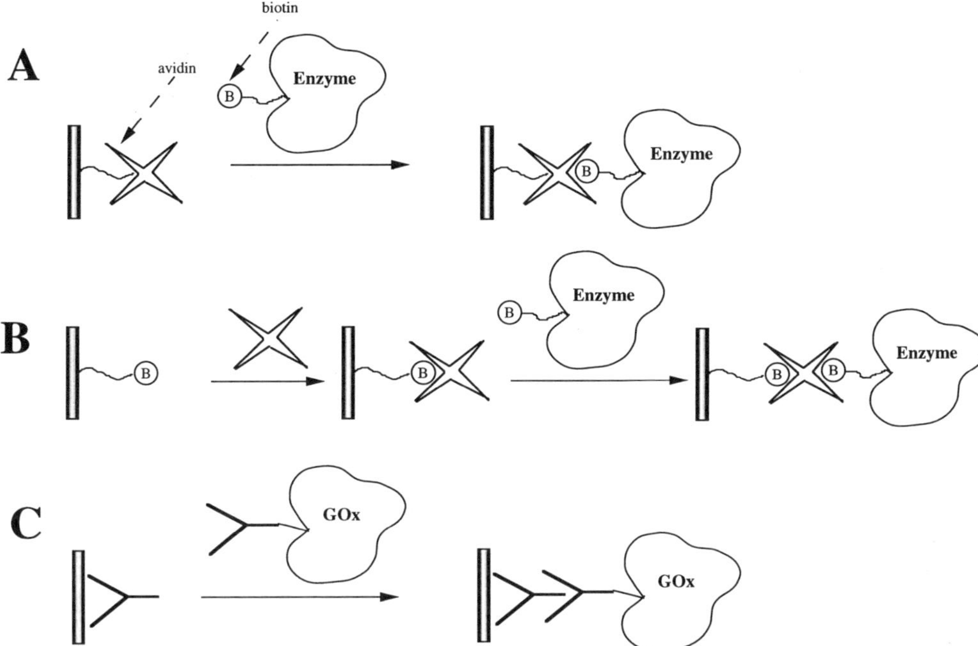

Figure 4.4. Monolayer enzyme immobilization based on affinity interactions: (A) interaction of biotinated enzyme molecules with an avidin-modified surface; (B) interaction of biotinated enzyme molecules with a biotin-modified surface via avidin located between them; (C) interaction of a conjugate of GOx and goat antirabbit IgG antibodies with an electrode surface functionalized with rabbit IgG antibodies.

was assembled as a monolayer on a Au electrode,[30,31] surface coverage 2×10^{-11} mole/cm^2 (Fig. 4.5A). The resulting protein was modified with the N-methyl-N'-carboxyalkyl-4,4'-bipyridinium (**1**) electron relay group. The relay units differ in the chain lengths that tether the electroactive group to the protein ($n = 2, 5, 11$). The resulting enzyme electrodes revealed bioelectrocatalytic features, and bioelectrocatalyzed reduction of oxidized glutathione (GSSG) was stimulated upon application of the potential characteristic to the reduction of the bipyridinium units ($E° = -0.58\,\mathrm{V}$ vs. SCE) (Fig. 4.5B).

The modification of the GR electrode by the bipyridinium units was performed in the presence of urea, a protocol that turned out to be an essential step in generating electrical contact in the resulting enzyme electrode. This was attributed to the partial unfolding of the protein by urea, which enabled the implanting of redox relay sites in inner protein positions, close to the enzyme redox center. Positioning of these relay groups at inner protein sites and at the enzyme periphery provides the pathway for electrical communication and activation of the enzyme. The effectiveness of the bioelectrocatalyzed reduction of GSSG is controlled by the chain length tethering the bipyridinium groups to the protein (Fig. 4.5B). The longer the tether, the more efficient the bioelectrocatalytic reduction of GSSG. This was attributed to improved electrical contact between the redox enzyme and the electrode surface with

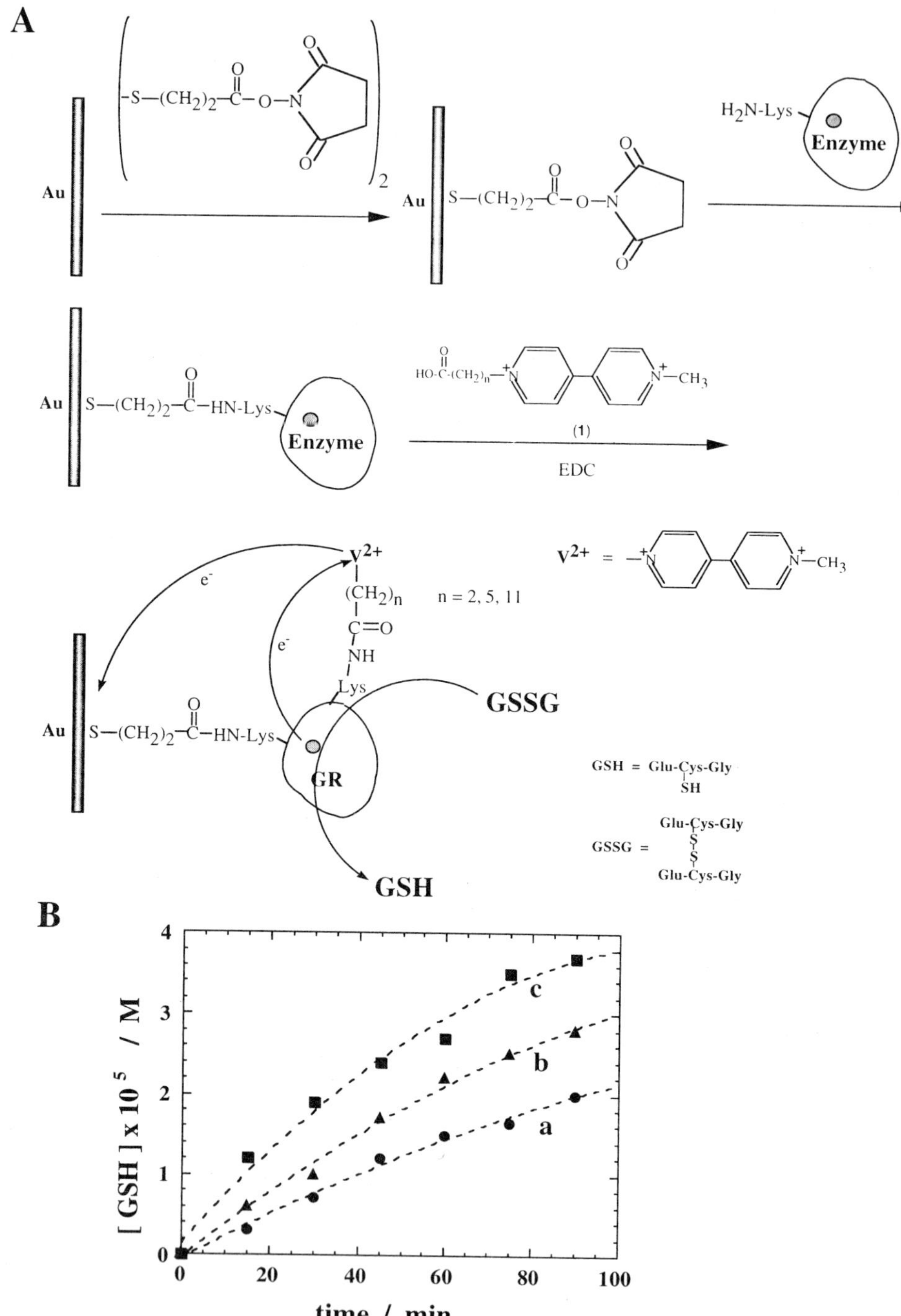

Figure 4.5.   (A) Schematic configuration of an electrically contacted GR monolayer electrode. (B) Rate of reduced glutathione (GSH) formation using GR-monolayer electrodes with bipyridinium tethers consisting of different chain lengths: [(a) $n = 2$; (b) $n = 5$; (c) $n = 11$]. For all systems [GSSG] = 1×10^{-2} M and the applied potential corresponded to $E = -0.72$ V vs. SCE.

longer tethers. With the longer tethers, the flexibility of the chains linking the redox relay to the protein enables the formation of shorter redox relay-site distances, which enhances electrical communication.

The protein content in a single monolayer is low, so that the electrical contact of the electrode can be detected by the steady-state accumulation of the product or by a very low amperometric response of the enzyme electrode. Enhanced amperometric responses of the electrode, and consequently higher sensitivities of the biosensor device, can be achieved either by increasing the protein content on the electrode surface or by improving the electrical contact between the redox site and the electrode.

It is possible to increase the enzyme content on the electrode by depositing multilayers of the enzyme. The thin enzyme layer can be organized on the electrode in a random, nonordered, configuration[91] or, alternatively, by a tailored approach with a controllable number of protein layers.[107,108] GOx and ferrocene ethylamine (**2**) were cross-linked in the presence of a Au electrode functionalized with a cystamine monolayer using glutaric dialdehyde as the coupling reagent.[91] The cross-linking leads to the coupling of **2** to the enzyme, to the interprotein coupling, and to the binding of the ferrocene-functionalized cross-linked array to the base cystamine monolayer associated with the electrode (Fig. 4.6). The resulting non-ordered layered enzyme electrode stimulates the bioelectrocatalyzed oxidation of glucose and reveals linear amperometric responses in the range of glucose concentrations corresponding to 0–10 mM.

Ordered layered assemblies on the electrode support can be organized by the stepwise construction of the enzyme layers by covalent[93,107,108] or noncovalent interactions.[102,105,106] The covalent organization of the enzyme net is exemplified here by the structure of a GOx multilayer electrode (Fig. 4.7).[107] GOx was covalently linked to a base cystamine monolayer associated with a Au electrode. By a two-step sequence of reactions with the bifunctional reagent *trans*-stilbene-(4,4′-diisothiocyanate)-2,2′-disulfonic acid and then with GOx, layers of the biocatalyst were assembled on the base enzyme film. The number of repeating cycles of this sequence of reactions controls the number of enzyme layers associated with the electrode, and enzyme electrodes with a predetermined number of layers can be prepared. The resulting layered enzyme electrodes were reacted with *N*-aminomethylferrocene caproic acid (**3**) in the presence of urea, to yield an electrically contacted GOx network.

The structure of the enzyme network shown in Fig. 4.7 is certainly over-simplified, and enzyme units within a layer or between layers can be cross-linked by the coupling reagent. However, analysis of the enzyme content in the multilayer assembly revealed an almost identical biocatalyst coverage per layer (7.5×10^{-12} mole/cm^2) up to a network that consists of eight GOx layers. Figure 4.8A shows the amperometric responses of the layered GOx electrodes consisting of variable numbers of enzyme layers in the presence of glucose. The catalytic anodic currents occur at the oxidation potential of ferrocene ($E° = 0.32$ V vs. SCE) implying that the enzyme networks are electrically contacted with the conducting support by the covalently linked ferrocene relay groups. This electrical communication activates the bioelectrocatalyzed oxidation of glucose. It has been demonstrated

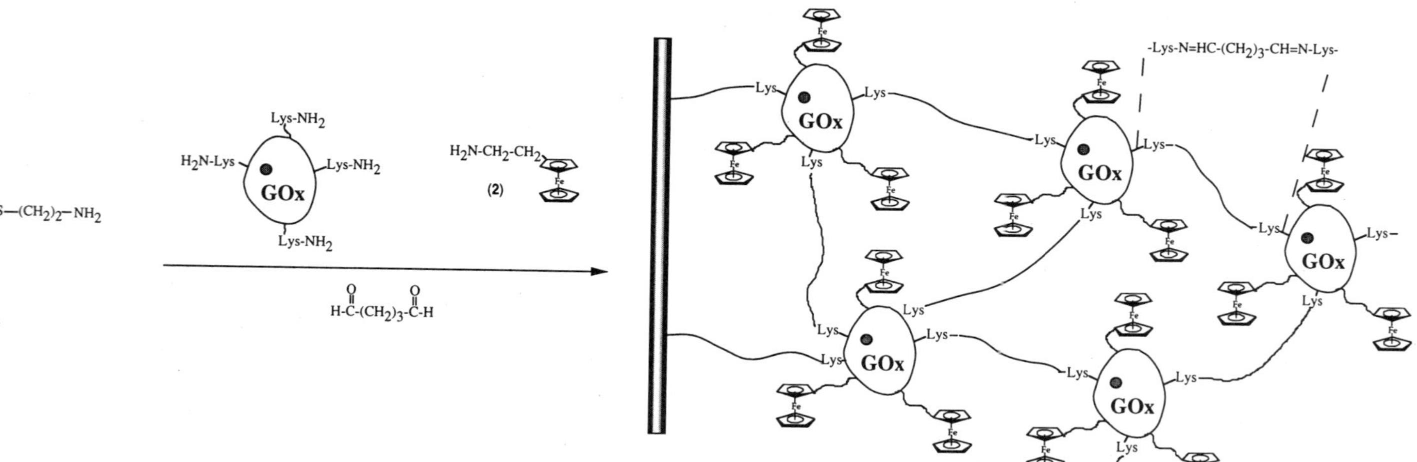

Figure 4.6. A nonordered enzyme electron relay multilayer electrode prepared by polycondensation of GOx aminoethylferrocene and glutaric dialdehyde at a cystamine-modified Au electrode surface.

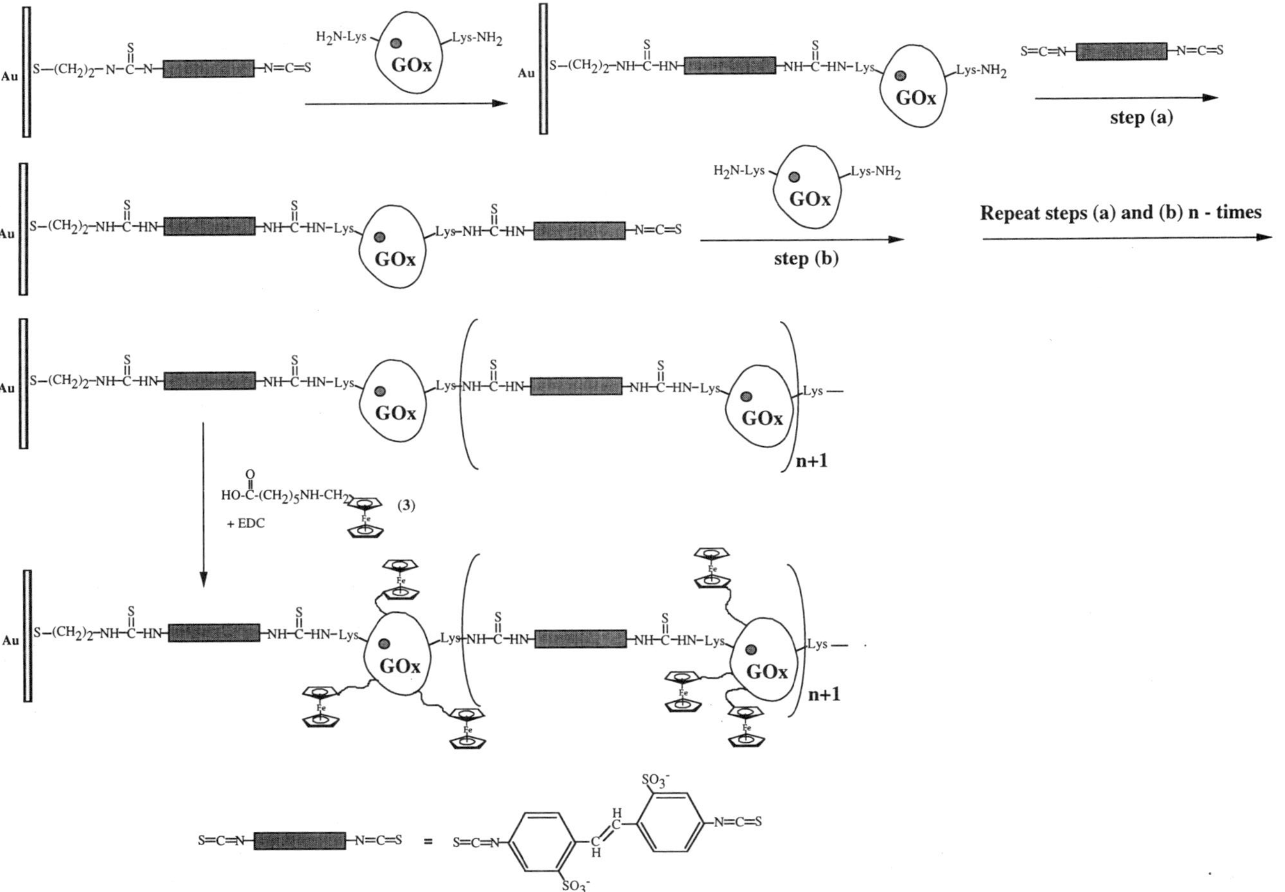

Figure 4.7. Assembly of an electrically contacted GOx ordered multilayer array on a base cystamine monolayer associated with a Au electrode.

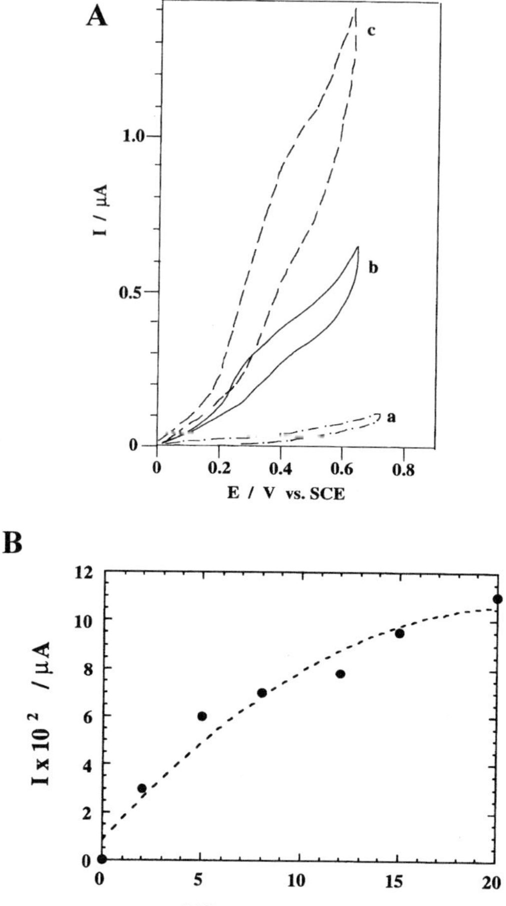

Figure 4.8. (A) Electrocatalytic currents developed by the electrically contacted layered GOx–Au electrode in the presence of 20 mM glucose [(a) 1, (b) 4, (c) 8 layers of GOx]; potential scan rate, 2 mV/s. Electrode area, ca. $0.2\,cm^2$, roughness coefficient, ca. 1.3]. (B) Amperometric responses of four-layer GOx array Au electrode as a function of glucose concentration. Current values were recorded by maintaining the enzyme electrode at a fixed potential (E = +0.4 V vs. SCE) and sequential injection of glucose into the electrochemical cell; the electrolyte consists of 0.1 M phosphate buffer, pH = 7.3.

that the transduced current is enhanced with an increase in the number of enzyme layers. This indicates that the sensitivity of layered enzyme electrodes can be tuned by the number of biocatalytic layers associated with the electrode support. Figure 4.8B shows the derived calibration curve that corresponds to the amperometric responses of a four-layer GOx electrode at different glucose concentrations.

The organization of ordered enzyme multilayer arrays opens the possibility of assembling multilayers of two or more different enzymes,[109] one of which is a redox biocatalyst electrically contacted with the electrode surface. The other enzyme(s) turns the analyte substrate into a product that is the substrate for the redox biocatalyst. The resulting transduced current is proportional to the concentration of the analyte. A multilayer electrode, consisting of choline oxidase (ChO) and acetylcholine esterase (AChE) was tailored according to this concept (Fig. 4.9A).[109] The analyte acetylcholine is hydrolyzed to choline in the presence of AChE, and the resulting product is bioelectrocatalytically oxidized by ChO to betaine, in the

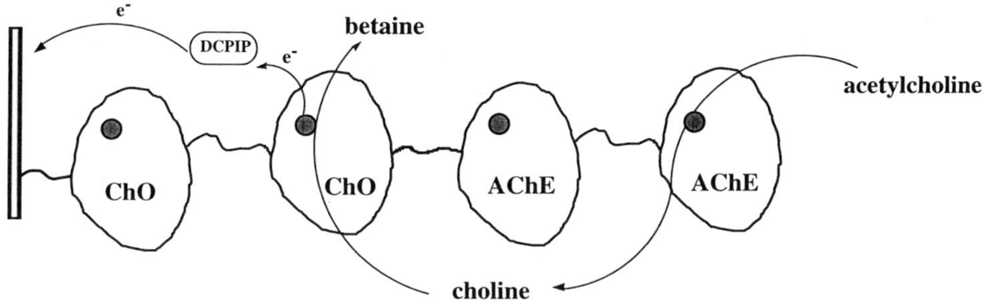

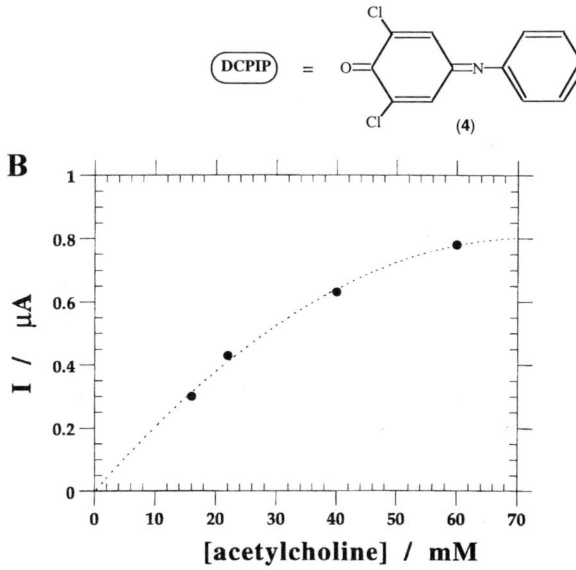

Figure 4.9. (A) Assembly of a bifunctional multilayer network consisting of ChO and AChE for the amperometric detection of acetylcholine. (B) Concentration dependence of the electrocatalytic current produced by the bifunctional ChO–AChE-modified electrode in the presence of acetylcholine; the electrode operated at an applied potential $E = +0.3\,V$ vs. SCE in the presence of $4 \times 10^{-5}\,M$ 2,6-dichlorophenol indophenol as a diffusional electron mediator.

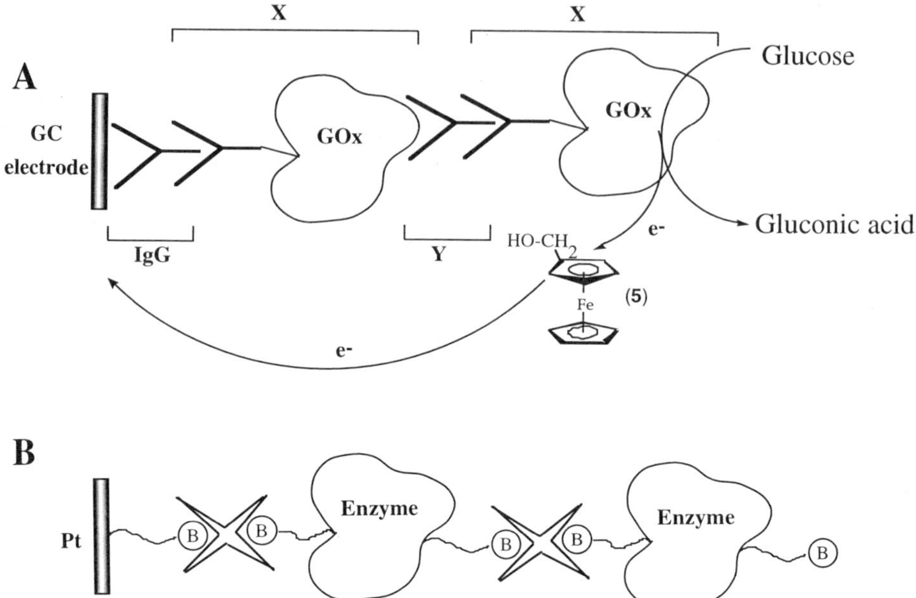

Figure 4.10. Assembly of a multilayer enzyme network using affinity interactions: (A) application of antigen/antibody/antiantibody interactions. (B) Application of biotin-avidin interactions.

presence of 2,6-dichlorophenol indophenol (DCPIP) (**4**) as diffusional electron mediator. Figure 4.9B shows the resulting currents at different concentrations of acetylcholine.

Substrate–receptor or antigen–antibody affinity interactions can be used for the noncovalent assembly of enzyme multilayers. For example,[105,106] a mouse IgG antibody was deposited onto a glassy carbon electrode and a conjugate (X) consisting of antimouse IgG antibody and GOx was linked to the base antibody. An IgG antibody produced in mouse (Y) against GOx was then linked to the enzyme antigen. The number of surface-deposited enzyme layers was controlled by repeated association of the antibody–GOx conjugate (X) and the GOx–antibody (Y) (Fig. 4.10A). The bioelectrocatalyzed oxidation of glucose occurs in the presence of ferrocene methanol (**5**), as the diffusional electron mediator. The magnitude of the amperometric response of the layered enzyme electrode is controlled by the number of layers associated with the electrode surface.

A related approach to assembling multilayer enzyme electrodes utilizes the high-affinity interactions between biotin and the tetradentate avidin receptor. This is exemplified[102] here with the assembly of alcohol oxidase (ALOx) on a Pt electrode (Fig. 4.10B). Avidin was deposited onto the Pt support and biotin-labeled ALOx was interacted with the modified electrode to yield an enzyme monolayer. The number of layers deposited on the electrode was controlled by the sequential treatment of the electrode with avidin and biotinylated ALOx. The multilayer enzyme electrode oxidizes alcohols with the concomitant generation of hydrogen peroxide [Eq. (5)]:

$$CH_3CH_2OH + O_2 \xrightarrow{\text{ALOx}} CH_3\overset{\overset{\displaystyle O}{\displaystyle \|}}{C}H + H_2O_2 \tag{5}$$

The current resulting from the oxidation of the H_2O_2 reflects the concentration of the analyte alcohol. The current resulting from the multilayer ALOx electrode is controlled by the number of biocatalytic layers associated with the electrode. A similar approach was used to develop lactate- and glucose-sensing electrodes in the presence of lactate oxidase (LOx) and GOx multilayer electrodes, respectively.[110–112]

The architecture of multilayer electrically contacted enzyme electrodes reveals two features that are important for biosensor devices: (1) It enables the assembly of a sensing electrode of tunable sensitivity based on the number of protein layers associated with the electrode. (2) It allows the organization of composite enzyme multilayers consisting of two (or more) biocatalysts, which broadens the possibilities of sensing enzyme electrodes as it allows the secondary electrochemical detection of a chemical component that is the product of a primary biocatalytic process.

Increasing the surface density of the bioactive material on the electrode support is essential for enhancing the sensor sensitivity. Another way to achieve this is the use of roughened large-surface-area electrodes (Fig. 4.11).[109] Treatment of Au surfaces with Hg results in roughening of the conducting support by the generation of an Au amalgam and its further dissolution.[113] Typically, base Au surfaces with a characteristic roughness factor of 1.2–1.5 can be roughened to surfaces exhibiting a roughness factor of 15–25. Multilayers of GOx were linked to smooth and rough Au electrodes by the coupling of GOx to cystamine-functional-ized surfaces, and electrical-contacting of the multilayer enzyme electrodes with ferrocene carboxylic acid (**6**) as the diffusional electron mediator.[109] The transduced

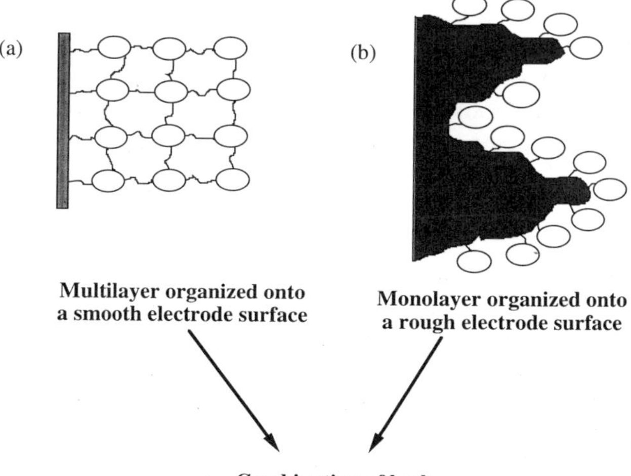

Figure 4.11. Increase of protein content of enzyme electrodes by: (a) multilayer enzyme network; (b) application of rough electrode support.

currents from a four-layer GOx electrode assembled on rough and smooth electrode surfaces at different concentrations of glucose are shown in Figs. 4.12A and B, respectively. The resulting currents of the four-layer roughened GOx electrodes are ca. six-fold enhanced as compared to the respective GOx multilayer on smooth Au supports (Fig. 4.12C).

4.4. ELECTRICALLY CONTACTED RECONSTITUTED ENZYME ELECTRODES

Another possibility for increasing the transduced currents of enzyme electrodes involves improving the electrical contact between the enzyme and the electrode surface. The electrical communication of the layered enzyme electrodes was accomplished by the random functionalization of proteins with electron relay groups. In fact, the detailed analysis of the transduced currents of GOx, modified by tethered ferrocene units at variable loading degrees enabled the estimation of the electron-transfer turnover in the various bioelectrocatalytic compositions.[29] The bioelectrocatalytic features of the ferrocene-functionalized GOx are delicately controlled by the degree of loading of the enzyme with the ferrocene units. An increase in relay loading improves the electron-transfer turnover rate owing to the shortening of the electron-transfer distances. However, high loading of the enzyme by the electrical relay components adversely affects the enzyme, and the bioelectrocatalytic properties of the enzyme are deactivated. This was attributed to the partial denaturation of the enzyme at high loadings. With the optimal bioelectrocatalyst, the electron-transfer turnover was only ca. 10% of the turnover of GOx with molecular oxygen, implying that the electrical communication between the enzyme redox site and the electrode limits the bioelectrocatalytic functions of enzyme electrodes. This relatively poor electrical communication was attributed to the fact that the electron relay units are randomly linked to the protein. Their position on the protein is not optimal for electrical communication, and extensive loading of the protein adversely affects the enzyme activity owing to partial denaturation.

Enhanced electrical contact between the redox enzymes and the electrodes has several important implications in biosensor technology. Enzyme electrodes exhibiting improved electrical communication would yield higher amperometric responses and therefore superior sensitivities. There are, however, other important aspects to such enzyme electrodes specifically related to their application as invasive sensors. For example, there has been an ongoing scientific and technological demand for the development of an invasive glucose sensor that is coupled to an insulin-containing pump by an electronic circuit for the continuous adjustment of blood-glucose levels.[25,26,114] As the pain accompanying the penetration of a needle electrode is proportional to its diameter, it is expected that microelectrodes will be used for invasive sensing. Redox enzymes with improved electrical contact are expected to yield physically detectable current responses even with small-surface-area microelectrodes, owing to their high electron-transfer turnover rates. A further aspect of enzyme electrodes with improved electrical communication relates to the specificity of the sensing interfaces. The activation of bioelectrocatalyzed oxidation or reduction processes by enzyme electrodes is often accompanied by nonspecific oxidation (or

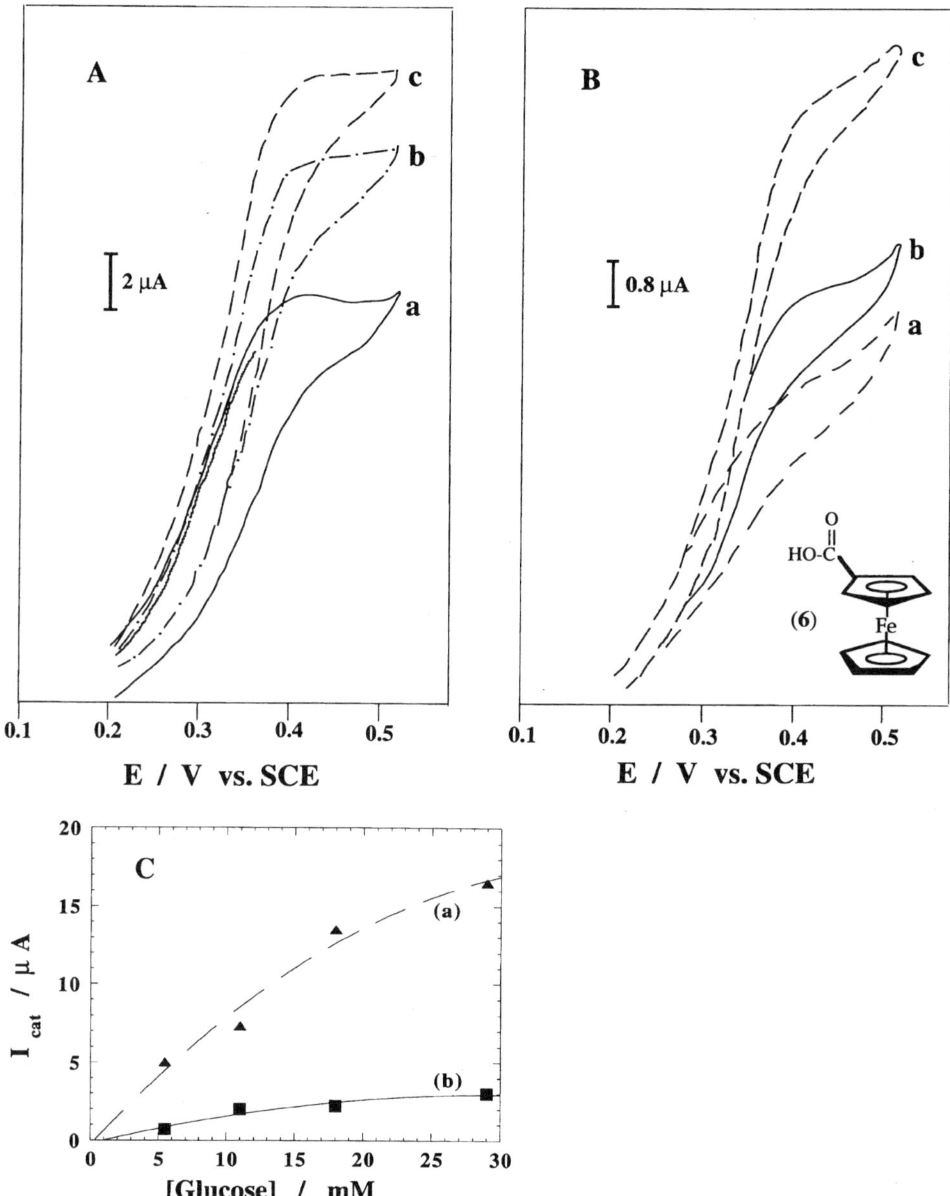

Figure 4.12. Electrochemical responses of a four-layer GOx electrode in the presence of 2.9×10^{-4} M ferrocenecarboxylic acid as a diffusional electron mediator and different concentrations of glucose: [(a) 0, (b) 6.8×10^{-3}, (c) 36×10^{-3} M]: (A) in the presence of a rough Au electrode (geometrical area $0.4\,\text{cm}^2$, roughness factor ca. 30); (B) in the presence of a smooth Au-electrode (geometrical area $0.4\,\text{cm}^2$, roughness factor ca. 1.2); (C) calibration curve of the amperometric response of a four-layer GOx electrode as a function of glucose concentration [(a) rough Au electrode, (b) smooth Au electrode]. All experiments were recorded in 0.1 M phosphate buffer, pH = 7.3, $35 \pm 0.5°$C, under Ar; potential scan rate, $2\,\text{mV/s}$.

reduction) of electroactive chemical ingredients available in the analyte samples.[26] For example, glucose- , lactate- , or bilirubin-sensing electrodes, based on the electrobiocatalyzed oxidation of analyte substrates by enzyme electrodes consisting of GOx, LOx, or bilirubin oxidase (BOx), is accompanied by the nonspecific oxidation of ascorbic acid, uric acid, and *p*-acetamole at the electrode interface. The latter compounds act as interferants, and the current originating from the interfering components perturbs the net amperometric responses originating from the bioactive sensing electrodes.[26] With effective electrically contacted bioactive electrodes, the current transduced by the electrobiocatalyzed reaction is expected to be overwhelmingly higher than the current contribution from the noncatalyzed redox reaction of the interfering compounds. Hence, such enzyme electrodes are expected to reveal high selectivity for the protein substrate. In addition, many oxidases, e.g., GOx, LOx, and BOx use molecular oxygen (O_2) as the oxidizing agent of the respective substrates. Thus, the currents transduced by the enzyme electrodes (especially in *in vivo* applications), are sensitive to the oxygen content in the surroundings.[25] With enzyme electrodes exhibiting efficient electrical communication, the electron-transfer turnover between the redox center and the electrode surface competes with the oxidation of the active site by oxygen. Consequently, little effect of oxygen on the resulting currents of such electrodes is expected.

It was suggested that the organization of monolayer enzyme electrodes by a reconstitution method (Fig. 4.13) could lead to enhanced electron transfer between the biocatalyst and the conductive support.[115] By this approach, an electron relay

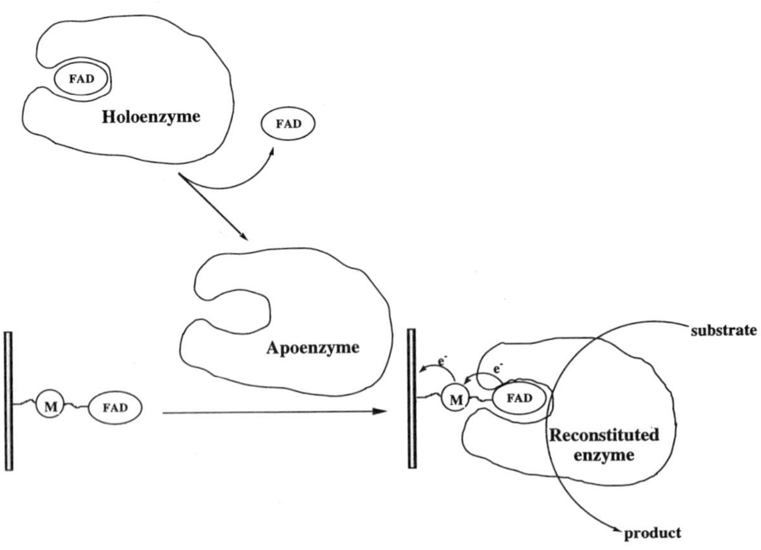

Figure 4.13. Obtaining an apo-flavoenzyme and its reconstitution on a relay-FAD monolayer linked to an electrode support.

is assembled as a monolayer on the electrode, and the native enzyme cofactor, i.e., the FAD cofactor, is covalently linked to the relay. The apo-redox enzyme (the protein lacking its native cofactor) is reconstituted onto the surface-bound FAD monolayer, which results in the alignment of the redox biocatalyst on the electrode and mediated electron transfer between the active site and the electrode occurs. Design of such electrodes signals a new level of molecular architecture of biomolecules on surfaces, as it enables the optimization of the electrical contact of the resulting enzyme electrode by systematic alteration of the distances separating the electrode–catalyst–FAD components.

The organization of an aligned reconstituted enzyme on an electron-relay–FAD monolayer was achieved by the reconstitution of the GOx apo-enzyme on a surface functionalized with a relay-FAD monolayer.[115] Figure 4.14 shows the method of assembly of the enzyme electrode. Pyrroloquinoline quinone (PQQ), was covalently linked to a base cystamine monolayer,[113] and N^6-(2-aminoethyl)-FAD was bound to the PQQ redox-relay units. Apo-GOx was then reconstituted[115] onto the semisynthetic FAD unit to yield an immobilized biocatalyst on the electrode with a surface coverage corresponding to ca. 2×10^{-12} mole/cm^2. The resulting reconstituted enzyme reveals bioelectrocatalytic properties.

Figure 4.15A shows the cyclic voltammograms of the enzyme electrode in the absence and presence of glucose. With the substrate, an electrocatalytic anodic current is observed, implying electrical contact between the reconstituted enzyme and the electrode surface leading to the bioelectrocatalyzed oxidation of glucose. Control experiments revealed that reconstituted GOx on the FAD monolayer lacking the PQQ component does not exhibit direct electron-transfer communication with the electrode surface. Thus, it was concluded that the redox relay unit consisting of PQQ is a key component in mediating electron transfer between the biocatalyst and the electrode. The PQQ site, located at the protein periphery, is constantly oxidized by the electrode, and PQQ-mediated oxidation of the FAD center activates the bioelectrocatalytic oxidation of glucose. The resulting electrical current is controlled by the recycling rate of the reduced FAD by the substrate or the concentration of glucose. Figure 4.15B shows the derived calibration curve for the amperometric responses of the reconstituted enzyme electrode at different concentrations of glucose. The resulting current densities are unprecedentally high in the area of biosensors and bioelectronic devices ($300\,\mu$A/cm^2 at 80 mM of glucose).

The electron-transfer turnover rate of GOx with molecular oxygen as the electron acceptor corresponds to ca. $600\,$s^{-1} at 25°C. Using the activation energy of $7.2\,$kcal/mole, the electron-transfer turnover rate of GOx at 35°C is estimated[115] to be ca. $900\,$s^{-1}. A densely packed monolayer of GOx (ca. 2×10^{-12} mole/cm^2) that exhibits the theoretical electron-transfer turnover rate is expected to yield an amperometric response of ca. $300\,\mu$A/cm^2. This indicates that reconstituted GOx on the PQQ–FAD monolayer exhibits an electron-transfer turnover with the electrode that is as effective as that observed with the enzyme and oxygen as the mediator. Apart from the high sensitivity of the resulting enzyme electrode, the unprecedented efficiency of electrical contact yields a selective, oxygen-insensitive sensing electrode. Figure 4.16 shows the amperometric responses of the

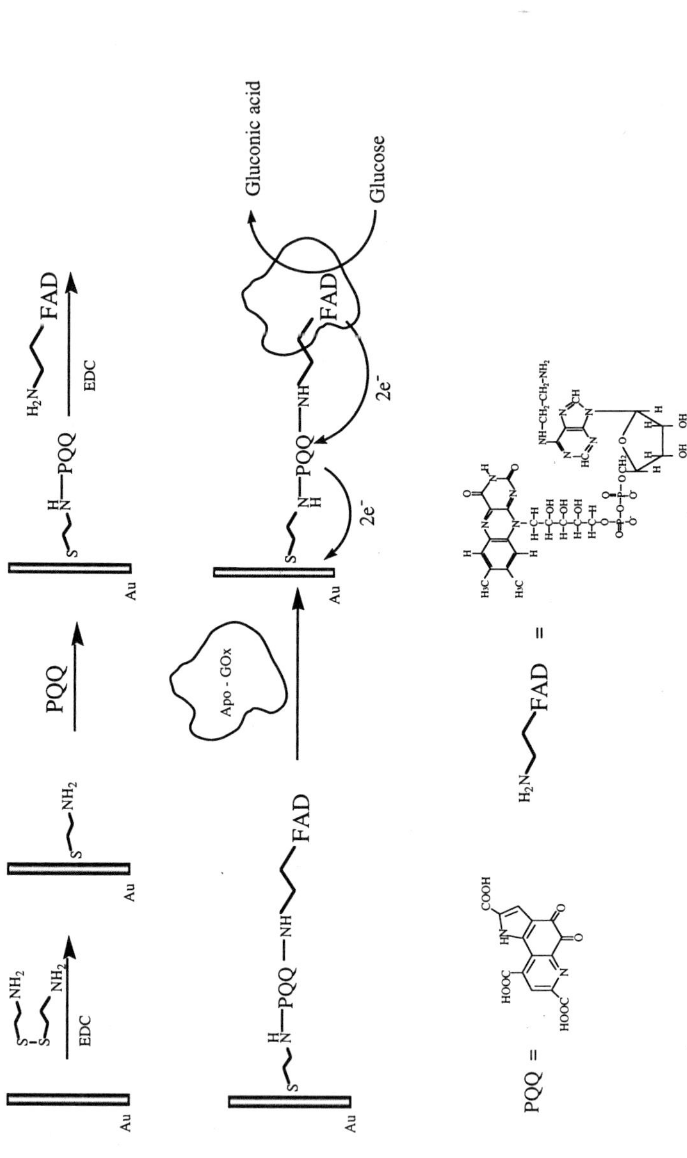

Figure 4.14. Assemblage of PQQ–FAD monolayer on a Au electrode and reconstitution of a GOx apoenzyme on the PQQ–FAD monolayer.

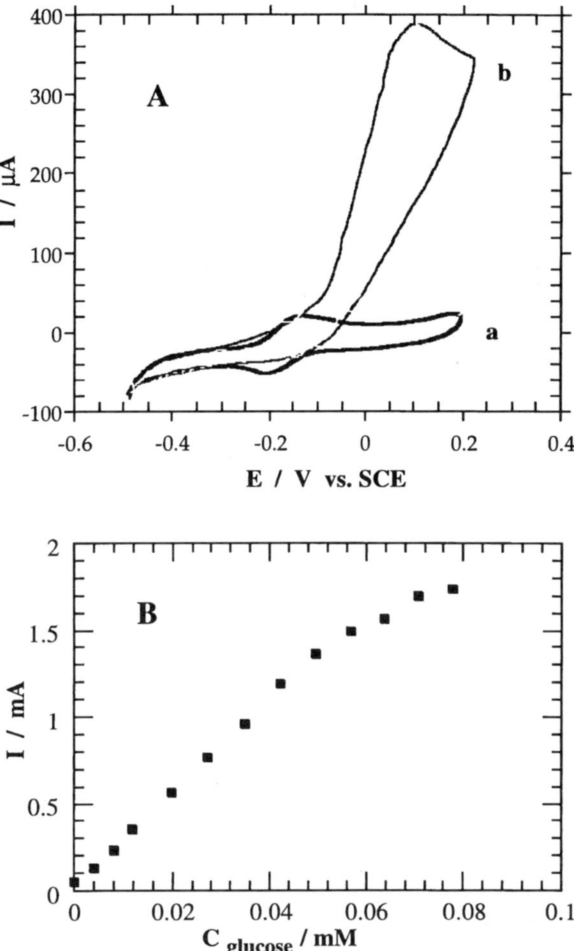

Figure 4.15. (A) Cyclic voltammograms of the PQQ-FAD-surface-reconstituted GOx electrode in the absence of glucose (curve a) and in the presence of glucose, 80 mM (curve b); data recorded under argon, 35°C, 0.01 M phosphate buffer and 0.1 M sodium sulfate, pH = 7.0, scan rate 5 mV/s. Geometrical electrode area, 0.4 cm^2, roughness factor 15 ± 5. (B) Amperometric responses of the reconstituted enzyme electrode at different glucose concentrations; data recorded by chronoamperometric measurements at a final potential +0.2 V vs. SCE, 35°C.

reconstituted GOx monolayer electrode in the presence of glucose upon the addition of ascorbic acid or the monitoring of the transduced amperometric signal in the absence or presence of oxygen. The transduced currents are almost identical in their magnitudes, implying that the sensing enzyme electrode is insensitive to these interferants.

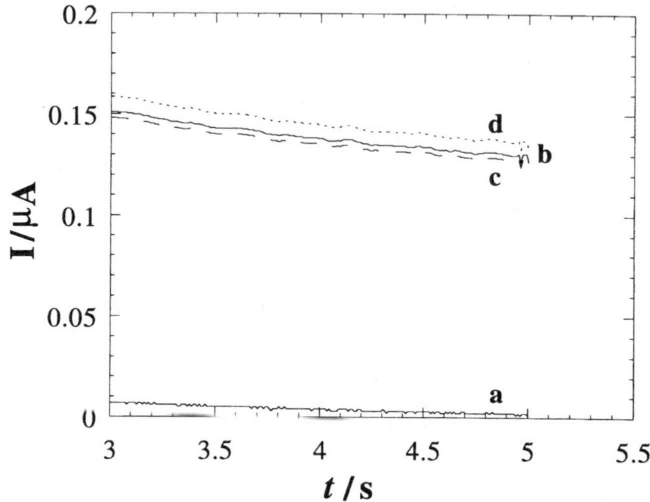

Figure 4.16. Amperometric responses produced by the GOx reconstituted with a PQQ–FAD mono-layer in the absence of glucose (a), in the presence of glucose, 50 mM, in the absence of O_2 (b), in the presence of glucose, 50 mM, in a solution saturated with air (c), in the presence of glucose, 50 mM, and ascorbic acid, 0.1 mM, in the absence of O_2 (d). Currents determined by chronoamperometry at final potentail 0.0 V vs. SCE; electrolyte composed of 0.01 M phosphate buffer and 0.1 M sodium sulfate, pH = 7.0, 35 ± 0.5°C; electrode geometrical area, 0.4 cm^2; roughness factor, ca. 15 ± 5.

4.5. INTEGRATED LAYERED NAD(P)$^+$-DEPENDENT ENZYME ELECTRODES

The development of amperometric biosensor devices has concentrated to date on applications for redox enzymes capable of electrical contact with electrodes by using synthetic electron mediators.[18] These enzymes usually include molecular cofactors embedded in the proteins, such as FAD, PQQ, or Fe–S clusters. The vast majority of redox enzymes however, are dependent on the diffusional cofactors: NAD$^+$ (**7**), and NADP$^+$ (**8**). These NAD(P)$^+$ cofactors stimulate the oxidation of the enzyme redox site by a diffusional pathway, and the resulting reduced cofactors, NAD(P)H, that diffuse out of the protein are oxidized by another reductive biocatalytic transformation. Substitution of the native oxidized cofactor, NAD(P)$^+$, or the reduced cofactor, NAD(P)H, by artificial electron carriers is usually impos-sible.[18] Only in a few cases were synthetic electron mediators successfully applied to activate redox processes with NAD(P)$^+$-dependent enzymes, i.e., the *N,N'*-dimethyl-4,4'-bipyridinium radical cation mediated reduction of nitrate,[34,35] or oxidized glutathione[30,31] in the presence of nitrate reductase or glutathione reductase, respectively, that belong to the series of viologen-accepting pyridine nucleotide oxidoreductases.[116]

Electrochemical activation of NAD(P)$^+$-dependent enzymes and their applica-tion as bioactive sensing materials is shown schematically in Fig. 4.17. Electroreduction

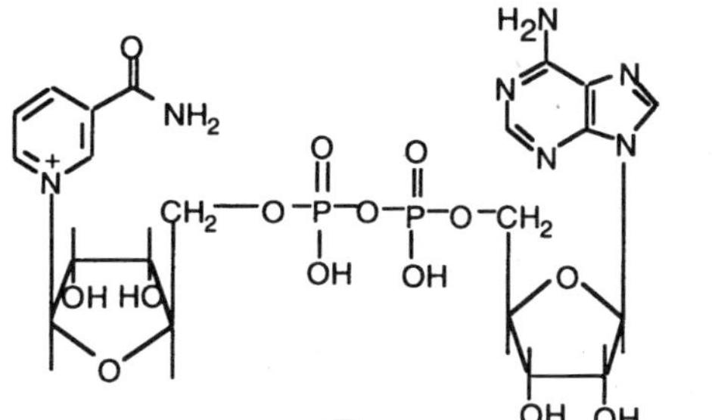

(7)

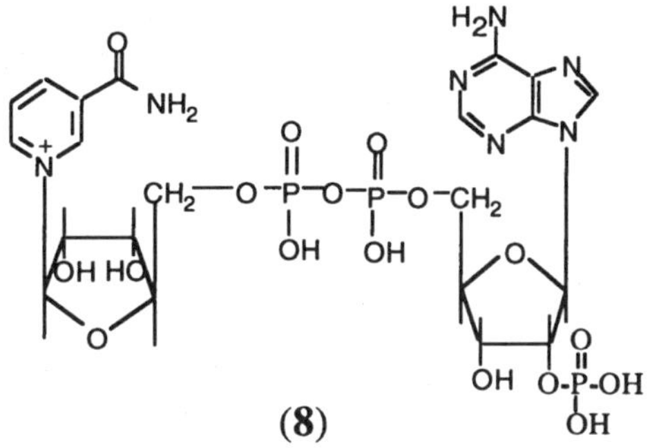

(8)

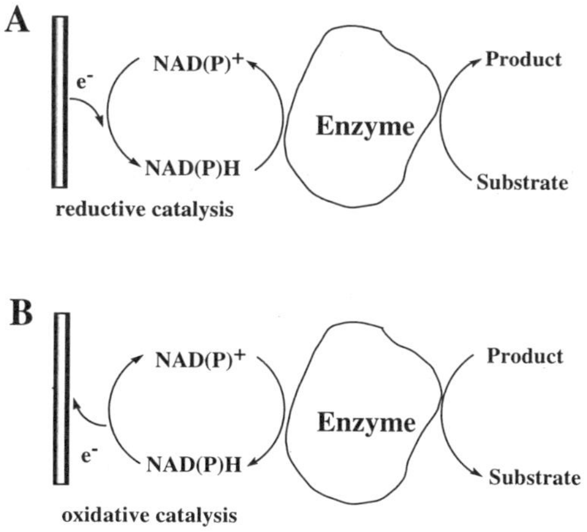

Figure 4.17. Electrochemical activation of NAD(P)$^+$-dependent enzymes: reductive (A) and oxidative (B) bioelectrocatalyzed processes.

of the NAD(P)$^+$-cofactor yields NAD(P)H, which stimulates the biocatalyzed reduction of the substrate (Fig. 4.17A).[117,118] Provided the formation of NAD(P)H is not rate-limiting, the resulting current is controlled by the enzyme substrate. Alternatively, the enzyme can be used in the oxidative pathway (Fig. 4.17B). Biocatalyzed oxidation of the substrate is accompanied by the formation of NAD(P)H. Electrochemical oxidation of NAD(P)H regenerates the cofactor and the transduced current is controlled by the substrate analyte.

These approaches to tailoring NAD(P)$^+$/NAD(P)H-dependent enzyme-based biosensors suffer from several limitations. Direct, noncatalyzed oxidation of NAD(P)H and reduction of NAD(P)$^+$ proceed with large overpotentials at very positive and very negative potentials, respectively.[119] Noncatalyzed reduction of NAD(P)$^+$ also results in the formation of radical products, and dimerization of these products results in the production of enzymatically nonactive dimers.[18,118] Thus, in order to apply NAD(P)$^+$/NAD(P)H-dependent enzymes as active biointerfaces in electrochemical biosensors, the introduction of appropriate catalysts for reduction of NAD(P)$^+$ and oxidation of NAD(P)H is important.[18,120] For example, Rh complexes[117,118] were applied to catalyze the reduction of NAD(P)$^+$ to NAD(P)H. Many different materials, including *o*-quinones,[121–125] *p*-quinones,[126,127] phenazine, phenoxazine and phenothiazine derivatives,[128–134] ferrocenes,[135] or Os complexes[136] were applied as electrocatalysts for the oxidation of NAD(P)H and the regeneration of bioactive NAD(P)$^+$ cofactors. For the practical application of NAD(P)$^+$-dependent enzymes as bioactive sensing matrices, integration of the electrocatalyst, the cofactor, and the enzyme into a united assembly that allows the electrochemical activation of the biocatalyst is essential. This originates from the fact that the cost of NAD(P)$^+$/NAD(P)H cofactors is high and their use as diffusional components in biosensor devices is economically impractical. Different configurations that employ polymeric NAD(P)$^+$ cofactors[137–143] entrapped with the respective biocatalysts in cross-linked polymers or the trapping of the polymeric NAD(P)$^+$ and enzyme by means of a membrane were used to confine the cofactor and biocatalyst at the electrode surface.[144–146] Covalent coupling of NAD$^+$ derivatives to NAD$^+$/NADH-dependent enzymes was also applied to activate them.[147–151]

Recent advances in the functionalization of electrodes by electroactive monolayers[83,84] and specifically electrocatalyst monolayers for the oxidation of NAD(P)H,[66,120–136] have provided new means of fabricating integrated electrodes with NAD(P)$^+$-dependent enzymes. PQQ was assembled as a monolayer on Au electrodes by its covalent coupling to a cystamine monolayer.[113] The quinone monolayer revealed electrocatalytic properties for NAD(P)H oxidation,[125,152] particularly in the presence of added Ca^{2+}-ions.[125] The electrocatalytic features of the PQQ monolayer were used to develop an integrated, electrically controlled PQQ–NAD$^+$ enzyme-layered electrode.[19] This will be illustrated here by describing an integrated lactate-sensing electrode using the NAD$^+$-dependent lactate dehydrogenase (LDH) as the bioactive interface. A bifunctional redox-active PQQ–NAD$^+$ monolayer was assembled on a Au electrode by the covalent linkage of N^6-(2-aminoethyl)-NAD$^+$ (**9**) to the PQQ monolayer (Fig. 4.18). Interaction of the PQQ–NAD$^+$ monolayer electrode with the NAD$^+$-dependent LDH led to the

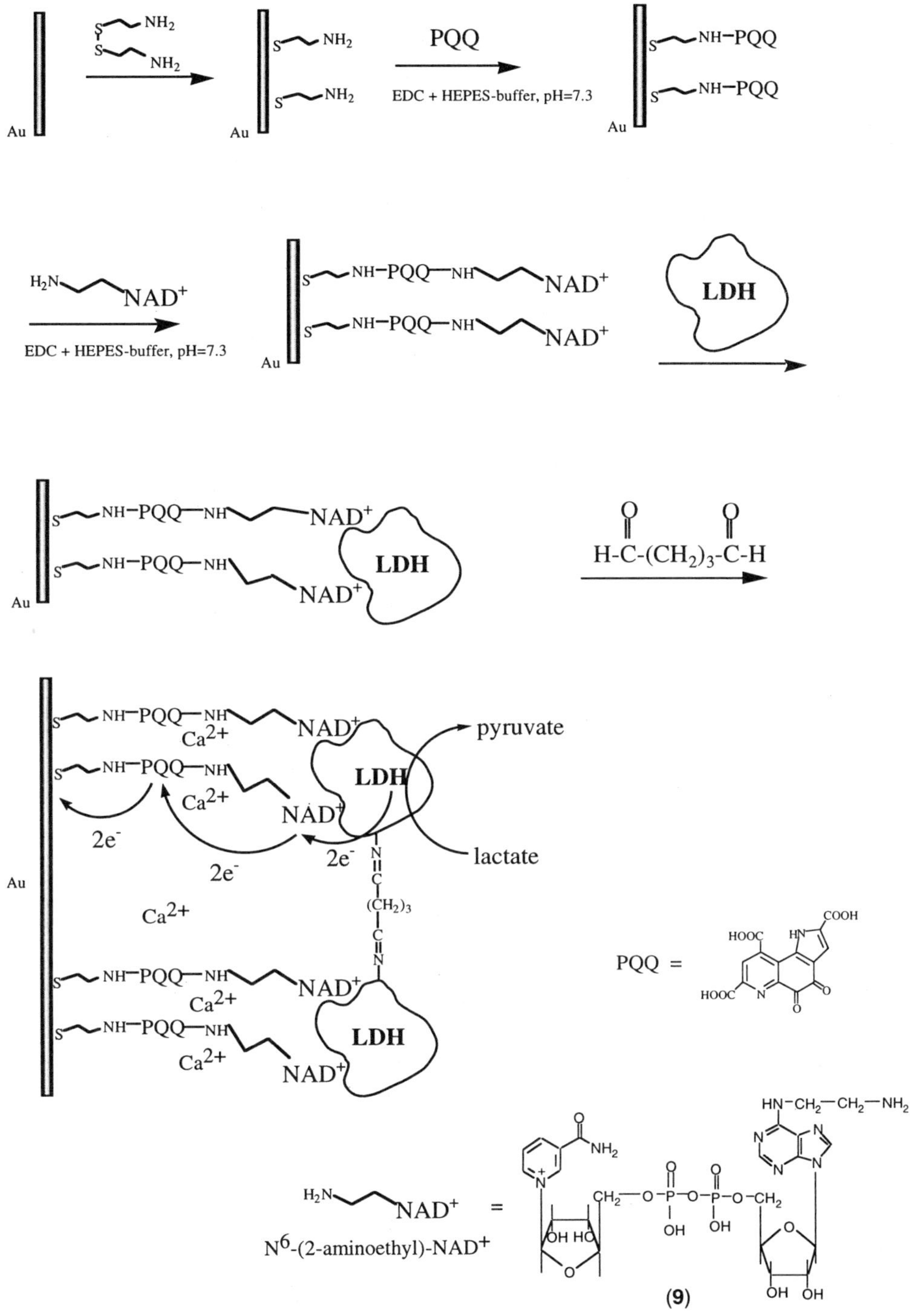

Figure 4.18. Stepwise organization of an integrated PQQ–NAD$^+$/LDH electrode and bioelectrocatalyzed oxidation of lactate.

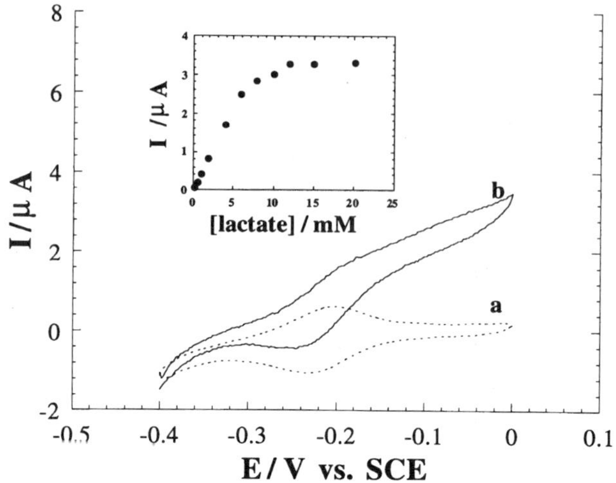

Figure 4.19. Cyclic voltammograms of the PQQ–NAD$^+$/LDH-modified Au electrode (roughness factor, ca. 15): (a) with the background solution only, (b) in the presence of 20 mM lactate [Background electrolyte: 0.1 M tris-buffer and 10 mM CaCl$_2$, pH 8.0; potential scan rate, 2 mV/s]. Inset: Currents generated by the integrated PQQ–NAD$^+$/LDH-modified Au electrode in the presence of different concentrations of lactate; applied potential $E = 0.1$ V vs. SCE.

selective association of the biocatalyst to the NAD$^+$ cofactor units by affinity associative interactions. Microgravimetric quartz crystal microbalance experiments revealed that the coverage of the surface-associated LDH by the respective affinity interactions is 3.5×10^{-12} mole/cm^2. The PQQ–NAD$^+$-bound LDH exhibited bioelectrocatalytic features for the oxidation of lactate. The surface-bound LDH revealed temporary stability on the electrode surface, and only ca. 10% of the biocatalyst is dissociated from the electrode within 30 min. However, the surface-bound LDH was immediately detached from the surface upon addition of diffusional NAD$^+$ to the electrolyte solution, implying the labile association of LDH to the monolayer electrode. The temporary stability of the monolayer of LDH associated with the PQQ–NAD$^+$ interface enabled the cross-linking of the enzyme layer with glutaric dialdehyde (Fig. 4.18). The resulting enzyme-monolayer electrode represents an electrically integrated NAD$^+$-dependent enzyme electrode. The enzyme is not dissociated from the electrode either by itself or upon the addition of diffusional NAD$^+$. The resulting PQQ–NAD$^+$-cross-linked enzyme assembly is electroactive for the oxidation of lactate (Fig. 4.19). The inset in the figure shows the calibration curve that corresponds to the amperometric response of the electrode at different lactate concentrations. This approach appears to be general, and a similar pathway was used to assemble an alcohol-sensing electrode using the NAD$^+$-dependent alcohol dehydrogenase as the bioactive interface for the analysis of ethanol.[19]

4.6. *LAYERED ANTIGEN MONOLAYER ELECTRODES FOR ELECTROCHEMICAL PROBING OF ANTIGEN–ANTIBODY INTERACTIONS*

In the previous sections, the electrical contact of redox proteins and electrode supports was used to probe the analyte substrate by the corresponding electrobio-catalyzed oxidation (or reduction) of the substrate. Now the electron transfer at the electrode/solution interface will be used to probe the formation of antigen–antibody complexes in a layered configuration on the electrode surface.[153–158] Figure 4.20A shows a basic configuration for the electrochemical sensing of an antibody.[159,160] A low-molecular-weight antigen is assembled as a monolayer on the electrode surface. A redox couple (R^+/R) solubilized in the electrolyte solution generates an amperometric response upon application of the appropriate potential on the electrode. Complexation of the antigen monolayer electrode with the complementary antibody results in the electrical insulation of the electrode with respect to the redox probe.

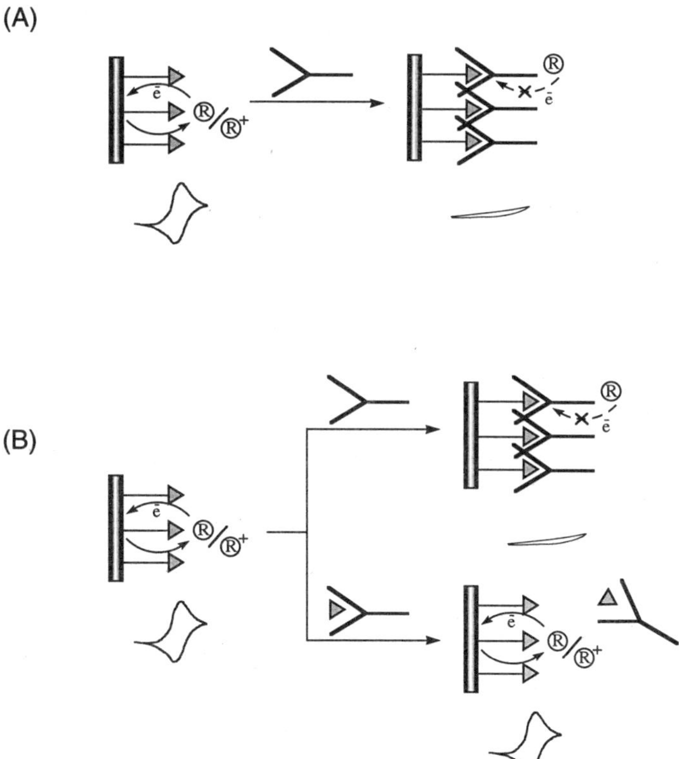

Figure 4.20. (A) Electrochemical analysis of an antibody by an antigen-monolayer-modified electrode, using a redox probe in the electrolyte solution. (B) Electrochemical analysis of an antigen by an antigen-monolayer-modified electrode in the presence of a constant antibody concentration.

As a result, the generated amperometric response decreases. The extent of electrode insulation correlates with the antibody concentration in the solution, and hence the extent of decrease in the amperometric response reflects the concentration of the analyte antibody. This approach was successfully applied[159,160] to sense the dinitrophenyl antibody (DNP-Ab). Figure 4.21 shows the assembly of the Nε-2,4-dinitrophenyl lysine antigen monolayer on a Au electrode. The extent of insulation of the electrode surface by the DNP-Ab was monitored by following the decrease (ΔI) in the amperometric responses of the electrode at various antibody concentrations, using $Fe(CN)_6^{3-}/Fe(CN)_6^{4-}$ as redox probe (Fig. 4.22A). Figure 4.22B shows the derived calibration curve that corresponds to the transduced currents of the antigen monolayer electrode at different DNP-Ab concentrations.

The antigen monolayer electrode can also be used as an active sensing interface for the antigen itself (see Fig. 4.20B). The sample is interacted with a fixed amount of the antibody for a constant incubation time, and then the antigen-monolayer-functionalized electrode is treated with the antibody sample mixture. A sample that includes the antigen results in the formation of the antibody–antigen complex. As the antibody is occupied by the antigen in the sample, no insulation of the electrode will occur. A sample that lacks the antigen will result in a vacant antibody in the probe solution, and in insulation of the electrode. The sensitivity range is controlled by the concentration of the antibody in the probe solution. Although this method was used successfully for the electrochemical analysis of dinitrophenol by the DNP-Ab and the Nε-2,4-dinitrophenyl lysine monolayer shown in Fig. 4.21, using $Fe(CN)_6^{3-}/Fe(CN)_6^{4-}$ as the redox indicator, it suffers from several limitations: (1) The differences in the transduced currents, owing to insulation of the electrode, are quite small, and hence an amplification in the transduced currents is desirable. (2) The assembly of the antigen monolayer is accompanied by the formation of "pinhole" surface defects that lack the antigen. A low-molecular-weight redox probe yields an electrical response at these microscopic defects. As a result, the real extent of surface insulation is not transduced by the electrode.

To amplify the effect of the formation of an antigen–antibody complex at the electrode surface, an electrocatalyst was coimmobilized with the antigen as a mixed- monolayer assembly on the electrode surface.[161] The redox probe in the electrolyte solution does not undergo its thermodynamically feasible electrical process owing to overpotential limitations, but the electrocatalyst associated with the

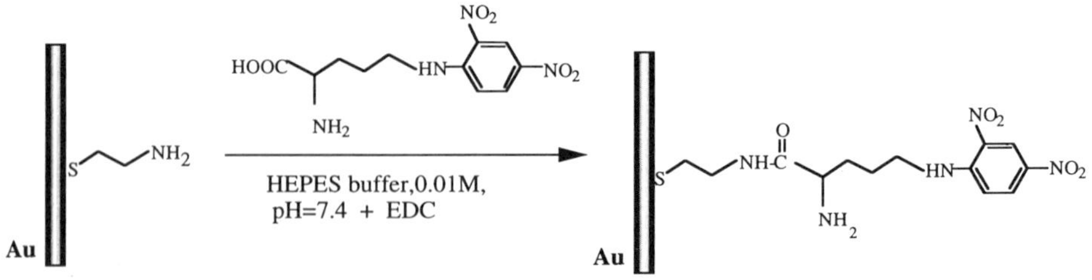

Figure 4.21.   Assembly of a Nε-2,4-dinitrophenyl lysin monolayer onto a Au electrode.

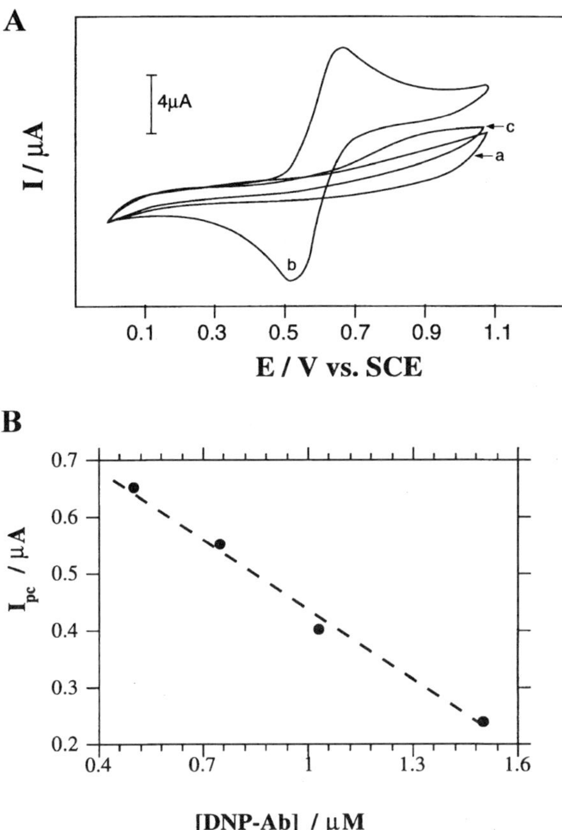

Figure 4.22. (A) Cyclic voltammograms of Nε-2,4-dinitrophenyl lysine monolayer-modified Au electrode: (a) in the presence of the background electrolyte composed of 0.1 M phosphate buffer, pH = 7.0; (b) upon addition of 1.1×10^{-3} M $K_3[Fe(CN)_6]$ to the electrolyte solution; (c) upon treatment of the electrode with DNP-Ab, 15 µM, for 15 min and in the presence of $K_3[Fe(CN)_6]$, at 35°C; potential scan rate, 100 mV/s. (B) Amperometric responses of the Nε-2,4-dinitrophenyl lysine monolayer-modified Au electrode at different DNP-Ab concentrations used for the electrode treatment; applied potential $E = 0.8$ V vs. SCE.

layered assembly mediates the redox reaction of the probe. This yields an amplified current from the system owing to the electrocatalytic reaction at the electrode surface. Interaction of the electrode with the antibody and formation of the layered antigen–antibody complex perturbs the electrocatalyzed reaction, and the transduced current is blocked. This method was used to sense DNP-Ab by a mixed monolayer consisting of Nε-2,4-dinitrophenyl lysine and PQQ (Fig. 4.23).[161] Electrocatalyzed oxidation of NADH by the PQQ component of the monolayer and CaCl$_2$ as cocatalyst provides the electrochemical transduction signal. Figure 4.24A shows the cyclic voltammograms of the antigen and electrocatalyst mixed monolayer in the presence of NADH before (curve b) and after (curve c) treatment of the electrode with the DNP-Ab. Association of the antibody to the electrode blocks the

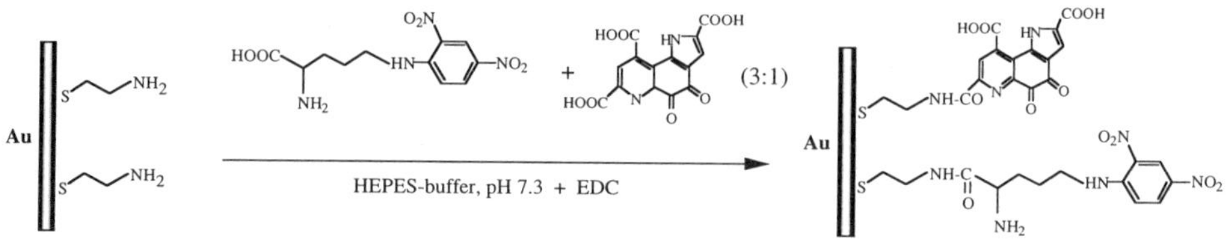

Figure 4.23.   Assembly of a mixed monolayer of Nε-2,4-dinitrophenyl lysine and PQQ on a Au electrode.

electrocatalyzed oxidation of NADH. The extent of the electrode insulation by the antibody is controlled by the bulk antibody concentration in the sample, and a typical calibration curve is shown in Fig. 4.24B. Other electrocatalysts and redox probes, such as microperoxidase-11 and $S_2O_8^=$ were used to probe the antigen–antibody complex at the antigen monolayer interface.[161]

Another way to amplify the electrical signal transduced by the formation of an antigen–antibody complex at the electrode surface includes the use of a redox enzyme.[160,162] Owing to the macromolecular structure of the protein, its redox response is insensitive to molecular defects in the antigen monolayer assembly. Figure 4.25 shows two configurations of antigen-monolayer-functionalized electrodes for the electrochemical analysis of the antibody by a redox enzyme acting as a probe.[160,162] One configuration includes the coassembly of the low-molecular-weight antigen and an electron mediator group as a mixed monolayer on the electrode (Fig. 4.25A). In the presence of the enzyme, mediated electrobiocatalyzed oxidation of the substrate proceeds and an electrocatalytic current is developed. Association of the antibody to the functionalized electrode blocks the electrical contact between the electron mediator and the redox enzyme, and the amperometric response of the electrode is inhibited. This method was used to sense the DNP-Ab by a mixed monolayer consisting of Nε-2,4-dinitrophenyl lysine antigen sites and ferrocene electron mediator groups (Fig. 4.26)[162]. GOx and its substrate glucose were used to probe the formation of the antigen–antibody complex. The extent of electrode insulation, reflected by the effectiveness of electrical contacting of the electrode and the probing redox enzyme (Fig. 4.27) points at the derived calibration curve that shows the progressive insulation of the electrode surface as the bulk concentration of DNP-Ab increases (Fig. 27 inset). At an antibody concentration of ca. 5 μg/mL the electrical contact between the enzyme and the electrode is almost entirely blocked.

An alternative method for sensing the antigen–antibody complex at the monolayer-functionalized electrode is shown in Fig. 4.25B. An "electrically-wired" enzyme that consists of a redox enzyme functionalized by electron relay groups is used to probe the antigen–antibody complex at the electrode surface.[160,162] In the presence of the antigen monolayer electrode, the enzyme probe is electrically contacted with the electrode and stimulates the electrobiocatalyzed oxidation of glucose. Association of the antibody with the antigen perturbs the electrical communication between the redox-functionalized enzyme and the electrode, and the electrobiocatalytic process is inhibited. This approach will be illustrated here with

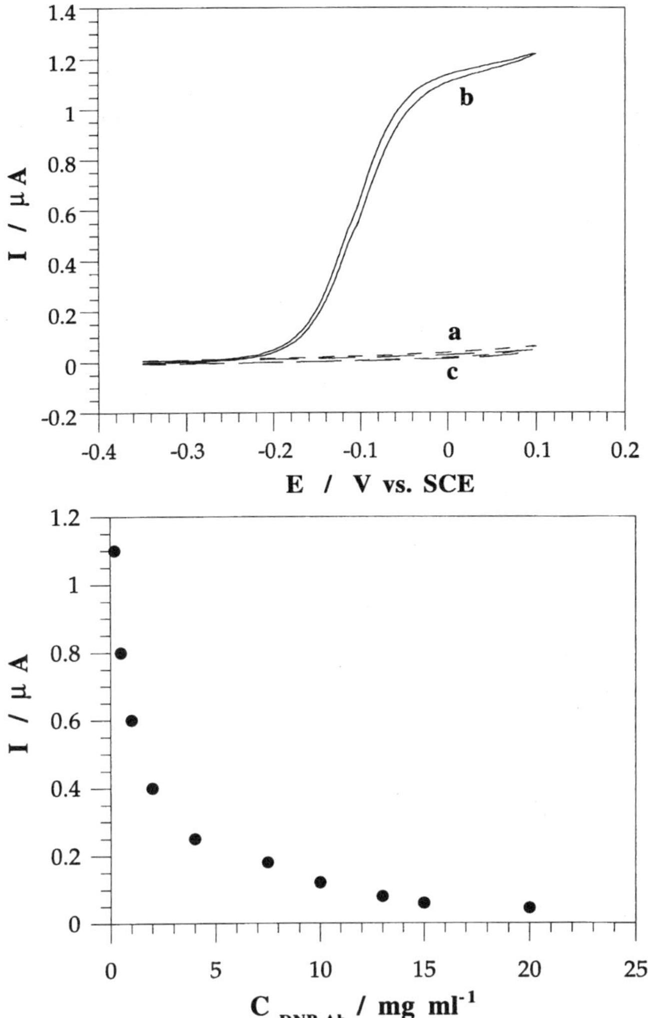

Figure 4.24. (A) Cyclic voltammograms of a mixed monolayer electrode composed of Nε-2,4-dini-trophenyl lysine and PQQ: (a) without added NADH; (b) with NADH, 10 mM; (c) after treatment of the electrode with DNP-Ab solution, 50 μg/mL, for 20 min, 35°C, and in the presence of NADH, 10 mM. In all experiments a 0.1 M tris-buffer solution, pH = 7.0, that included $CaCl_2$, 20 mM, was used as electrolyte; scan rate, 2 mV/s. (B) Calibration curve for the amperometric responses of the Nε-2,4-dini-trophenyl lysine–PQQ mixed monolayer electrode after treatment with different concentrations of DNP-Ab.

the amperometric detection of the DNP-Ab using glucose oxidase functionalized by tethered ferrocene units as the "electrically wired" enzyme redox probe. The electrical response of the Nε-2,4-dinitrophenyl lysine-functionalized monolayer electrode in the presence of ferrocene-tethered GOx and glucose is shown in Fig. 4.28, curve b. The electrocatalytic anodic currents imply efficient electrical contact between the

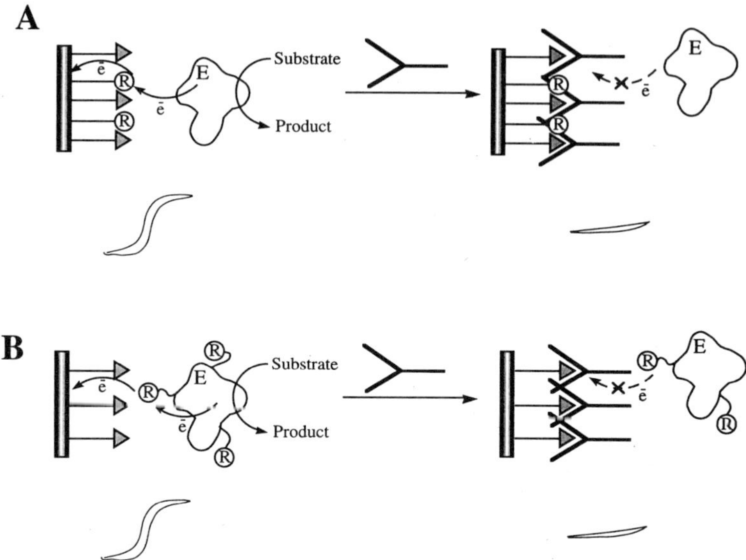

Figure 4.25. (A) Electrochemical analysis of an antibody by a mixed monolayer composed of the complementary antigen and an electron-transfer mediator assembled on an electrode and a solubilized redox enzyme. (B) Amperometric analysis of an antibody by an antigen monolayer electrode using a solubilized, "electrically wired" enzyme modified with relay groups.

enzyme and the electrode that stimulates the oxidation of glucose. Association of the antibody to the monolayer blocks the electron transfer between the enzyme and the electrode and the subsequent biocatalyzed oxidation of glucose (Fig. 4.28, curve c). Other antibodies, such as fluorescein-Ab, were similarly assayed at antigen-monolayer-functionalized electrodes.[160,162]

4.7. LAYERED PHOTOISOMERIZABLE ANTIGEN MONOLAYER ELECTRODES FOR REVERSIBLE PROBING OF ANTIGEN–ANTIBODY INTERACTIONS

A serious limitation of immunosensors is the fact that the sensing device is a single-cycle system. Tight antigen–antibody interactions result in blocking of the antigen monolayer by the analyte antibody. Extreme environmental conditions, such as acidification of the medium[163] or the addition of a chaotropic agent (glycine-NaCl or urea),[164,165] are known to dissociate antigen–antibody complexes, but the use of these methods is limited owing to the partial degradation/denaturation of the bioactive components. A method based on the use of photoisomerizable antigen monolayers on electrode surfaces was suggested for the tailoring of reversible, multicycle, immunosensor devices.[159,160] According to this approach (Fig. 4.29) a

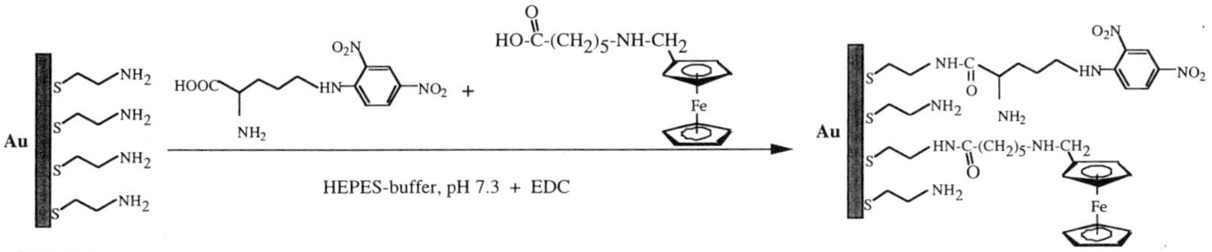

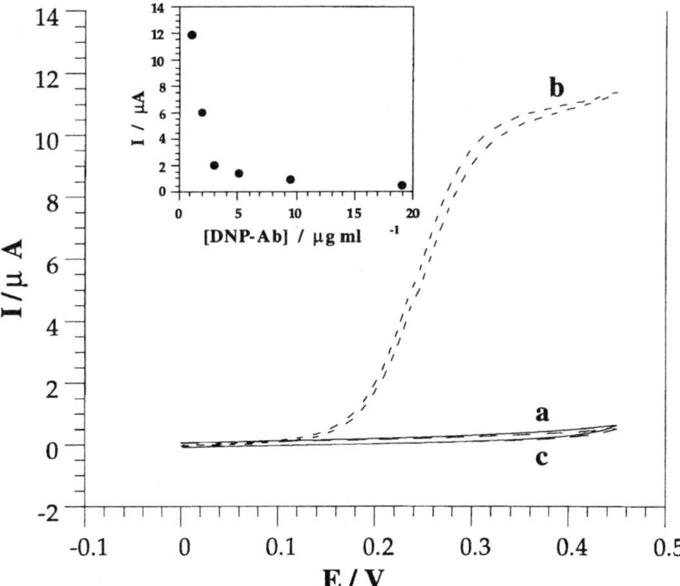

Figure 4.26. Assembly of a mixed monolayer of Nε-2,4-dinitrophenyl lysine (antigen) and ferrocene units (electron relay) on a Au electrode for the analysis of the DNP-Ab.

photoisomerizable antigen is assembled on the sensing interface (i.e., the electrode). In one photoisomer state (A) the monolayer exhibits affinity for the antibody. Challenging of the sensing interface with the complementary antibody results in the formation of the antigen-monolayer–antibody complex. Formation of the complex is physically transduced — by electrochemical transduction using one of the redox-probe configurations discussed in Section 4.6. Completion of the analysis cycle is followed by the photoisomerization of the monolayer to state B, which lacks antigen features and affinity for the antibody. By rinsing the electrode, the antibody is washed off, and a second illumination cycle restores the antigen monolayer to its

Figure 4.27. Cyclic voltammograms of the ferrocene/Nε-2,4-dinitrophenyl lysine mixed monolayer electrode: (a) in the background electrolyte composed of 0.1 M phosphate buffer, pH 7.0; (b) in the presence of added GOx (5 mg/mL) and glucose (0.05 M); (c) after the electrode treatment with a DNP-Ab solution (50 μg/mL) for 6 min and recording the electrochemical response in the presence of GOx (5 mg/mL) and glucose (0.05 M); potential scan rate, 2 mV/s; temperature, 35°C. Inset: Electrochemical response after the electrode treatment with different concentrations of DNP-Ab.

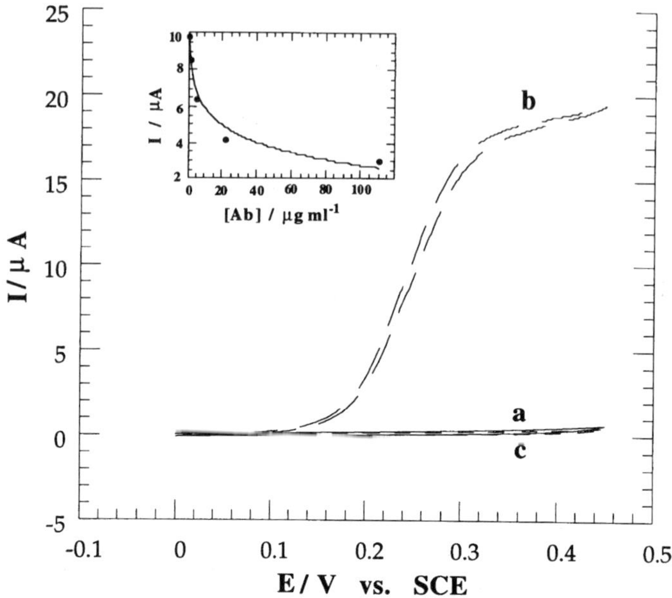

Figure 4.28. Cyclic voltammograms of Nε-2,4-dinitrophenyl lysine monolayer electrode: (a) in the presence of the background electrolyte composed of 0.1 M phosphate buffer, pH 7.0; (b) upon addition of ferrocene-modified GOx (Fc-GOx, 5 mg/mL) and glucose (0.05 M) to the electrolyte; (c) upon treatment of the electrode with the DNP-Ab (50 μg/mL) for 6 min and recording the response in the presence of Fc-GOx (5 mg/mL) and glucose (0.05 M). Potential scan rate, 2 mV/s, temperature, 35°C. Inset: Electrochemical responses after the electrode treatment with different concentrations of DNP-Ab.

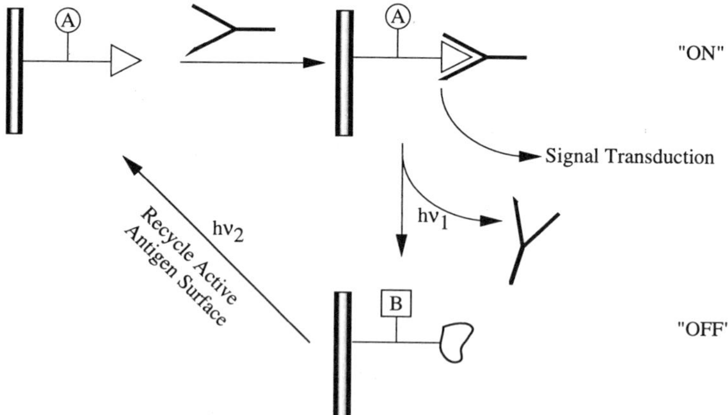

Figure 4.29. Schematic configuration of a reversible immunosensor device using a photoisomerizable antigen monolayer assembled on a transducing support as a functionalized interface for sensing the Ab and photochemical regeneration of the sensing surface.

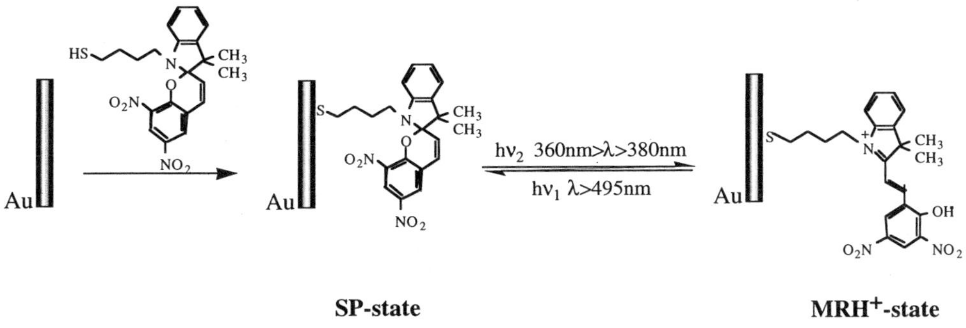

SP-state   **MRH$^+$-state**

Figure 4.30. Assembly of mercaptobutyl dinitrospiropyran onto a Au electrode to generate a photo-isomerizable antigen-monolayer electrode.

original state A. That is, by a two-step illumination cycle, the antibody sensed in the primary analysis step is washed off and the active sensing monolayer interface is regenerated.

This concept was proved experimentally with the organization of a photoisomerizable dinitrophenyl-antigen monolayer for the cyclic sensing of the DNP-Ab.[159,160] Figure 4.30 shows the assembly of the mercaptobutyl dinitro-spiropyran antigen monolayer on a Au electrode. The monolayer exhibits reversible photoisomerizable properties and irradiation of the dinitrospiropyran, SP state, with UV light yields, at neutral pH, the protonated dinitromerocyanine photoisomer monolayer, MRH$^+$ state. Further irradiation of the MRH$^+$ monolayer with visible light regenerates the SP monolayer state. The dinitros-piropyran monolayer acts as an antigen for the DNP-Ab, whereas the MRH$^+$ state lacks antigen features for the DNP-Ab. Figure 4.31 shows the electrochemical response of the SP monolayer electrode before (curve a) and after (curve b) treatment with DNP-Ab, using ferrocene-tethered GOx and glucose as the redox probe to analyze the formation of the antigen–antibody complex.[160] The antigen-monolayer electrode is completely insulated by the antibody with respect to the enzyme bioelectrocatalyzed transformation. This implies that the association of the DNP antibody to the antigen monolayer perturbs the electrical contact between the redox enzyme and the electrode support. Photoisomerization of the monolayer to the MRH$^+$ state followed by rinsing of the electrode yields the electrical response shown in Fig. 4.31 (curve c). The high electrocatalytic response indicates the dissociation of the DNP-Ab from the monolayer interface. Addition of the DNP-Ab to the electrode functionalized by the MRH$^+$ monolayer, generates the high amperometric response shown in curve d, indicating that the MRH$^+$ monolayer lacks antigen features for the association of the DNP-Ab. Further isomerization of the MRH$^+$ monolayer to the SP state generates an active sensing interface for the DNP-Ab. The inset in Fig. 4.31 shows the cyclic amperometric analysis of the DNP-Ab by the SP-monolayer-functionalized electrode.

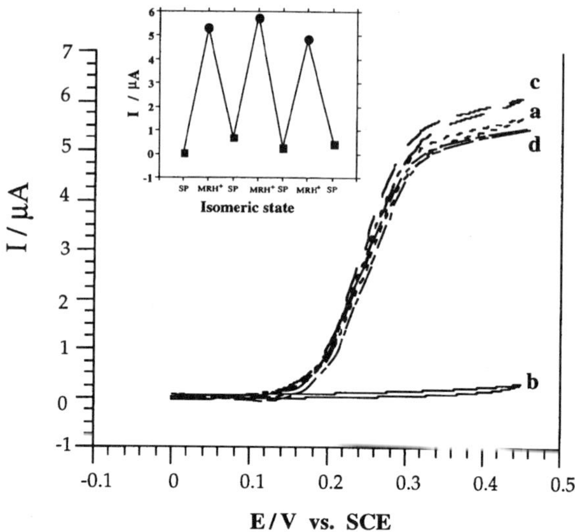

Figure 4.31. Cyclic voltammograms of the dinitrospiropyran photoisomerizable monolayer electrode in the presence of Fc–GOx, 5 mg/mL, and glucose, 50 mM: (a) in the presence of SP state; (b) in the presence of the SP state after the electrode treatment with the DNP-Ab solution, 0.15 mg/mL, for 6 min; (c) in the presence of MRH^+ state; (d) in the presence of MRH^+ state after the electrode treatment with DNP-Ab, 0.15 mg/L, for 6 min. All experiments were performed in 0.1 M phosphate buffer, pH 7.0, under argon, 35°C, scan rate, 5 mV/s. The SP state of the monolayer-modified electrode was generated by irradiation of the electrode, $\lambda > 495$ nm. The MRH^+ state of the monolayer-modified electrode was generated by irradiation of the electrode, 360 nm $< \lambda < 380$ nm. Inset: Cyclic amperometric responses of the dinitro-spiropyran photoisomerizable monolayer electrode: (■) in the presence of the SP state, after treatment with a DNP-Ab solution (0.15 mg/mL for 6 min); (●) in the presence of MRH^+ state, after rinsing with water. Net electrocatalytic anodic currents at $E = 0.4$ V vs. SCE are presented.

4.8.   *LAYERED OLIGONUCLEOTIDE ELECTRODES FOR ELECTROCHEMICAL PROBING OF DNA*

Biosensors for DNA analysis have evoked substantial recent research efforts within the broad search for means of gene detection. The development of sensitive means to identify precise mutations in DNA could lead to diagnostic methods to assess genetic disorders or infectious diseases. A variety of electrochemical DNA sensors have been reported in recent years.[166-168] These include devices that incorporate an oligonucleotide probe, complementary to the sensed single-stranded (ss) DNA, onto the conductive electrode support. Formation of the double-stranded (ds) assembly upon hybridization with complementary DNA was sensed by probing the intercalation of electroactive dyes or transition metal complexes with the ds-DNA. For example, the oligonucleotide 5′-TGCAGTTCCGGTGGCTGATC-3′ that is complementary to the oncogene-*v-myc* was adsorbed onto a basal plane pyrolytic graphite electrode. The hybridization with pVM 623 carrying the Pst fragment of *v-myc* was identified by following the anodic potential shift of acridine orange that is intercalated in the ds-DNA. Similarly, carbon-paste electrodes modified by the inclusion of either octadecylamine or stearic acid were used as a solid phase for the covalent attachment of the oligonucleotide. The formation of the ds-DNA by hybridization of the complementary DNA was sensed by following the electrical response of the Co(II)/Co(III)-*tris*-phenanthroline complex that is intercalated in the ds assembly.

The advantages associated with bioactive monolayer electrodes recently led to the architecture of oligonucleotide-probe monolayer assemblies on electrode sup-

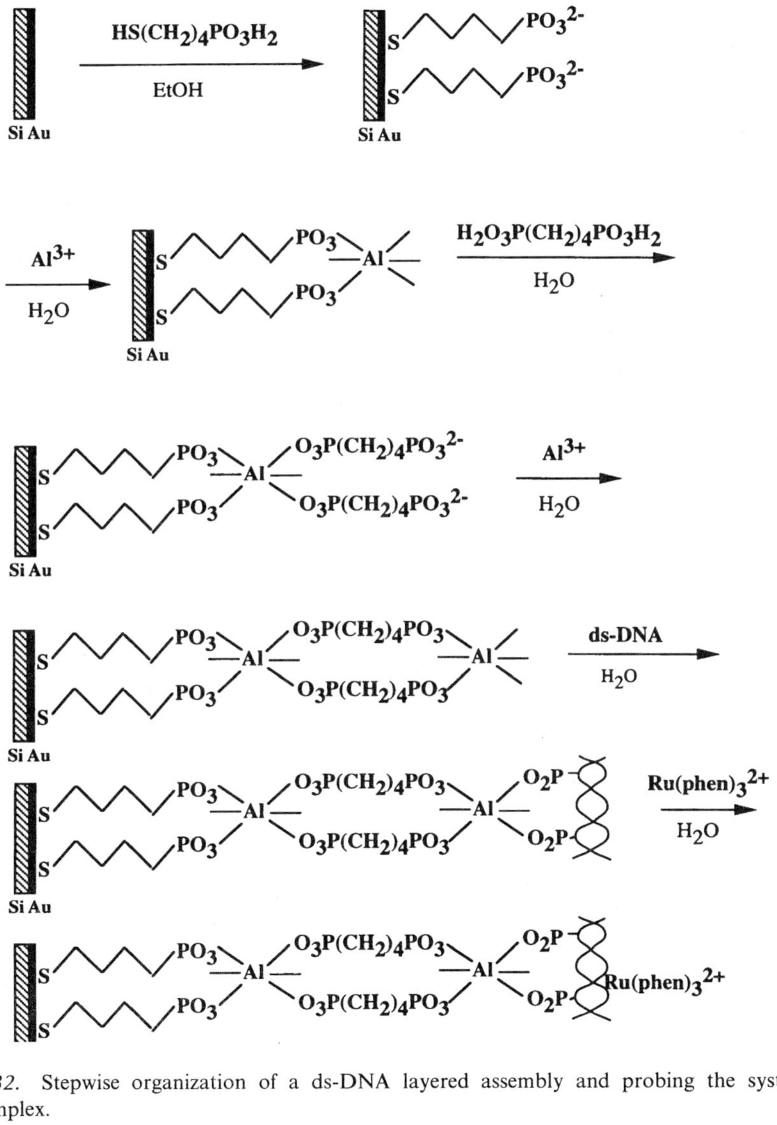

Figure 4.32. Stepwise organization of a ds-DNA layered assembly and probing the system by a Ru(II)-complex.

ports and the direct electrochemical or indirect electrogenerated chemoluminescent (ECL) detection of hybridized ds-DNA on the conducting support. Figure 4.32 shows the assembly of an oligonucleotide layer on a Au electrode supported on a Si wafer by the association of the oligonucleotide phosphate groups to Al(III) and the organization of an Al(III)-phosphonate layer for the binding of the oligonucleotide.[169] A thiol phosphonate was linked to a Au layer associated with the Si wafer, and Al(III) was linked to this base layer. An alkane bisphosphonate was then linked to the base metal centers and a second Al(III) ion layer was bound to the layered assembly. Phosphate groups of ss-DNA or ds-DNA were bound to the Al(III) sites. The existence of ds-DNA was confirmed by the intercalation of Ru(II)-*tris*-phenanthroline [$Ru(phen)_3^{2+}$] to the ds-DNA, and identification of the ds-associated metal complex by ECL in the presence of a tertiary amine, e.g., tripropylamine. The mechanism of ECL is summarized in Eqs. (6)–(10) and includes the electrooxidation of $Ru(phen)_3^{2+}$ and the tertiary amine:

$$ds\text{-}DNA\text{-}Ru(phen)_3^{2+} + e^- \rightarrow ds\text{-}DNA\text{-}Ru(phen)_3^{3+} \tag{6}$$

$$CH_3CH_2CH_2NPr_2 + e^- \rightarrow CH_3CH_2CH_2N^{+\cdot}Pr_2 \tag{7}$$

$$CH_3CH_2CH_2N^{+\cdot} \rightarrow CH_3CH_2CHNPr_2 + H^+ \tag{8}$$

$$ds\text{-}DNA\text{-}Ru(phen)_3^{3+} + CH_3CH_2CHNPr_2 \rightarrow ds\text{-}DNA\text{-}{}^*Ru(phen)_3^{2+}$$

$$+ CH_3CH_2CH{=}N^+Pr_2 \tag{9}$$

$$ds\text{-}DNA\text{-}{}^*Ru(phen)_3^{2+} \rightarrow ds\text{-}DNA\text{-}Ru(phen)_3^{2+} + h\nu \tag{10}$$

The resulting amine radical cation is deprotonated to the alkyl radical that is a powerful reductant. Back electron transfer between $Ru(phen)_3^{3+}$ and the radical yields the excited metal complex that emits light. As $Ru(phen)_3^{2+}$ binds only to

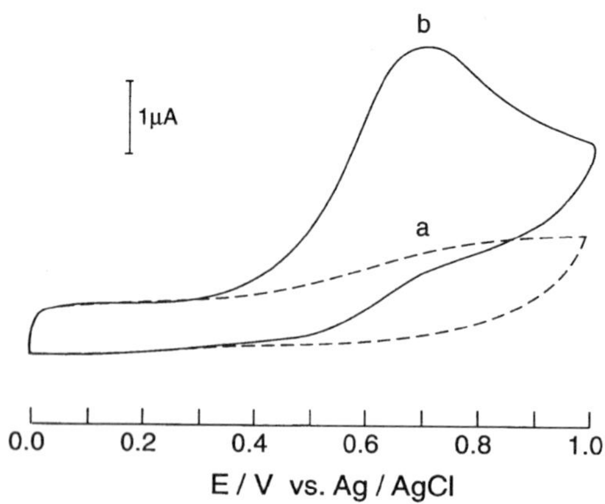

Figure 4.33. Cyclic voltammograms of ethydium bromide (**11**) in the presence of: (a) ss-DNA-electrode; (b) ds-DNA-electrode (data taken from Hashimoto et al.[170]

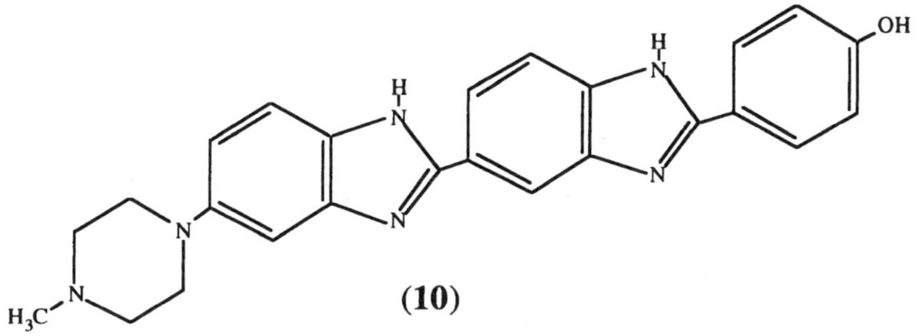

(10)

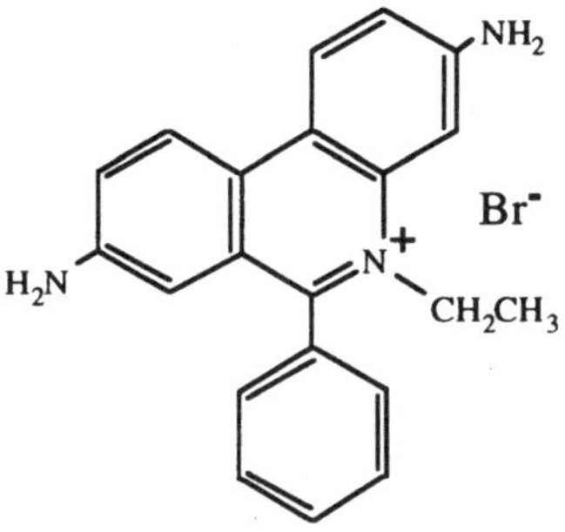

(11)

ds-DNA, ECL provides a sensitive method for following the formation of ds-helices between a probe oligonucleotide and its complementary analyte ss-DNA.

Layered oligonucleotide electrodes for the electrochemical detection of electroactive intercalator that associates to ds-helix assemblies, formed upon hybridization of the complementary DNA with the probe oligonucleotide, were tailored on different conductive supports and using different intercalating substances.[170,171] A thiol-functionalized 20-mer oligonucleotide probe, HS-5′pd TGCAGTTCCGGTG-GCTGATC-3′, which is complementary to an oncogene *v-myc*, was assembled on a Au electrode. Hybridization of the modified electrode with the ss-pVM 623 that contains a complementary fragment of *v-myc*, yields the ds-DNA that was identified by the amperometric response of 2′-(4-hydroxyphenyl)-5-(4-methyl-1-piperazinyl)-2,5′-bi-1H-benzimidazole trihydrochloride (Hoechst 33258) (**10**). The amperometric response of **10** relates to the amount of complementary target DNA in the analyte sample. The probe was found to be effective in the determination of the sequence-specific DNA in the range 10^{-7}–10^{-13} g/mL.

An oxidized graphite electrode was silanized with propyl amino triethoxysilane and the 24-mer deoxyoligonucleotide 5′-dGCCCCTTCCCTATGAGAATTCGGG-3′ was covalently linked in the presence of 1-ethyl-3-(3-dimethylaminopropyl)-carbodiimide (EDC).[171] Hybridization of the complementary 24-mer oligonucleotide was probed by the electrochemical response of ethydium bromide (**11**). Figure 4.33 shows the anodic currents of ethydium bromide in the presence of the ss-oligonucleotide (curve a) and the ds-assembly (curve b). Concentration of the redox probe by intercalation to the ds-helical assembly results in the high amperometric response of ethydium bromide. The anodic currents related linearly to the concentration of the complementary oligonucleotide within the range 1×10^{-7}–1×10^{-9} mg/mL. Related studies were performed also by the immobilization of ss-DNA on Au electrodes and the use of Co(III)-*tris*-bipyridine as redox probe for the hybridized ds-DNA.[172]

An enzyme-amplified amperometric detection of the formation of hybridized ds-oligonucleotide assemblies was recently illustrated in a layered hydrogel assembly.[173] This system includes basic elements for the future development of electrochemical enzyme-linked DNA sensors. A vitreous carbon electrode was functionalized with a hydrogel consisting of the copolymer polyacrylamide, polyacrylhydrazide, and Os(II)-bis-dimethylbipyridine polyvinylimidazole (**12**) and the copolymer polyacrylamide-polyacrylhydrazide (**13**), which are cross-linked by poly(ethylene glycol diglycidyl ether) (PEGDGE) (**14**). Poly(deoxythymidine)-5'-phosphate [pd $(T)_{25-30}$], was covalently linked as a probe oligonucleotide to the hydrazide functionalities by EDC coupling (Fig. 4.34A). The enzyme horseradish peroxidase (HRP) was used as the enzyme label for the complementary oligonucleotide pd $(A)_{25-30}$. Labeling of the complementary oligonucleotide was accomplished as outlined in Fig. 4.34B. The phosphate groups of the complementary oligonucleotide were functionalized by hydrazine and the glycoprotein HRP was activated by $NaIO_4$ to yield aldehyde protein functions. Reaction between the modified oligonucleotide and the activated protein generates the Schiff base, which, after reduction with $NaBH_4$, leads to the enzyme-labeled oligonucleotide, pd$(A)_{25-30}$-HRP. Hybridization of pd$(A)_{25-30}$-HRP with the functionalized electrode yields the ds-HRP-labeled oligonucleotide. The enzyme is electrically contacted by the redox hydrogel and stimulates the bioelectrocatalyzed reduction of hydrogen peroxide. Current densities as high as $1.4 \pm 0.1 \, \mu A/cm^2$ were detected. With noncomplementary HRP-labeled oligonucleotides, e.g., pd$(G)_{12-18}$-HRP, no significant amperometric signal is transduced ($\sim 0.07 \pm 0.02 \, \mu A/cm^2$), implying that the amperometric response originates from the specific complementarity of the enzyme-labeled oligonucleotide and the probe nucleotide. Thus, by primary assaying of an analyte sample with the electrode and subsequent treatment of the electrode with the enzyme-labeled oligonucleotide, a competitive analysis of a sample in the presence of the labeled oligonucleotide using the functionalized electrode, the amplified electrochemical sensing of the complementary DNA is feasible.

4.9. CONCLUSIONS AND PERSPECTIVES

The assembly of bioactive layers on conductive supports provides a novel method of organizing biosensor devices. The present study has emphasized the organization of sensing interfaces on electrodes for the electrochemical transduction of analysis events at the electrode surface. Similar layered structures on solid supports can be used to develop other biosensor systems with different transduction

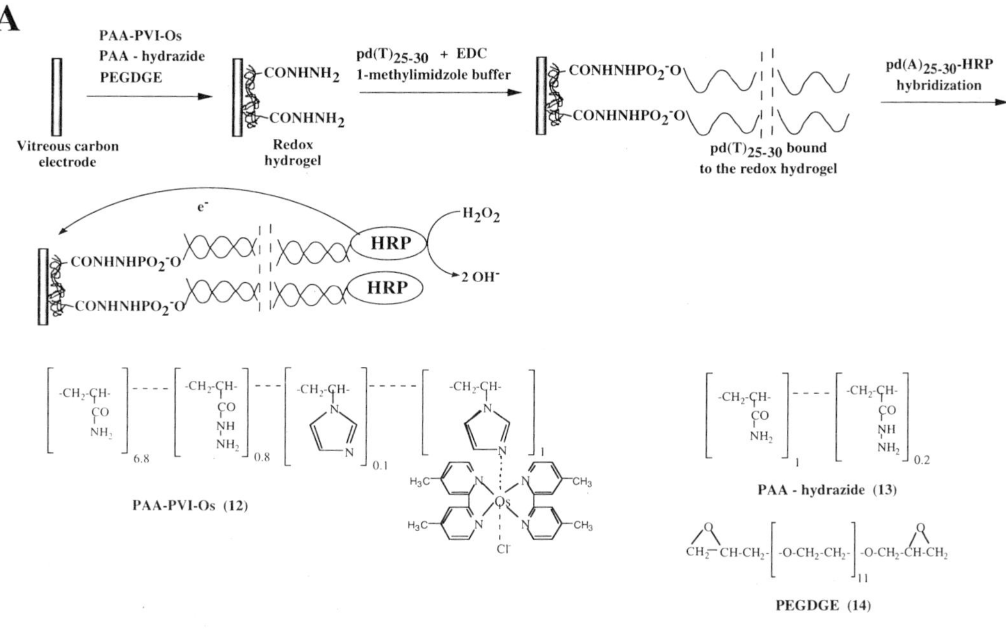

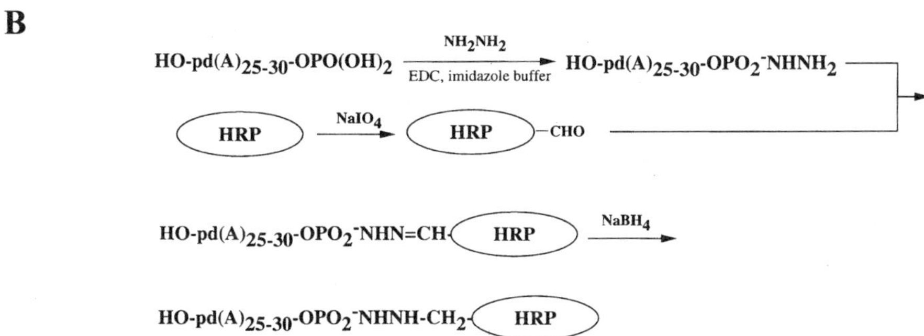

Figure 4.34. (A) Organization of a DNA-sensing electrode by immobilization of an oligonucleotide onto a hydrogel and using a HRP-labeled oligonucleotide for the detection of ds-DNA formation. (B) Synthesis of HRP-labeled oligonucleotide.

signals, e.g., surface plasmon resonance, SPR,[97,174] piezoelectric transduction,[175,176] ellipsometric[177] electrochemiluminescence detection,[178] and impedance measurements.[179,180] The organization of bioactive layers on the electrode supports revealed several important new concepts in biosensor technology:

1. The sensing processes at a monolayer in thin multilayer assemblies are diffusion-controlled and thus the recognition of the analyte by the sensing interface is not accompanied by transport barriers.[30,31]
2. The assembly of multilayers of enzymes on electrode supports allows the design of enzyme electrodes of tunable and controlled sensitivities.[107,108]

Also, the possibility of combining two (or more) enzymes in the multilayer array[109] opens the way to the tailoring of a variety of sensing interfaces through the activation of an enzyme cascade at the electrode surface.

3. Reconstitution of apo-flavoenzymes in the form of monolayers on functionalized electrodes yields aligned enzyme units of unprecedentally high electrical contact efficiency.[115] This novel method of assembling enzyme electrodes represents a new concept in the tailoring of invasive and specific electrochemical biosensors.

4. Affinity interactions between cofactor/enzyme components occurring at the electrode surface provide a general means of organizing integrated stable electrodes by cross-linking the macromolecular complexes onto the conducting supports. This was illustrated by cross-linking a $PQQ–NAD^+/$ enzyme complex layer assembled on the electrode. The resulting cross-linked cofactor–enzyme polymer layer is electrically contacted with the electrode surface. The method represents a new approach to tailoring integrated biosensors using $NAD(P)^+$-dependent enzymes.[153] In fact, the method of generating integrated bioelectrocatalytic interfaces by cross-linking of biomolecular complexes exhibiting affinity associative interactions seems to be of broad scope and applicability in bioelectronics and biosensor devices. Recently, it was shown that cross-linking of hemoproteins that form affinity complexes with microperoxidase-11 assembled as a monolayer on a Au electrode yields integrated, electrically contacted, bioelectrocatalytically active electrodes.[181,182] For example, Co(II)-protoporphyrin IX-reconstituted myoglobin forms an affinity associative complex with microperoxidase assembled on a Au electrode. Cross-linking of the affinity complex on the electrode surface yields a stable, integrated, bioactive electrode that stimulates the electrobiocatalyzed hydrogenation of acetylenes.

5. A series of configurations of antigen-monolayer-functionalized electrodes for electrochemical immunosensor systems was addressed. In order to enhance the sensitivity of the immunosensors, it is important to develop further means of amplifying the electrochemical signal transduced by the electrode as a result of the antigen–antibody recognition event at the electrode.[160–162].

6. The use of photoisomerizable antigen-layered electrodes to tailor reusable, reversible, sensing interfaces represents a novel concept in immunosensor design.[159,160] The photostimulated and optically controlled association of antibodies to photoisomerizable antigen monolayers associated with solid supports is not only important for organizing reversible immunosensors, but it allows surface photolithographic patterning with two-dimensional monolayer antibody domains.[183] This opens the possibility of linking other bioactive materials (e.g., enzymes, neurons, and oligonucleotides) to the antibody, and using it as a "directing arrow" to position the bioactive sensing material on the photoisomerizable "target support."

ACKNOWLEDGMENT: The support of the Israel Ministry of Science and the Bundesministerium für Bildung und Forschung (BMBF), Germany, is gratefully acknowledged.

REFERENCES

1. Schmidt H-L, Schuhmann W. Reagentless oxidoreductase sensors. *Biosens Bioelectron* 1996;11:127–135.
2. Bartlett PN. Applications of enzymes in amperometric sensors: problems and possibilities. In: Buck RP, Hatfield WE, Umaña M, Bowden EF, eds. *Biosensor Technology: Fundamentals and Applications.* New York: Marcel Dekker, 1990, Ch 7, pp 95–115.
3. Hill HAO, Sanghera GS. Mediated amperometric enzyme electrodes. In: Cass AEG, ed. *Biosensors: A Practical Approach.* Oxford: Oxford University Press, 1990, Ch 2, pp 19–46.
4. Katz E, Heleg-Shabtai V, Willner B, et al. Electrical contact of redox enzymes with electrodes: novel approaches for amperometric biosensors. *Bioelectrochem Bioenerg* 1997;42:95–104.
5. Bourdillon C, Bourgeois JP, Thomas D. Chemically modified electrodes bearing grafted enzymes. *Biotech Bioeng* 1979;21:1877–1879.
6. Bourdillon C, Bougeois JP, Thomas D. Covalent linkage of glucose oxidase on modified glassy carbon electrodes. Kinetic phenomena. *J Am Chem Soc* 1980;102:4231–4235.
7. Wieck HJ, Shea C, Yacynych AM. Reticulated vitreous carbon electrode materials chemically modified with immobilized enzyme. *Anal Chim Acta* 1982;142:277–279.
8. Urban G, Jobst G, Kohl F, et al. Miniaturized thin-film biosensors using covalently immobilized glucose oxidase. *Biosens Bioelectron* 1991;6:555–562.
9. Ianniello RM, Yacynych AM. Immobilized enzyme chemically modified electrode as an amperometric sensor. *Anal Chem* 1981;53:2090–2095.
10. Clark LC Jr. The hydrogen peroxide sensing platinum anode as an analytical enzyme electrode. In: Fleischer S, Packer L, eds. *Methods in Enzymology.* New York: Academic Press, 1979, vol 56, pp 448–479.
11. Ianniello RM, Yacynych AM. Chemically modified graphite electrode with immobilized enzyme as a potentiometric sensor for some L-amino acids. *Anal Chim Acta* 1981;131:123–132.
12. Kumar SD, Kulkarni AV, Dhaneshwar RG, et al. Potentiometric studies at the glucose oxidase enzyme electrode. *Bioelectrochem Bioenerg* 1994;34:195–198.
13. Clark Jr LC, Lyons C. Electrode systems for continuous monitoring in cardiovascular surgery. *Ann NY Acad Sci* 1962;102:29–45.
14. Bardeletti G, Séchaud F, Coulet PR. Amperometric enzyme electrodes for substrate and enzyme activity determinations. In: Blum LJ, Coulet PR, eds. *Biosensor Principles and Applications.* New York: Marcel Dekker, 1991, Ch 2, pp 7–45.
15. Heller A. Electrical wiring of redox enzymes. *Acc Chem Res* 1990;23:128–134.
16. Marcus RA, Sutin N. Electron transfer in chemistry and biology. *Biochim Biophys Acta* 1985;811:265–322.
17. Dryhurst G, Kadish KM, Scheller F, et al. *Biological Electrochemistry, Vol 1,* New York: Academic Press, 1982:459.
18. Bartlett PN, Tebbutt P, Whitaker RG. Kinetic aspects of the use of modified electrodes and mediators in bioelectrochemistry. *Prog React Kinet* 1991;16:55–155.
19. Bardea A, Katz E, Bückmann AF, et al. NAD$^+$-dependent enzyme electrodes: electrical contact of cofactor-dependent enzymes and electrodes. *J Am Chem Soc* 1997;119:9114–9119.
20. Brunori M. Control of electron transfer in metalloproteins. *Biosens Bioelectron* 1994;9:633–636.
21. Tarasevich MR. Way of using enzymes for acceleration of electrochemical reactions. *Bioelectrochem Bioenerg* 1979;6:587–597.
22. Williams DL, Doig Jr AP, Korosi A. Electrochemical-enzymatic analysis of blood glucose and lactate. *Anal Chem* 1970;42:118–121.

23. Cass AEG, Davis G, Green MJ, et al. Ferricinium ion as an electron acceptor for oxido-reductases. *J Electroanal Chem* 1985;190:117–127.
24. Ianniello RM, Lindsay TJ, Yacynych AM. Immobilized xanthine oxidase chemically modified electrode as a dual analytical sensor. *Anal Chem* 1982;54:1980–1984.
25. Zhang Y, Wilson GS. In vitro and in vivo evaluation of oxygen effects on a glucose oxidase based implantable glucose sensor. *Anal Chim Acta* 1993;281:513–520.
26. Csöregi E, Quinn CP, Schmidtke DW, et al. Design, characterization and one-point in vivo calibration of a subcutaneously implanted glucose electrode. *Anal Chem* 1994;66:3131–3138.
27. Degani Y, Heller A. Direct electrical communication between chemically modified enzymes and metal electrodes: 1. Electron transfer from glucose oxidase to metal electrodes via electron relays bound covalently to the enzyme. *J Phys Chem* 1987;91:1285–1289.
28. Degani Y, Heller A. Direct electrical communication between chemically modified enzymes and metal electrodes: 2. Method for bonding electron-transfer relays to glucose oxidase and D-amino-acid oxidase. *J Am Chem Soc* 1988;110:2615–2620.
29. Badia A, Carlini R, Fernandez A, et al. Intramolecular electron-transfer rates in ferrocene-derivatized glucose oxidase. *J Am Chem Soc* 1993;115:7053–7060.
30. Willner I, Katz E, Riklin A, et al. Mediated electron transfer in glutathione reductase organized in self-assembled monolayers on Au-electrodes. *J Am Chem Soc* 1992;114:10965–10966.
31. Willner I, Lapidot N, Riklin A, et al. Electron transfer communication in glutathione reductase assemblies: Electrocatalytic, photocatalytic and catalytic systems for the reduction of oxidized glutathione. *J Am Chem Soc* 1994;116:1428–1441.
32. Degani Y, Heller A. Electrical communication between redox centers of glucose oxidase and electrodes via electrostatically and covalently bound redox polymers. *J Am Chem Soc* 1989;111:2357–2358.
33. Heller A. Electrical connection of enzyme redox centers to electrodes. *J Phys Chem* 1992;96:3579–3587.
34. Willner I, Katz E, Lapidot N, et al. Bioelectrocatalysed reduction of nitrate utilizing polythiophene bipyridinium enzyme-electrodes. *Bioelectrochem Bioenerg* 1992;29:29–45.
35. Cosnier S, Innocent C, Jouanneau Y. Amperometric detection of nitrate via a nitrate reductase immobilized and electrically wired at the electrode surface. *Anal Chem* 1994;66:3198–3201.
36. Schuhmann W. Electron-transfer pathways in amperometric biosensors. Ferrocene-modified enzymes entrapped in conducting-polymer layers. *Biosens Bioelectron* 1995;10:181–193.
37. Lin AWC, Yeh P, Yacynych AM, et al. Cyanuric chloride as a general linking agent for the attachment of redox groups to pyrolytic graphite and metal oxide electrodes. *J Electroanal Chem* 1997;84:411–419.
38. Yacynych AM, Kuwana T. Cyanuric chloride as a general linking agent for modified electrodes: attachment of redox groups to pyrolytic graphite. *Anal Chem* 1978;50:640–645.
39. Ianniello RM, Lindsay TJ, Yacynych AM. Differential pulse voltammetric study of direct electron transfer in glucose oxidase chemically modified graphite electrodes. *Anal Chem* 1982;54:1098–1101.
40. Abruña HD. Coordination chemistry in two dimensions: chemically modified electrodes. *Coord Chem Rev* 1988;86:135–189.
41. Murray RW. Chemically modified electrodes. *Acc Chem Res* 1980;13:135–141.
42. Albery WJ, Hillman AR. Modified electrodes. *Annu Rep Prog Chem C* 1981;C78:377–437.
43. Murray RW. Chemically modified electrodes. In: Bard AJ, ed. *Electroanalytical Chemistry.* New York: Marcel Dekker, 1984, Vol. 13, 191–368.
44. Yoshida S, Kanno H, Watanabe T. Glutamate sensors carrying glutamate oxidase/peroxidase bienzyme system on tin oxide electrode. *Anal Sci* 1995;11:251–256.
45. Tarasevich MR. *Electrochemistry of Carbon Materials.* Moscow: Nauka, 1984.
46. Ianniello RM, Wieck HJ, Yacynych AM. Characterization of chemically modified carbonaceous electrode materials by diffuse reflectance Fourier transform infrared spectrometry. *Anal Chem* 1983;55:2067–2070.
47. Koval CA, Anson FC. Electrochemistry of the ruthenium (3+, 2+) couple attached to graphite electrodes. *Anal Chem* 1978;50:223–229.
48. Rocklin RD, Murray RW. Chemically modified carbon electrodes: XVII. Metallation of immobilized

tetra(aminophenyl)porphyrin with manganese, iron, cobalt, nickel, copper and zinc, and electrochemistry of diprotonated tetraphenyl porphyrin. *J Electroanal Chem* 1979;100:271–282.

49. Bourdillon C, Hervagault C, Thomas D. Increase in operational stability of immobilized glucose oxidase by the use of an artificial cosubstrate. *Biotech Bioeng* 1985;27:1619–1622.

50. Osborn JA, Ianniello RM, Wieck HJ, et al. Use of chemically modified activated carbon as a support for immobilized enzymes. *Biotech Bioeng* 1982;24:1653–1669.

51. Bianco P, Haladjian J, Bourdillon C. Immobilization of glucose oxidase on carbon electrodes. *J. Electroanal Chem* 1990;293:151–163.

52. Schuhmann W, Lammert R, Uhe B, et al. Polypyrrole: a new possibility for covalent binding of oxidoreductases to electrode surfaces as a base for stable biosensors. *Sens Act* 1990;B1:537–541.

53. Schuhmann W, Wohlschlager H, Lammert R, et al. Leaching of dimethylferrocene: a redox mediator in amperometric enzyme electrodes. *Sens Act* 1990;B1:571–575.

54. Razumas VJ, Jasaitis JJ, Kulys JJ. Electrocatalysis on enzyme-modified carbon materials. *Bioelectrochem Bioenerg* 1984;12:297–322.

55. Laval JM, Boudillon C. Modified glassy carbon electrode with immobilized enzyme. NAD/NADH lactic dehydrogenase. *J Electroanal Chem* 1983;152:125–141.

56. Evans JF, Kuwana T. Introduction of functional groups onto carbon electrodes via treatment with radio-frequency plasmas. *Anal Chem* 1979;51:358–365.

57. Oyama N, Brown AP, Anson FC. Introduction of amine functional groups on graphite electrode surfaces and their use in the attachment of ruthenium(II) to the electrode surface. *J Electroanal Chem* 1978;87:435–441.

58. Oyama N, Anson FC. Attachment of the EDTA complex of ruthenium(III) to the surface of graphite electrodes. Electrochemistry and ligand substitution chemistry with the attached complex. *J Electroanal Chem* 1978;88:289–297.

59. Nowak R, Schultz FA, Umaña M, et al. Chemically modified electrodes: XIV. Attachment of reagents to oxide-free glassy carbon surfaces: electroactive RF polymer films on carbon and platinum electrodes. *J Electroanal Chem* 1978;94:219–225.

60. Kamin RA, Wilson GS. Rotating ring-disk enzyme electrode for biocatalysis kinetic studies and characterization of the immobilized enzyme layer. *Anal Chem* 1980;52:1198–1205.

61. Lane RF, Hubbard AT. Electrochemistry of chemisorbed molecules: 1. Reactants connected to electrodes through olefinic substituents. *J Phys Chem* 1973;77:1401–1410.

62. Lane RF, Hubbard AT. Electrochemistry of chemisorbed molecules: 2. The influence of charged chemisorbed molecules on the electrode reactions of platinum complexes. *J Phys Chem* 1973;77:1411–1421.

63. Sharp M, Petersson M, Edstrom K. Preliminary determinations of electron transfer kinetics involving ferrocene covalently attached to a platinum surface. *J Electroanal Chem* 1979; 95:123–130.

64. Hupp JT, Weaver MJ. Utility of surface reaction entropies for examining reactant-solvent interactions at electrochemical interfaces. Ferricinium–ferrocene attached to platinum electrodes. *J Electrochem Soc* 1984;131:619–622.

65. Brown AP, Anson FC. Molecular anchors for the attachment of metal complexes to graphite electrode surfaces. *J Electroanal Chem* 1977;83:203–206.

66. Jaegfeldt H, Torstensson A, Gorton L, et al. Catalytic oxidation of reduced nicotinamide adenine dinucleotide by graphite electrodes modified with adsorbed aromatics containing catechol functionalities. *Anal Chem* 1981;53:1979–1982.

67. Jonsson G, Gorton L, Pettersson L. Mediated electron transfer from glucose oxidase at a ferrocene-modified graphite electrode. *Electroanalysis* 1989;1:49–55.

68. Katz E. Application of bifunctional reagents for immobilization of proteins on a carbon electrode surface: oriented immobilization of photosynthetic reaction centers. *J Electroanal Chem* 1994;365:157–164.

69. Black AJ, Wooster TT, Geiger WE, et al. Synthesis of a rigid dimethoxynaphthacene-spacer-dithiol which spontaneously attaches to Au and Pt electrodes: properties of monolayer films in nonaqueous solvents. *J Am Chem Soc* 1993;115:7924–7925.

70. Bruant MA, Joa SL, Pemberton JE. Raman scattering from monolayer films of thiophenol and 4-mercaptopyridine at Pt surfaces. *Langmuir* 1992;8:753–756.

71. Katz E, Solov'ev AA. Chemical modification of platinum and gold electrodes by naphthaquinones using amines containing sulphydryl or disulphide groups. *J Electroanal Chem* 1990;291:171–186.

72. Walczak MM, Chung C, Stole SM, et al. Structure and interfacial properties of spontaneously adsorbed *n*-alkanethiolate monolayers on evaporated silver surfaces. *J Am Chem Soc* 1991;113:2370–2378.

73. Tang X, Schneider T, Buttry DA. A vibrational spectroscopic study of the structure of electroactive self-assembled monolayers of viologen derivatives. *Langmuir* 1994;10:2235–2240.

74. Laibinis PE, Fox MA, Folkers JP, et al. Comparisons of self-assembled monolayers on silver and gold: mixed monolayers derived from $HS(CH_2)_{21}X$ and $HS(CH_2)_{10}Y$ (X,Y$=CH_3$, CH_2OH) have similar properties. *Langmuir* 1991;7:3167–3173.

75. Laibinis PE, Bain CD, Whitesides GM. Attenuation of photoelectrons in monolayers of *n*-alkanethiols adsorbed on copper, silver and gold. *J Phys Chem* 1991;95:7017–7021.

76. Laibinis PE, Whitesides GM, Allara DL, et al. Comparison of the structures and wetting properties of self-assembled monolayers of *n*-alkanethiols on the coinage metal surfaces, Cu, Ag, Au. *J Am Chem Soc* 1991;113:7152–7167.

77. Laibinis PE, Whitesides GM. ω-Terminated alkanethiolate monolayers on surfaces of copper, silver and gold have similar wettabilities. *J Am Chem Soc* 1992;114:1990–1995.

78. Bain CD, Troughton EB, Tao Y-T, et al. Formation of monolayer films by the spontaneous assembly of organic thiols from solution onto gold. *J Am Chem Soc* 1989;111:321–335.

79. Bain CD, Evall J, Whitesides GM. Formation of monolayers by the coadsorption of thiols on gold: variation in the head group, tail group and solvent. *J Am Chem Soc* 1989;111:7155–7164.

80. Bain CD, Biebuyck HA, Whitesides GM. Comparison of self-assembled monolayers on gold: coadsorption of thiols and disulfides. *Langmuir* 1989;5:723–727.

81. Sheen CW, Shi J-X, Martensson J, et al. A new class of organized self-assembled monolayers: alkanethiols on GaAs (100). *J Am Chem Soc* 1992;114:1514–1515.

82. Gu Y, Lin Z, Butera RA, et al. Preparation of self-assembled monolayers on InP. *Langmuir* 1995;11:1849–1851.

83. Finklea HO, Electrochemistry of organized monolayers of thiols and related molecules on electrodes. In: Bard AJ, Rubinstein I, eds. *Electroanalytical Chemistry*, New York: Marcel Dekker, 1996, Vol. 19, pp 109–335.

84. Xu J, Li H. The chemistry of self-assembled long-chain alkanethiol monolayers on gold. *J Colloid Interf Sci* 1995;176:138–149.

85. Cheng Q, Brajter-Toth A. Permselectivity and high sensitivity at ultrathin monolayers: effect of film hydrophobicity. *Anal Chem* 1995;67:2767–2775.

86. Katz E. A chemically modified electrode capable of a spontaneous immobilization of amino compounds due to its functionalization with succinimidyl groups. *J Electroanal Chem* 1990;291:257–260.

87. Katz E, Schlereth DD, Schmidt H-L, et al. Reconstitution of the quinoprotein glucose dehydrogenase from its apoenzyme on a gold electrode surface modified with a monolayer of pyrroloquinoline quinone. *J Electroanal Chem* 1994;368:165–171.

88. Katz E, Riklin A, Willner I. Application of stilbene-(4,4'-diisothiocyanate)-2,2'-disulfonic acid as bifunctional reagent for the organization of organic materials and proteins onto electrode surfaces. *J Electroanal Chem* 1993;354:129–144.

89. Williams RA, Blanch HW. Covalent immobilization of protein monolayers for biosensor applications. *Biosens Bioelectron* 1994;9:159–167.

90. Creager SE, Olsen KG. Self-assembled monolayers and enzyme electrodes: progress, problems and prospects. *Anal Chim Acta* 1995;307:277–289.

91. Kuwabata S, Okamoto T, Kajiya Y, et al. Preparation and amperometric glucose sensitivity of covalently bound glucose oxidase to (2-aminoethyl)-ferrocene on an Au electrode. *Anal Chem* 1995;67:1684–1690.

92. McRipley MA, Linsenmeier RA. Fabrication of a mediated glucose oxidase recessed microelectrode for the amperometric determination of glucose. *J Electroanal Chem* 1996;414:235–246.

93. Jin W, Bier F, Wollenberger U, et al. Construction and characterization of a multi-layer enzyme electrode: covalent binding of quinoprotein glucose dehydrogenase onto gold electrodes. *Biosens Bioelectron* 1995;10:823–829.

94. Sawaguchi T, Matsue T, Uchida I. Catalytic capability of diaphorase bound to a self-assembled thiol monolayer at a gold electrode. *Bioelectrochem Bioenerg* 1992;29:127–133.

95. Prime KL, Whitesides GM. Adsorption of proteins onto surface containing end-attached oligo(ethylene oxide): a model system using self-assembled monolayers. *J Am Chem Soc* 1993;115:10714–10721.

96. Kinnear KT, Monbouquette HG. Direct electron transfer to *Escherichia coli* fumarate reductase in self-assembled alkanethiol monolayers on gold electrodes. *Langmuir* 1993;9:2255–2257.

97. Mrksich M, Sigal GB, Whitesides GM. Surface plasmon resonance permits in situ measurement of protein adsorption on self-assembled monolayers of alkanethiolates on gold. *Langmuir* 1995;11:4383–4385.

98. Nahir TM, Bowden EF. The distribution of standard rate constants for electron transfer between thiol-modified gold electrodes and adsorbed cytochrome c. *J Electroanal Chem* 1996;410:9–13.

99. Naumann R, Jonczyk A, Kopp R, et al. Incorporation of membrane proteins in solid-supported lipid layers. *Angew Chem Int Ed Engl* 1995;34:2056–2058.

100. Tarlov MJ, Bowden EF. Electron-transfer reaction of cytochrome c adsorbed on carboxylic acid terminated alkanethiol monolayer electrodes. *J Am Chem Soc* 1991;113:1847–1849.

101. Hoshi T, Takeshita H, Anzai J, et al. Use of an electrodeposited avidin film for the preparation of lactate and choline sensors. *Anal Sci* 1995;11:311–312.

102. Du X, Anzai J, Osa T, et al. Amperometric alcohol sensors based on protein multilayers composed of avidin and biotin-labeled alcohol oxidase. *Electroanalysis* 1996;8:813–816.

103. Pantano P, Morton TH, Kuhr WG. Enzyme-modified carbon-fiber microelectrodes with millisecond response times. *J Am Chem Soc* 1991;113:1832–1833.

104. Bourdillon C, Demaille C, Gueris J, et al. A fully active monolayer enzyme electrode derivatized by antigen–antibody attachment. *J Am Chem Soc.* 1993;115:12264–12269.

105. Bourdillon C, Demaille C, Moiroux J, et al. Step-by-step immunological construction of a fully active multilayer enzyme electrode. *J Am Chem Soc* 1994;116:10328–10329.

106. Bourdillon C, Demaille C, Moiroux J, et al. From homogeneous electroenzymatic kinetics to antigen–antibody construction and characterization of spatially ordered catalytic enzyme assemblies on electrodes. *Acc Chem Res* 1996;29:529–535.

107. Willner I, Riklin A, Shoham B, et al. Development of novel biosensor enzyme electrodes: glucose oxidase multilayer arrays immobilized onto self-assembled monolayers on electrodes. *Adv Mater* 1993;5:912–915.

108. Shoham B, Migron Y, Riklin A, et al. A bilirubin biosensor based on a multilayer network enzyme electrode. *Biosens Bioelectron* 1995;10:341–352.

109. Riklin A, Willner I, Glucose and acetylcholine sensing multilayer enzyme electrodes of controlled enzyme layer thickness. *Anal Chem* 1995;67:4118–4126.

110. He P-G, Takahashi T, Anzai J, et al. A facile method to regulate enzyme load on biosensor electrode based on avidin/biotin complexation. *Pharmazie* 1994;49:621–622.

111. Hoshi T, Anzai J, Osa T. Controlled deposition of glucose oxidase on platinum electrode based on an avidin/biotin system for the regulation of output current of glucose sensors. *Anal Chem* 1995;67:770–774.

112. Anzai J. Takeshita H, Hoshi T, et al. Regulation of output current of L-lactate sensors based on alternate deposition of avidin and biotinylated lactate oxidase on electrode surface through avidin/biotin complexation. *Chem Pharm Bull* 1995;43:520–522.

113. Katz E, Schlereth DD, Schmidt H-L. Electrochemical study of pyrroloquinoline quinone covalently immobilized as monolayer onto a cystamine modified gold electrode. *J Electroanal Chem* 1994;367:59–70.

114. Wang J. Electroanalysis and biosensors. *Anal Chem* 1995;67:487R–492R.

115. Willner I, Heleg-Shabtai V, Blonder R, et al. Electrical wiring of glucose oxidase by reconstitution of FAD-modified monolayers assembled onto Au-electrodes. *J Am Chem Soc* 1996;118:10321–10322.

116. Günther H, Paxinos AS, Schultz M. Direct electron transfer between carbon electrode, immobilized mediator and an immobilized viologen-accepting pyridine nucleotide oxidoreductase. *Angew Chem Int Ed Engl* 1990;29:1053–1055.

117. Wienkamp R, Steckhan E. Indirect electrochemical regeneration of NADH by a bipyridiner-hodium(I) complex as electron-transfer agent. *Angew Chem Int Ed Engl* 1982;21:782–783.

118. Rupert R, Herrmann S, Steckhan E. Efficient indirect electrochemical in-situ regeneration of NADH:

electrochemically driven enzymatic reduction of pyruvate catalyzed by D-LDH. *Tetrahedron Lett.* 1987;28:6583–6586.

119. Moiroux J, Elving PJ. Mechanistic aspects of the electrochemical oxidation of dihydronicotinamide adenine dinucleotide (NADH). *J Am Chem Soc* 1980;102:6533–6538.

120. Gorton L, Persson B, Hale PD, et al. Electrocatalytic oxidation of nicotinamide adenine dinucleotide cofactor at chemically modified electrodes. In: Edelman PG, Wang J, eds. *Biosensors and Chemical Sensors.* ACS Symp. Ser. 487. Washington, DC: American Chemical Society, 1992, pp 56–83.

121. Tse DC-S, Kuwana T. Electrocatalysis of dihydronicotinamide adenosine diphosphate with quinones and modified quinone electrodes. *Anal Chem* 1978;50:1315-1318.

122. Ueda C, Tse DC-S, Kuwana T. Stability of catechol modified carbon electrodes for electrocatalysis of dihydronicotinamide adenine dinucleotide and ascorbic acid. *Anal Chem* 1982;54:850–856.

123. Jaegfeldt H, Kuwana T, Johansson G. Electrochemical stability of catechols with a pyrene side chain strongly adsorbed on graphite electrodes for catalytic oxidation of dihydronicotinamide adenine dinucleotide. *J Am Chem Soc* 1983;105:1805–1814.

124. Degrand C, Miller LL. An electrode modified with polymer-bound dopamine which catalyzes NADH oxidation. *J Am Chem Soc* 1980;102:5728–5732.

125. Katz E, Lötzbeyer T, Schlereth DD, et al. Electrocatalytic oxidation of reduced nicotinamide coenzymes at gold and platinum electrode surfaces modified with a monolayer of pyrroloquinoline quinone: effect of Ca^{2+} cations. *J Electroanal Chem* 1994;373:189–200.

126. Huck H, Schmidt H-L. Chloranil as catalyst for the electrochemical oxidation of NADH to NAD^+. *Angew Chem* 1981;93:421–422.

127. Schuhmann W, Huber J, Wohlschlager H, et al. Electrocatalytic oxidation of NADH at mediator-modified electrode surfaces. *J Biotech* 1993;27:129–142.

128. Torstensson A, Gorton L. Catalytic oxidation of NADH by surface-modified graphite electrodes. *J Electroanal Chem* 1981;130:199–207.

129. Gorton L, Torstensson A, Jaegfeldt H, et al. Electrocatalytic oxidation of reduced nicotinamide coenzymes by graphite electrodes modified with an adsorbed phenoxazinium salt, meldola blue. *J Electroanal Chem* 1984;161:103–120.

130. Persson B, Lan HL, Gorton L, et al. Amperometric biosensors based on electrocatalytic regeneration of NAD^+ at redox polymer-modified electrodes. *Biosens Bioelectron* 1993;8:81–88.

131. Kulys J, Gleixner G, Schuhmann W, et al. Biocatalysis and electrocatalysis at carbon paste electrodes doped by diaphorase-methylene green and diaphorase-meldola blue. *Electroanalysis* 1993;5:201–207.

132. Schlereth DD, Katz E, Schmidt H-L. Toluidine blue covalently immobilized onto gold electrode surfaces: an electrocatalytic system for NADH oxidation. *Electroanalysis* 1994;6:725–734.

133. Schlereth DD, Katz E, Schmidt H-L. Surface-modified gold electrodes for electrocatalytic oxidation of NADH based on the immobilization of phenoxazine and phenothiazine derivatives on self-assembled monolayers. *Electroanalysis* 1995;7:46–54.

134. Ohtani M, Kuwabata S, Yoneyama H. Electrochemical oxidation of reduced nicotinamide coenzymes at Au electrodes modified with phenothiazine derivative monolayers. *J Electroanal Chem* 1997;422:45–54.

135. Vering T, Schuhmann W, Seiwald D, et al. A potentiostatic multi-pulse method using redox polymers for potentiometric measurements of enzymic redox-reactions. *J Electroanal Chem* 1994;364:277–279.

136. Marko-Varga G, Appelqvist R, Gorton L. A glucose sensor based on glucose dehydrogenase adsorbed on a modified carbon electrode. *Anal Chim Acta* 1986;179:371–379.

137. Schmidt H-L, Grenner G. Coenzyme properties of NAD^+ bound to different matrices through the amino group in the 6-position. *Eur J Biochem* 1976;67:295–302.

138. Zappelli P, Rossodivita A, Re L. Synthesis of coenzymically active soluble and insoluble macromolecularized NAD^+ derivatives. *Eur J Biochem* 1975;54:475–482.

139. Zappelli P, Rossodivita A, Prosperi G, et al. New coenzymically-active soluble and insoluble macromolecular NAD^+ derivatives. *Eur J Biochem* 1976;62:211–215.

140. Zappelli P, Pappa R, Rossodivita A, et al. Preparation and coenzymic activity of soluble polyethyleneimine-bound $NADP^+$ derivatives. *Eur J Biochem* 1977;72:309–315.

141. Furukawa S, Katayama N, Iizuka T, et al. Preparation of polyethylene glycol-bound NAD$^+$ and its application in a model enzyme reactor. *FEBS Lett* 1980;121:239–242.

142. Yamazaki Y, Maeda H. The synthesis of new polymer derivatives of NAD$^+$ by radical copolymerization and their coenzymic activity. *Agr Biol Chem* 1981;45:2277–2288.

143. Bückmann AF. A new synthesis of coenzymically active water-soluble macromolecular NAD$^+$ and NADP$^+$ derivatives. *Biocatalysis* 1987;1:173–186.

144. Yamazaki Y, Maeda H. The co-immobilization of NAD$^+$ and dehydrogenases and its application to bioreactors for synthesis and analysis. *Agr Biol Chem* 1982;46:1571–1581.

145. Eguchi T, Iizuka T, Kagotani T, et al. Covalent linking of poly(ethylene-glycol)-bound NAD$^+$ with *Thermus thermophilus* malate dehydrogenase. NADH regeneration unit for a coupled second-enzyme reaction. *Eur J Biochem* 1986;155:415–421.

146. Montagné, M, Marty J-L. Bi-enzyme amperometric D-lactate sensor using macromolecular NAD$^+$. *Anal Chim Acta* 1995;315:297–302.

147. Månsson M-O, Larsson P-O, Mosbach K. Covalent binding of an NAD$^+$ analogue to liver alcohol dehydrogenase resulting in an enzyme–coenzyme complex not requiring exogeneous coenzyme for activity. *Eur J Biochem* 1978;86:455–463.

148. Månsson M-O, Larsson P-O, Mosbach K. Recycling by a second enzyme of NAD$^+$ covalently bound to alcohol dehydrogenase. *FEBS Lett* 1979;98:309–313.

149. Woenckhaus C, Koob R, Burkhard A, et al. Preparations of holodehydrogenases by covalent fixation of NAD$^+$-analogs to alcohol and lactate dehydrogenase. *Bioorg Chem* 1983;12:45–57.

150. Kovár J, Šimek K, Kucera I, et al. Steady-state kinetics of horse-liver alcohol dehydrogenase with covalently bound coenzyme analogue. *Eur J Biochem* 1984;139:585–591.

151. Goulas P. Covalent binding of an NAD$^+$ analogue to horse liver alcohol dehydrogenase in a ternary complex with pyrazole. *Eur J Biochem* 1987;168:469–473.

152. Willner I, Riklin A. Electrical communication between electrode and NAD(P)$^+$-dependent enzymes using pyrroloquinoline quinone-enzyme electrodes in a self-assembled monolayer configuration: design of a new class of amperometric biosensors. *Anal Chem* 1994;66:1535–1539.

153. Anderson JL, Bowden EF, Pickup PG. Dynamic electrochemistry: methodology and application. *Anal Chem* 1996;68:379R–444R (Section: Immunological and recognition-based electrochemistry, 422R).

154. Ho WO, Atehy D, McNeil CJ. Amperometric detection of alkaline phosphatase activity at a horseradish peroxidase enzyme electrode based on activated carbon: potential application to electrochemical immunoassay. *Biosens Bioelectron* 1995;10:683–691.

155. McNeil CJ, Atehy D, Ho WO. Direct electron transfer bioelectronic interfaces: application to clinical analysis. *Biosens Bioelectron* 1995;10:75–83.

156. Ivnitski D, Rishpon J. A one-step, separation-free amperometric enzyme immunosensor. *Biosens Bioelectron* 1996;11:409–417.

157. Wittstock G, Emons H, Heineman WR, Electron transfer through an immunoglobulin layer via an immobilized redox mediator. *Electroanalysis* 1996;8:143–146.

158. Huang S-C, Caldwell KD, Lin J-N, et al. Site-specific immobilization of monoclonal antibodies using spacer-mediated antibody attachment. *Langmuir* 1996;12:4292–4298.

159. Willner I, Blonder R, Dagan A. Application of photoisomerizable antigenic monolayer-electrodes as reversible amperometric immunosensors. *J Am Chem Soc* 1994;116:9365–9366.

160. Blonder R, Levi S, Tao G, et al. Development of amperometric and microgravimetric immunosensors and reversible immunosensors using antigen and photoisomerizable antigen monolayer electrodes. *J Am Chem Soc* 1997; 119: 10467–10478.

161. Katz E, Willner I. Amperometric amplification of antigen–antibody association at monolayer interfaces: design of immunosensor electrodes. *J Electroanal Chem* 1996;418:67–72.

162. Blonder R, Katz E, Cohen Y, et al. Application of redox enzymes for probing the antigen–antibody association at monolayer interfaces: development of amperometric immunosensor electrodes. *Anal Chem* 1996;68:3151–3157.

163. Ilchmann D, Helbig D, Göhler H, et al. Regeneration of adsorbed and covalently immobilized antibodies on solid phases for immunoassay. *J Clin Chem Clin Biochem* 1990;28:677–681.

164. Sibley DET, Ramsay G, Lubrano GJ, et al. Stability and reusability of enzyme-linked immunosorbent assay (ELISA) plates. *Anal Lett* 1993;26:1623–1634.

165. Joyeux C, Chrzavzez E, Boquien CY, et al. Reusable solid phase immunoassay for the detection of citrate lyase. *Anal Chim Acta* 1996;320:77–86.
166. Hashimoto K, Ito I, Ishimori Y. Novel DNA sensor for electrochemical gene detection. *Anal Chim Acta* 1994;286:219–224.
167. Millan KM, Saraullo A, Mikkelsen SR. Voltammetric DNA biosensor for cystic fibrosis based on a modified carbon paste electrode. *Anal Chem* 1994;66:2943–2948.
168. Hashimoto K, Miwa K. Goto M,, et al. DNA-sensor: a novel electrochemical gene detection method using carbon electrode immobilized DNA-probes. *Supramolecular Chem* 1993;2:265–270.
169. Xu X-H, Yang HC, Mallouk TE, et al. Immobilization of DNA on an aluminium(III) alkane bisphosphonate thin film with electrogenerated chemiluminescent detection. *J Am Chem Soc* 1994;116:8386–8387.
170. Hashimoto K, Ito K. Ishimori Y. Sequence-specific gene detection with a gold electrode modified with DNA probes and an electrochemically active dye. *Anal Chem* 1994;66:3830–3833.
171. Liu S, Ye J, He P, et al. Voltammetric determination of sequence-specific DNA by electroactive intercalator on graphite electrodes. *Anal Chim Acta* 1996;335:239–243.
172. Pang D-W, Zhang M, Wang Z-L, et al. Modification of glassy carbon and gold electrodes with DNA. *J Electroanal Chem* 1996;403:183 188.
173. de Lumley-Woodyear T, Campbell CN, Heller A. Direct enzyme-amplified electrical recognition of a 30-base model oligonucleotide. *J Am Chem Soc* 1996;118:5504–5505.
174. Liedberg B, Nylander C, Lundström I. Biosensing with surface plasmon resonance—how it all started. *Biosens Bioelectron* 1995;10:i–ix.
175. Lion-Dagan M, Ben-Dov I, Willner I. Microgravimetric quartz-crystal-microbalance analysis of cytochrome c interactions with pyridine and pyridine-nitrospiropyran monolayer electrodes and characterization of interprotein complexes at the functionalized surfaces. *Colloids Surf B: Biointerfaces* 1997; 8:251–260.
176. Ben-Dov I, Willner I, Zisman E. Piezoelectric immunosensors for urine specimens of *Chlamydia trachomatis* employing quartz-crystal-microbalance microgravimetric analyses. *Anal Chem* 1997; 69:3506–3512.
177. Nygren H. Real-time recording of antigen–antibody reactions at surfaces: interpretation of data and a statistical model. *Colloids Surf B: Biointerfaces* 1995;4:243–250.
178. Wilson R, Kremeskötter J, Schiffrin DJ, et al. Electrochemiluminescence detection of glucose oxidase as a model for flow injection immunoassays. *Biosens Bioelectron* 1996;11:805–810.
179. Rickert J, Göpel W, Beck W, et al. A "mixed" self-assembled monolayer for an impedimetric immunosensor. *Biosens Bioelectron* 1996;11:757–768.
180. Mirsky VM, Krause C, Heckmann DK. Capacitive sensor for lipolytic enzymes. *Thin Solid Films* 1996;284:939–941.
181. Heleg-Shabtai V, Katz E, Willner I. Assembly of microperoxidase-11 and Co(II)-protoporphyrin IX reconstituted myoglobin monolayers on Au-electrodes: integrated bioelectrocatalytic interfaces. *J Am Chem Soc* 1997; 119:8121–8122.
182. Heleg-Shabtai V, Katz E, Levi S, et al. Microperoxidase-11 functionalized electrodes: an active monolayer interface for the electrocatalyzed reduction of Co(II)-protoporphyrin IX reconstituted myoglobin and for the generation of integrated protein electrodes for bioelectrocatalyzed hydrogenation of acetylenes. *J Chem Soc: Perkin Trans 2* 1997; 2645–2651.
183. Willner I, Blonder R. Patterning of surfaces by photoisomerizable antibody–antigen monolayers. *Thin Solid Films* 1995;266:254–257.

5

Biosensors Based on "Wired" Peroxidases

Qiang Chen, Adam Heller, and Gregory L. Kenausis

5.1. INTRODUCTION

Electrochemical and optical hydrogen peroxide sensors based on H_2O_2 producing enzyme-catalyzed reactions are commonly used in medical diagnostic assays.[1-5] Electrochemical sensing usually allows use of small-volume and light-absorbing samples, such as whole blood, as well as facile integration into flow systems.[6,7] Many assays involve an oxidase-catalyzed oxidation of a substrate by molecular oxygen in their first step, a reaction whereby an equimolar amount of H_2O_2 is formed:

$$\text{Substrate} + O_2 \xrightarrow{\text{Oxidase}} \text{Product} + H_2O_2 \tag{1}$$

The H_2O_2 is then amperometrically measured in the second step through its electrooxidation to oxygen. This reaction proceeds at its transport-limited rate at a potential that depends on the nature of the surface of the electrode. On platinum, H_2O_2 is electrooxidized at $+700\,\text{mV}$ (SCE).[8,9]

At this potential, some endogenous compound species commonly present in biological samples, such as ascorbate and urate, as well as exogenous compounds, such as acetaminophen, are also electrooxidized, and these compounds interfere with the assay.[10,11] Furthermore, fouling of the platinum electrode causes a baseline drift and extends the reequilibration time required after even a minor change in the system.[12] In cleaning of the platinum electrodes through electrooxidation of the adsorbed impurities, a platinum oxide film is formed. Reducing this oxide often leads to the formation of a platinum hydride film. The residual hydride film, being oxidized

Qiang Chen • Cygnus Inc, Redwood City, California 94063-4719.   Adam Heller • Department of Chemical Engineering, University of Texas at Austin, Austin, Texas 78712.   Gregory L. Kenausis • Laboratory for Surface Science and Engineering, ETH Zurich, Zurich CH-8092, Switzerland.

Biosensors and Their Applications, edited by Yang and Ngo, Kluwer Academic/Plenum Publishers, New York, 1999.

at the operating potential, is a source of electrochemical noise. By using very small particles of platinum-group metals on carbon, it is possible to increase the surface area of the H_2O_2 oxidizing electrodes and shift the potential at which H_2O_2 is electrooxidized at the mass-transport-limited rate to more reducing potentials. Such a shift, however, makes the electrodes more sensitive to contaminants, increasing their drift. To eliminate the effect of electrooxidizable interferants, the platinum electrodes have been overcoated with polymers having pores small enough to allow H_2O_2 transport, but preventing transport of interfering electrooxidizable organic species. These electrodes, when operated at $+700\,mV$ (SCE), are selective for H_2O_2. In an alternative approach to the amperometric assays of H_2O_2, the peroxide is electroreduced. On peroxidase-modified cathodes, H_2O_2 is electroreduced to water:

$$H_2O_2 + 2H^+ + 2e^- \xrightarrow{\text{Peroxidase}} 2H_2O \qquad (2)$$

Amperometric peroxidase-containing H_2O_2 sensors based on mediation by fast and reversible redox couples have been reported.[13-15] In these, H_2O_2 is reduced by a mediator,

$$2\,\text{Med}_{red} + H_2O_2 + 2H^+ \xrightarrow{\text{Peroxidase}} 2\,\text{Med}_{ox}^+ + 2H_2O \qquad (3)$$

which is subsequently regenerated through its electroreduction, usually at $0\,mV$ (SCE), at its mass-transport-limited rate. This electrode does not require a clean, catalytic metal (e.g., platinum) surface and is not subject to fouling, while its sensitivity is more stable than that of the platinum anodes. However, the problem of oxidizable interferants persists, because the peroxidase, when oxidized by H_2O_2, oxidizes urate, ascorbate, and acetaminophen.

Horseradish peroxidase (HRP) is the most commonly used peroxidase for biosensor fabrication. Other peroxidases, being less easily available, are used less frequently.[16] Table 5.1 shows the operating potential, the sensitivity and the linear range for H_2O_2 sensors based on peroxidases.

5.2. "WIRING" OF HORSERADISH PEROXIDASE

Peroxidases are small ($\approx 40\,kD$) glycoproteins, and their heme-containing reaction centers are near the surface.[17] As a result, direct electron transfer between peroxidases and electrodes is possible.[18-20] Again, the resulting cathodes, unlike the platinum anodes, do not foul. Because electron transfer from the heme to carbon or metal electrodes takes place only when the peroxidase is so oriented that its heme function approaches the electrode, only some of the adsorbed enzyme molecules are electroactive, and, with multiple enzyme layers, only reaction centers of the first layer are electrically "connected" to the electrode. For this reason, diffusional mediators, particularly redox dyes, are used to connect multiple enzyme layers to electrodes. However, such mediators are often leached out of the electrodes, which precludes their use *in vivo*, in flow systems, and in microelectrodes. To avoid leaching, HRP electrodes, where electron-conducting redox hydrogels "wire" the enzyme to the electrode, were constructed in 1992.[21] One redox hydrogel was formed on the surface of a glassy carbon electrode by cross-linking HRP and poly(4-vinylpyridine), which was partly complexed with osmium bipyridine redox centers and partly quaternized with ethylamine groups (Fig. 5.1). The osmium centers served to relay electrons from the electrode surface to the reactive centers of the peroxidase (POD) as shown schematically in Fig. 5.2. The ethylamine groups enabled the cross-linking of the film with polyethylene glycol diglycidyl ether (PEGDGE). With sensitivities as high as 1 A/cm^2 M, these "wired" HRP electrodes reached a limiting current that was two orders of magnitude higher than that of HRP electrodes that were similarly made, but without the redox polymer that formed an electron-conducting hydrogel upon cross-linking.

This hydrogel offered some of the advantages of both of the diffusional-mediator-containing systems. It had no leachable mediator, yet connected not only those enzyme molecules that were adsorbed to the electrode surface, but also those remote from it. At the same time, there was no need to add mediator to the sample, and the mediator could not leach out or contaminate the sample.

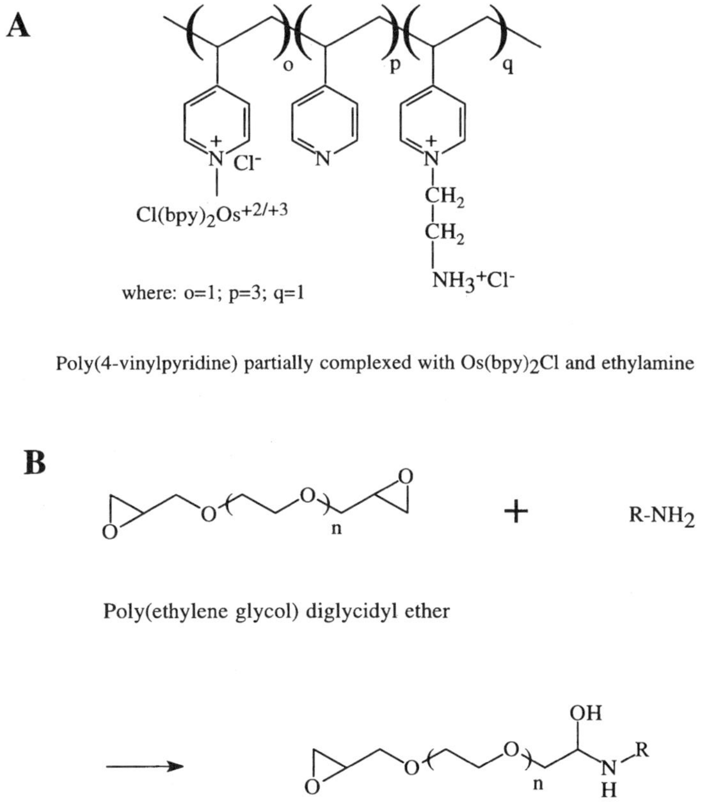

A

where: o=1; p=3; q=1

Poly(4-vinylpyridine) partially complexed with Os(bpy)$_2$Cl and ethylamine

B

Poly(ethylene glycol) diglycidyl ether

Figure 5.1. (A) The structure of the osmium containing redox polymer. (B) the cross-linker, PEGDGE, and its reaction with the primary amines of the redox polymer and the enzyme.

Table 5.1. Amperometric H_2O_2 Sensors Based on HRP-Modified Electrodes[21]

Electrode surface	Mediator or redox matrix	Electrode potential (V)	Sensitivity ($A/cm^2\ M^{-1}$)	Linear range (mM)	Comments
Glassy carbon	None	0.00	0.01		HRP covalently bound to a hydrophilic epoxy network; polyvinylpyridine-derived polyamine cross-linked with PEGDGE
Glassy carbon	Polymer I	0.00	1.00	0.1–100	HRP covalently bound to a hydrophilic epoxy network; polymer I cross-linked with PEGDGE
Spectrographic graphite	None[a]	0.05	0.18	0.1–500	BSA with glutaraldehyde cross-linking
Carbon paste	o-Phenylenediamine[c]	−0.15	na[b]	3.1–200	Butanone peroxide used as the substrate
Platinum	Hexacyanoferrate (0.01 M)[c]	−0.05	[d]	5–1700	HRP immobilized onto nylon net
NMP⁺TCNQ⁻	None[e]	0.05	0.17		HRP entrapped with a dialysis membrane
SnO₂	Ferrocenecarboxylic acid[c]	0.20	0.04	0.01–1	HRP immobilized with glutaraldehyde
Carbon paste	Ferrocene[c]	0.05	na[b]	0.1–10	Nafion coating applied to electrode to prevent loss of mediator
Graphite foil	Potassium Hexacyanoferrate(II)[c]	−0.02[g]	0.03	<600	Electrolyte was dioxane with 15% aqueous buffer
Carbon fiber[f]	None	[h]		40–5000	Biotin–avidin complex used to obtain a surface layer of HRP

Platinum, organic metal or glassy carbon	Potassium ferrocyamide	[i]			Membrane with albumin and glutaraldehyde
Spectrographic graphite or carbon film	Hexacyanoferrate(II)[c]	0.00	[d]	0.1–1000	HRP immobilized on arylamino-derivatized controlled-pore glass, packed into a flow-through reactor
Aminosilylated glassy carbon	Hexacyanoferrate(II)[c]	0.00	[i]		Glycerophosphate oxidase, HRP and BSA covalently cross-linked on the glassy carbon surface
Glassy carbon	Hexacyanoferrate(II)[c]	0.00	[i]		Albumin, glutaraldehyde, HRP and oxidase (xanthine, uricase, glucose) matrix contained at electrode with a dialysis membrane
Gold or graphite	Several[j]	[j]	2.0[k]	0.05–6[k]	HRP free in solution

[a] Uncertainty as to whether the surface species created during electrode pretreatment are mediating.

[b] Macroporous electrode where the true surface area is unknown.

[c] Freely diffusing mediator.

[d] Flow-through system.

[e] Probably mediated by the soluble component of the organic metal or the reaction product of the organic metal.

[f] Microelectrode.

[g] Potential vs. SCE.

[h] Cyclic voltametry used to provide selective detection of oxygen generated by autocatalytic decomposition of hydrogen peroxide.

[i] HRP incorporated into a bienzyme system.

[j] Mediators used and redox potential: $[Ru(NH_3)_5py](ClO_4)_3 = +0.028$ V, $CpFeC_2B_9H_{11} = -0.008$ V, 1,1'-dimethyl-3-(2-aminoethyl)ferrocene $= +0.075$ V, (2-aminoethyl)ferrocene $= +0.185$ V, ferrocenemonocarboxylic acid $= +0.275$ V, (aminomethyl)ferrocene $= +0.309$ V.

[k] Best reported result for ferrocenemonocarboxylic acid.

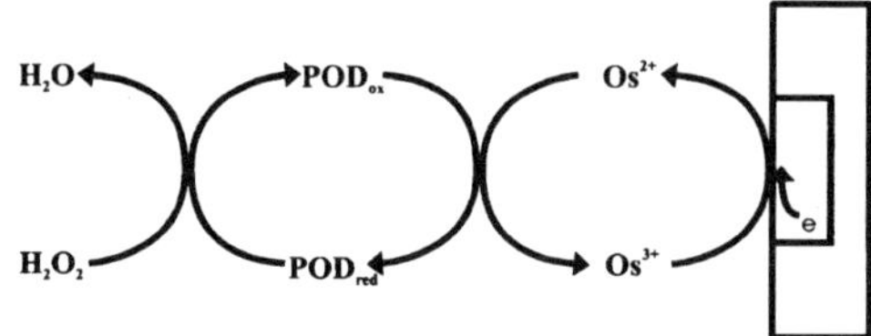

Figure 5.2. Schematic of the electron transfer from the electrode surface to the reactive center of peroxidase (POD) via the osmium bipyridine redox center.

5.3. THERMOSTABLE SOYBEAN PEROXIDASE

The life of the HRP-based sensor was limited by the life of its enzyme. The rate of HRP deactivation, which may involve loss of heme, increases with temperature. It is also conceivable, though not proven, that H_2O_2 itself reacts with the enzyme irreversibly at high temperatures. In the case of the HRP electrodes "wired" with the redox hydrogels, a 10% loss in sensitivity was observed after 3 days of continuous operation at room temperature. However, at the physiological temperature of 37°C, "wired" HRP electrodes showed a 10% loss in sensitivity after only 2 h. Although such stability was sufficient for *ex vivo* use of one week or less, better stability was needed in sensors designed for *in vivo* monitoring, which have to operate continuously for weeks or even months.

In 1995, a thermostable peroxidase isolated from soybean became commercially available.[22] Soybean peroxidase (SBP) was found to have a more tightly bound heme than HRP. Improved retention of the prosthetic group was found to aid in maintaining the catalytically active conformation of the protein, and the rate of SBP deactivation in solution at high temperatures was substantially slower than that of HRP.[22,23]

"Wired" soybean peroxidase electrodes were subsequently built and were found to be thermostable. In continuous operation at 45°C, the "wired" SBP electrodes lost only 0.05% of their sensitivity per hour, while "wired" HRP electrodes lost 5.6%. Even at a temperature as high as 65°C, the SBP-based electrodes lost less than 2% of their sensitivity per hour, compared to a 50% loss per hour in the HRP-based electrodes. Table 5.2 compares the thermostabilities of the "wired" SBP electrodes and the "wired" HRP electrodes.

5.4. BIENZYME SYSTEMS

The development of "wired" peroxidase electrodes opened the way to new and improved bienzyme and multienzyme electrodes. It allowed replacing of the platinum anodes with "wired" peroxidase cathodes as the H_2O_2 sensing element, substantially lessening its fouling. Figure 5.3 is a schematic representation of the process whereby the substrate flux is transduced into an electrical current in such a system.

Table 5.2. Loss in Sensitivity (%/h) at Different
Temperatures of the Electrodes with PEGDGE
Cross-Linked Films

Electrode	Sensitivity loss				
	25°C	45°C	55°C	65°C	75°C
Wired HRP	0.9	5.6	9.9	50	> 50
Wired SBP	0.1	0.05	0.15	1.7	13

Two types of bienzyme electrodes have been built. In the first, both an oxidase and a peroxidase were coimmobilized within the redox hydrogel. In 1981, Kulys et al.[24] introduced the first peroxide-containing bienzyme electrodes. These electrodes had a film consisting of HRP, an oxidase of glucose (lactate, choline, alcohol, amino acid), and bovine serum albumin (BSA), coimmobilized through glutaraldehyde cross-linking. The H_2O_2, formed in the GOx-catalyzed reaction of glucose and oxygen, oxidized HRP, which, in turn, oxidized dissolved $Fe(CN)_6^{-4}$ to $Fe(CN)_6^{-3}$. Electroreduction of the latter ion generated the signal. One problem of this single-layer system was the competing reverse reaction of $Fe(CN)_6^{-3}$ reduction to $Fe(CN)_6^{-4}$ by the glucose-reduced GOx ($FADH_2$) reaction centers. Such reduction, or short-circuiting, diminished the cathodic current. Furthermore, leaching of the mediator limited the life of the electrodes and contaminated the sample.

Ohara et al. reported single-layer bienzyme electrodes without a diffusional mediator based on the "wiring" of HRP in a redox hydrogel.[25] In these, the redox polymer, HRP, and the oxidase were cross-linked in one layer onto the surface of a glassy carbon electrode. Again, the oxidase-catalyzed substrate oxidation produced H_2O_2, which oxidized the HRP, which, in turn, was electroreduced by electron transfer from the electrode to the osmium relays of the redox hydrogel. The sensitivity of this system was high and leaching was no longer a problem. However, short-circuiting between the reduced oxidase and the oxidized osmium relays suppressed the current, particularly at high substrate concentrations. Furthermore, these electrodes had a half-life of only 15 h at 37°C.

A similar configuration combining "wired" HRP and a H_2O_2-producing oxidase has been used in carbon-paste electrodes. Vijayakumar et al.[26] constructed amperometric alcohol biosensors by coimmobilizing commercially available alcohol oxidase with HRP in a carbon-paste matrix.

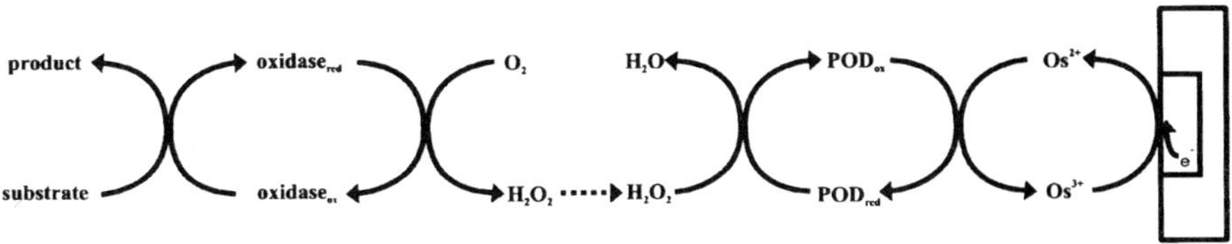

Figure 5.3. Cycle diagram of the "wired" bienzyme system.

Vreeke and Rocca[27] introduced an alternative bienzyme "wired" peroxidase single-layer electrode using biotin and avidin. Because the biotin–avidin complex has a dissociation constant as low as 10^{-15}, this conjugate is frequently used for immobilization of biological molecules onto polymers, including polymers on surfaces of electrodes. In this configuration, avidin, HRP, and the redox polymer were coimmobilized onto the surface of a glassy carbon electrode. In the following step, biotintylated GOx was bound to the avidin in the film. The resulting glucose electrodes behaved similarly to other single-layer bienzyme electrodes and were also subject to short-circuiting.

To eliminate the short-circuiting caused by the transport of electrons from GOx $FADH_2$ centers via the redox centers of the cross-linked hydrogel to the H_2O_2-oxidized heme centers of the peroxidases Ohara et al.[25] proposed the use of bilayer electrodes. In these, a first layer of redox polymer and HRP was cross-linked onto the surface of a glassy carbon electrode. Subsequently, a layer of cross-linked oxidase was immobilized over the preceding layer. Despite the application of separate layers, some short-circuiting was still observed, as evidenced by current suppression at high substrate concentrations. Thus, when the electrode was poised at $+500\,mV$ (SCE) in the presence of its substrate,[27] an electrooxidation current, rather than an electroreduction current, was observed.

Kenausis et al.[28] refined this design by introducing an insulating layer of cellulose acetate between the "wired" peroxidase layer and the oxidase layer, eliminating all short-circuiting without appreciably reducing the sensitivity (Fig. 5.4). Furthermore, by using thermostable soybean peroxidase (SBP) rather than HRP, the operational stability at 37°C was improved. These glucose-sensing bienzyme electrodes lost less than 20% of their initial current after 2 weeks of continuous operation at 37°C when an excess of GOx was used. Even after part of the enzyme degraded, enough activity remained to ensure the complete conversion of the glucose flux through its reaction with oxygen to H_2O_2 and gluconolactone (Fig. 5.5).

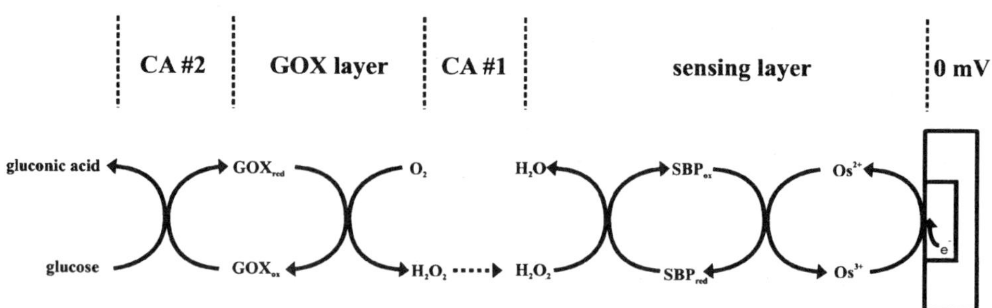

CA#1: prevents short-circuiting
CA#2: prevents GOX leaching

Figure 5.4. Cycle diagram of the four-layer glucose sensor (CA = cellulose acetate).

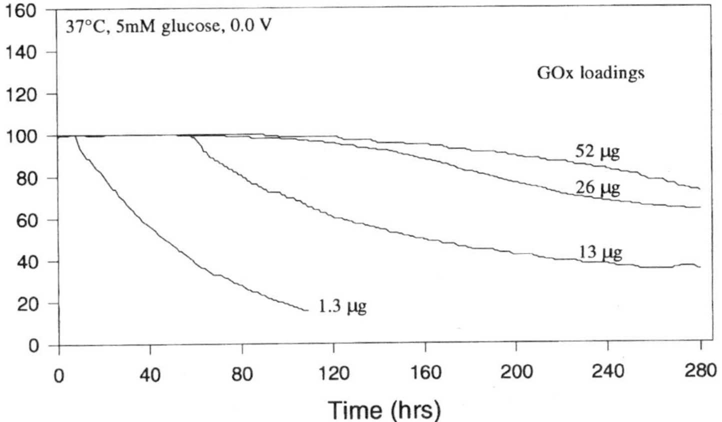

Figure 5.5. Stability of SBP-based glucose sensors at different GOx loadings.

"Wired" peroxidase electrodes have also been used to detect glucose, lactate, glutamate, and acetylcholine in systems with a liquid chromatographic column (LC) and a postcolumn-immobilized enzyme reactor (IMER).[12,29,30] In these systems, the test samples are run through the chromatographic column in order to separate the components of the mixture. The column effluent is subsequently passed through an immobilized enzyme reactor, where in the presence of oxygen, the analyte, i.e., the substrate of the oxidase, is converted to product and H_2O_2. The effluent of the IMER is then passed through a flow-injection cell where a "wired" peroxidase electrode measures the concentration of H_2O_2, which is then related to the concentration of analyte.

5.5. APPLICATIONS

5.5.1. NAD(P)H Sensing

NADH and NADPH are cofactors of dehydrogenases and reductases. Unlike $FADH_2$, NAD(P)H is not readily oxidized by molecular oxygen. However, Gorton, Kulys, and co-authors[8,31,32] showed, that the quinoid dye mediators catalyze the conversion of NAD(P)H and oxygen to $NAD(P)^+$ and H_2O_2. Thus, NAD(P)H was assayed by measuring the depletion of oxygen or by spectrophotometrically measuring the peroxidase-catalyzed oxidation of a leuco-dye by H_2O_2.[33] Vreeke et al.[21] showed that the *N*-methylphenazolium ion is rapidly reduced by NAD(P)H, and that the *N*-methylphenazine produced reacts rapidly with dissolved molecular oxygen. Thus, in the presence of $\sim 10^{-6}$ M *N*-methylphenazonium ion, the NAD(P)H is quantitatively air-oxidized to H_2O_2 and $NAD(P)^+$. The H_2O_2 produced is electroreduced at 0.0 V (SCE) by the "wired" peroxidase electrode. The sensitivity of this NAD(P)H electrode was 1 A/cm² M, and its linear range was 1×10^{-7} to 2×10^{-4} M.

5.5.2. Avidin and Biotin

Enzyme-linked immunosorbent assay (ELISA) and related affinity sensing methods, by which antigens and antibodies are commonly assayed, often involve attachment of an affinity reagent to a surface, to which a complementary reagent is selectively bound. Vreeke et al.[34] designed a heterogeneous electrochemical system, based on "wired" avidin and peroxidase-labeled biotin. Catalytic electroreduction of H_2O_2 was observed for as little as 1 μg/cm^2 of bound HRP, and sensitivities were as high as 1 A/cm^2 M. Competitive binding of free biotin resulted in a signal decrease (Fig. 5.6). Similarly, the presence of avidin brought about a decrease in signal. The competitive binding allowed use of the electrode as a sensor for both biotin and avidin. Although the sensitivity and detection limits of this sensor did not approach those of the most sensitive immunosensors, it had the advantages of being fast and easy to use, in that it did not require washing steps or separation of reagents. Recently, Lu et al.[35] reported a similar sensor design based on anti-HRP rather than avidin.

5.5.3. Oligonucleotide Sensing

De Lumley-Woodyear et al.[36] developed an amperometric oligonucleotide sensor, conceptually related to the avidin–biotin sensor. Figure 5.7 shows the chemical and electrical steps involved in observing the hybridization through the generation of an electrocatalytic H_2O_2 reduction current. First, a copolymer of acrylamide, (PAA), $[Os(bpy)^2Cl]^{+/2+}$ complexed vinylimidazole (PVI-Os), and acrylhydrazide (PAA-hydrazide), was cross-linked on the surface of a glassy carbon electrode. The 25–30 base assayed oligonucleotide $[p(dT)_{25-30}]$ was carbodiimide-coupled through reacting its terminal phosphate with the redox polymer's hydrazide.

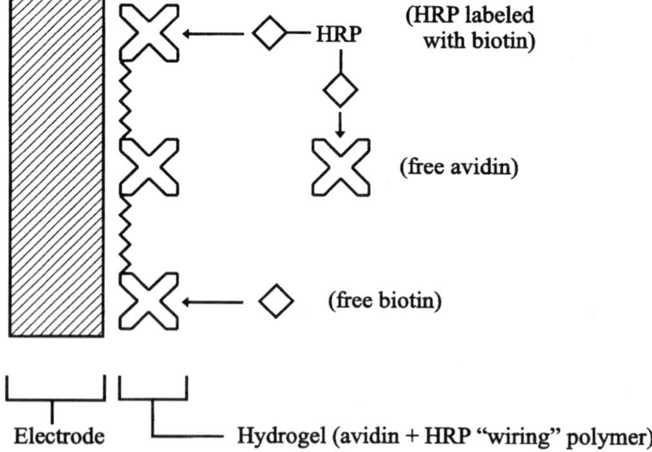

Figure 5.6. Schematic of the competitive binding processes determining the current output of the avidin/biotin/biotintylated HRP sensor.

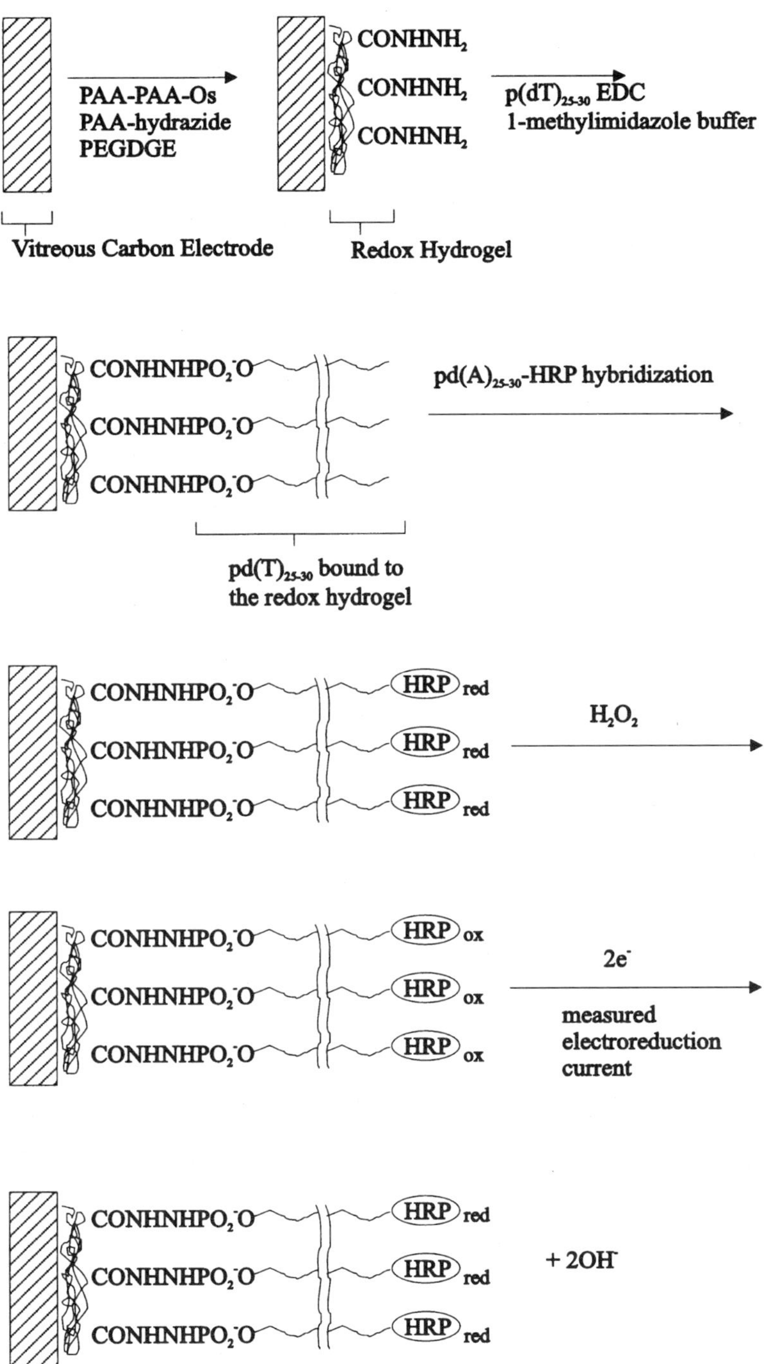

Figure 5.7. Design of the oligonucleotide probe.

When the electrode was incubated in a solution containing HRP-labeled probing oligonucleotide, the oligonucleotide–HRP was bound to the redox hydrogel and became "wired." The electrode now contained both HRP and its "wire," becoming catalytic for H_2O_2 electroreduction to water. Thus, hybridization resulted in a catalytic electroreduction current. In electrodes prepared identically, but without the target oligonucleotide attached to the hydrogel, only a fifteenfold smaller current was detected.

5.5.4.   Characterization of Electrodes Generating and Consuming H_2O_2

"Wired" peroxidase microelectrodes were applied in the locating of sites on electrodes where H_2O_2 was generated. On an indium tin oxide glass partially coated with the photocatalyst titanium dioxide, Sakai et al.[37] determined that H_2O_2 was produced mainly through the reduction of dissolved oxygen by photogenerated electrons. Horrocks et al.[38] analyzed the potential ranges where H_2O_2 was generated and where it was consumed on carbon, gold, and platinum surfaces.

5.5.5.   Organic-Phase Peroxide Sensors

An organic phase enzyme electrode (OPEE) was made with native and N-hydroxysuccinimide ester (NHS)-modified HRP electrostatically complexed, with an osmium bis(bipyridyl) poly(4-vinylpyridine) polymer.[39] The enzyme electrodes operated in organic (90% acetonitrile) phase. Chemical modification of the HRP by its NHS coupling to the redox polymer improves the peroxide sensitivity fivefold and the linear response range threefold.

ACKNOWLEDGMENTS: We acknowledge support of this work by the National Institutes of Health, the Welch Foundation, the National Science Foundation, and the Office of Naval Research.

REFERENCES

1. Kacaniklic V, Johnasson K, Marko-Varga G, et al. Amperometric biosensors for the detection of L- and D-amino acids based on coimmobilized peroxidase and L- and D-amino acid oxidases in carbon paste electrodes. *Electroanal* 1994;6:381–390.
2. Moussy F, Harrison D, O'Brien D, et al. Performance of subcutaneously implanted needle-type glucose sensors employing a novel trilayer coating. *Anal Chem* 1993;65:2072–2077.
3. Thome-Duret V, Gangneran M, Zhang Y, et al. Modification of the sensitivity of glucose sensor implanted into subcutaneous tissue, *Diab Metab* 1996;22:174–178.
4. Hlavay J, Haemmerli S, Guilbault G. G. Fibre-optic biosensor for hypoxanthine and xanthine based on a chemiluminescence reaction. *Biosens Bioelectron* 1994;9:189–195.
5. Papkovsky D, Olah J, Kurochkin I. Fiber-optic lifetime-based enzyme biosensor, *Sens Act* 1993;11:525–530.
6. Turner APF. Applications of amperometric biosensors. *Anal Proc* 1991;28:376–377.

7. Jacobs E, Vadasdi E, Sarkozi L, et al. Analytical evaluation of i-STAT portable clinical analyzer and use by nonlaboratory health-care professionals. *Clin Chem* 1993;39:1069–1074.

8. Gorton L, Domingnez E., Emneus J, et al. Selective detection in flow analysis based on the combination of immobilized enzymes and chemically modified electrodes. *Anal Chim Acta* 1991;250:203–248.

9. Vadgama P, Crump P. Biosensors: Recent trends, a review, *Analyst* 1992;117:1657–1670.

10. Wang J, Chen Q. Enzyme microelectrode array strips for glucose and lactate. *Anal Chem* 1994;66:1007–1011.

11. Wang J, Chen Q, Renschler C, et al. Ultrathin porous carbon films as amperometric transducers for biocatalytic sensors. *Anal Chem* 1994;66:1988–1992.

12. Yang L, Janle E, Huang T, et al. Applications of "wired" peroxidase electrodes for peroxide determination in liquid chromatography coupled to oxidase immobilized enzyme reactors. *Anal Chem* 1995;67:1326–1331.

13. Wang J, Frieha B, Naser N, et al. Amperometric biosensing of organic peroxides with peroxidase-modified electrodes. *Anal Chim Acta* 1991;254:81–88.

14. Cosgrove M, Moody GJ, Thomas JDR Chemically immobilized enzyme electrodes for hydrogen peroxide determination. *Analyst* 1988;113:1811–1815.

15. Tatsuma T, Okawa Y, Watanabe T. Enzyme monolayer- and bilayer-modified tin oxide electrodes for the determination of hydrogen peroxide and glucose. *Anal Chem* 1989;61:2352–2355.

16. Paddock RM, Bowden EF. Electrocatalytic reduction of hydrogen peroxide via direct electrical transfer from pyrolytic graphite electrodes to irreversibly adsorbed cytochrome c peroxidase, *J Electroanal Chem: Interf Electrochem* 1989;260:487–494.

17. Vreeke M, Heller A. In: Usmani AM, Akmal N, eds. *Diagnostic Biosensor Polymers*. ACS Symposium Series 556; American Chemical Society, Washington, DC, 1994, Ch 15.

18. Mulchandani A, Wang C, Weetall H. Amperometric detection of peroxides with poly(anilinomethyl-ferrocene)-modified enzyme electrodes. *Anal Chem* 1995;67:94–100.

19. Csoregi E, Gorton L, Marko-Varga G. Peroxidase-modified carbon fiber microelectrodes in flow through detection of hydrogen peroxide and organic peroxides. *Anal Chem* 1994;66:3604–3610.

20. Ruzgas T, Csoregi E, Emneus J, et al. Peroxidase-modified electrodes: fundamentals and application. *Anal Chim Acta* 1996;330:123–138.

21. Vreeke M, Maidan R, Heller A. Hydrogen peroxide and beta-nicotinamide adenine dinucleotide sensing amperometric electrodes based on electrical connection of horseradish peroxidase redox centers to electrodes through a three-dimensional electron relaying polymer network. *Anal Chem* 1992;64:3084–3090.

22. McEldoon J, Dordick J. Unusual thermal stability of soybean peroxidase. *Biotechnol Prog* 1996;12:555–558.

23. Vreeke M, Yong K, Heller A. A thermostable hydrogen peroxide sensor based on "wiring" of soybean peroxidase. *Anal Chem* 1995;67:4247–4249.

24. Kulys J, Pesliakinene M, Samalius A. The development of bienzyme glucose electrodes. *Bioelectrochem Bioenerg* 1981;8:81–88.

25. Ohara T, Vreeke M, Battaglin F, et al. Bienzyme sensors based on "electrically wired" peroxidase. *Electroanal* 1993;5:825–831.

26. Vijayakumar AR, Csoregi E, Heller A, et al. Alcohol biosensors based on coupled oxidase–peroxidase systems. *Anal Chim Acta* 1996;327:233–234.

27. Vreeke M, Rocca P. Biosensors based on cross-linking of biotinylated glucose oxidase by avidin. *Electroanal* 1996;8:55–60.

28. Kenausis G, Chen Q, Heller A. Electrochemical glucose and lactate sensors based on "wired" thermostable soybean peroxidase operating continuously and stably at 37°C. *Anal Chem* 1997;69:1054–1060.

29. Kato T, Liu J, Yamamoto K, et al. Detection of basal acetylcholine release in the microdialysis of rate frontal cortex by high performance liquid chromatography using horseradish peroxidase–osmium redox polymer electrode with pre-enzyme reactor. *J Chromatogr B: Biomed Appl* 1996;682:162–166.

30. Niwa O, Torimitsu K, Morita M, et al. Concentration of extracellular L-glutamate released from cultured nerve cells measured with a small volume online sensor. *Anal Chem* 1996;68:1865–1870.

31. Cenas N, Kanapieniene K, Kulys J. Electrocatalytic oxidation of NADH on carbon black electrodes. *J Electroanal Chem: Interf Electrochem* 1985;189:163–169.

32. Kulys J. Enzyme electrodes based on organic metals, *Biosensors* 1986;2:3–14.
33. Guilbault, G. G. *Analytical Uses of Immobilized Enzymes*. New York: Marcel Dekker, Inc, 1984, p 60.
34. Vreeke M, Rocca P, Heller A. Direct electrical detection of dissolved biotinylated horseradish peroxidase, biotin and avidin. *Anal Chem* 1995;67:303–306.
35. Lu B, Iwuoha E, Smyth M, et al. Development of an amperometric immunosensor for horseradish peroxidase (HRP) involving a non-diffusional osmium redox polymer co-immobilized with anti-HRP antibody. *Anal Comm* 1997;34:21–24.
36. De Lumley-Woodyear T, Campbell C, Heller A. Direct enzyme-amplified electrical recognition of a 30-base model oligonucleotide. *J Am Chem Soc* 1996;118:5504–5505.
37. Sakai H, Baba, R, Hashimoto K, et al. Local detection of photoelectrochemically produced H_2O_2 with a "wired" horseradish peroxidase sensor. *J Phys Chem* 1995;99:11896–11900.
38. Horrocks BR, Schmidtke D, Heller A, et al. Scanning electrochemical microscopy: 24. Enzyme ultramicroelectrodes for the measurement of hydrogen peroxide at surfaces. *Anal Chem* 1993;65:3605–3614.
39. Iwuoha EI, Leister I, Miland E, et al. Reactivities of organic phase biosensors enhancement of the sensitivity and stability of amperometric peroxidase biosensors using chemically modified enzymes. *Anal Chem* 1997,69.1674–1681.

Nonseparation Electrochemical Enzyme Immunoassay Using Microporous Gold Electrodes

M. W. Ducey, A. M. Smith, R. Smith, C. Duan, and Mark E. Meyerhoff

6.1. INTRODUCTION*

In recent years, the coupling of immunological reactions to electrochemical and optical detectors has led to a flurry of research in the area of immunosensor technology.[1-15] While there have been extensive efforts to devise so-called "direct" immunosensors, where antibodies are immobilized on the surface of a transducer and the immunological binding reaction is monitored via a change in, e.g., innate fluorescence, surface refractive index, or electrical capacitance, the most successful immunosensors from a practical analytical point of view are those that involve the use of labeled reagents (e.g., enzymes, fluorophores). Indeed, such methods are rather similar to their more classical immunoassay counterparts, except that the antibodies are immobilized on the surface of the transducer (e.g., electrode, optical fiber), rather than on the walls of microtiter plates, magnetic particles, or some other solid-phase material. Hence, in many of the immunosensor methods reported to date, where labeled reagents are utilized, extensive washing steps to remove unbound labeled

*ABBREVIATIONS: NEEIA, nonseparation electrochemical enzyme immunoassay; ALP, alkaline phosphatase; β-GAL, β-D-galactosidase; TBG, tris buffered gelatin (0.01 M tris(hydroxymethyl)aminomethane, pH 7.4, 1 mM $MgCl_2$, 0.1 mM $ZnCl_2$, 150 mM NaCl, and 0.1% gelatin); BSA, bovine serum albumin; hCG, human chorionic gonadotropin; PSA, prostate specific antigen; pAPP, *p*-aminophenyl phosphate; EDAC, 1-ethyl-3,3-dimethyl-aminopropyl carbodiimide; TBS, tris buffered saline (0.01 M tris(hydroxymethyl)aminomethane, pH 7.4, 1 mM $MgCl_2$, 0.1 mM $ZnCl_2$, 150 mM NaCl).

M. W. Ducey, A. M. Smith, R. Smith, C. Duan, and Mark E. Meyerhoff • Department of Chemistry, University of Michigan, Ann Arbor, Michigan 48109.

Biosensors and Their Applications, edited by Yang and Ngo, Kluwer Academic/Plenum Publishers, New York, 1999.

species are required.[6,13,14] Herein, we review recent efforts to devise a novel electrochemical enzyme immunoassay arrangement capable of detecting both large and small molecules at low concentrations without the need for any discrete separation or washing steps.

Relatively rapid homogeneous enzyme immunoassays, in which no separation of bound and free enzyme label is required, have already been devised for a large number of low-molecular-weight analytes. Two popular homogeneous enzyme immunoassay formats are the EMIT (enzyme multiplied immunoassay technique) and CEDIA (cloned enzyme donor immunoassay) methods. In the EMIT method,[16-20] an enzyme–analyte conjugate is mixed with a known amount of antibody and the sample or calibrator, resulting in competition between the antigen in the sample and the conjugate for the limited number of antibody binding sites. Antibody binding induces a conformational change in the enzyme or sterically hinders the diffusion of substrate to the active site of the enzyme, resulting in lower enzyme activity.[18] Malate dehydrogenase and glucose-6-phosphate dehydrogenase are the most commonly employed enzymes, owing to the high degree of inhibition of the enzyme activity (up to 90%) observed upon the formation of the ligand–enzyme conjugate/anti-ligand–antibody complex. Because the extent of antibody binding to the conjugate is dependent on analyte concentration in the sample, the observed enzyme activity is proportional to analyte concentration. Unfortunately, this method is generally limited to the detection of small molecules, although certain modifications (e.g., the use of macromolecular enzyme substrates[21]) have been proposed as a means of devising truly homogeneous assays of higher-molecular-weight protein analytes.

The CEDIA method relies on the use of a recombinant enzyme, commonly β-D-galactosidase (β-GAL), in which two inactive fragments of the enzyme are prepared.[22-25] Enzyme activity is restored by combining these fragments. In this assay, the smaller of these fragments (ED, or enzyme donor), is prepared by automated synthesis and is conjugated to the hapten in such a way that recombination of ED–hapten with the larger fragment (EA, or enzyme acceptor, a recombinant protein lacking the ED region) restores full enzyme activity. However, complexation of the ED–hapten with antibody inhibits reconstitution of the active enzyme. Thus, binding of the ED–hapten to the specific antibody is inversely proportional to the concentration of free hapten in the sample. As with EMIT, the CEDIA method is generally applicable only to the detection of low-molecular-weight analytes (>1000 Dal). Further, since both the EMIT and CEDIA methods are based on detecting enzymatic activity via spectrophotometric detection, it is difficult to adapt these assay methods for measurements in samples as complex as whole blood.

As an alternative to the homogeneous immunoassay formats described above, nonseparation immunoassays can also be devised by taking advantage of the enhanced surface concentration of labeled species that occurs after the formation of immune complexes on a solid-phase surface. If the surface is also a transducer, there is a potential for devising detection schemes to spatially discriminate labeled species bound to the surface from uncomplexed labeled reagents remaining free in solution. Immunoanalytical methods using the principles of total internal reflection fluorescence (TIRF)[5,26-30] are driven, in part, by taking advantage of such distance effects. In such methods, surface immunobound fluorophore labels are observed without removing unbound fluorophores via excitation by an evanescent wave of appropriate wavelength on the surface of either a fiber or planar waveguide. Although proposed more than 20 years ago,[28] these TIRF-based nonseparation immunosensors have not yet been successfully commercialized as a result of performance issues, especially when attempting to carry out assays in undiluted whole blood.

As an alternate approach to nonseparation immunoassay of both large and small molecules, we have recently described a novel system that employs electrochemical detection of enzyme labels at microporous gold electrodes onto which antibodies are immobilized, thus yielding a nonseparation electrochemical enzyme immunoassay (NEEIA).[31–34] Microporous gold electrodes are formed by vapor deposition of gold onto nylon membranes followed by immobilization of a binding protein or antibody via a chemisorbed layer of thioctic acid. Thus, the electrode serves as both the immunosorbent solid phase and the transduction element for the immunoassay. Initial NEEIA experiments were carried out using a noncompetitive sandwich-type immunoassay format and involved the detection of several proteins in serum as well as whole blood samples. Assays are performed by mounting the microporous electrode between two chambers of a diffusion cell and adding sample containing the analyte of interest and an excess of a second enzyme labeled antibody (reporter antibody) to the chamber facing the gold.

After completion of the binding reaction, a potential sufficient to oxidize or reduce the product of the enzymatic reaction is applied to the modified electrode. Substrate is then added to the backside of the diffusion cell (see Fig. 6.1) and allowed to diffuse through the membrane, where it encounters surface-bound reporter antibody and is converted to an electroactive product. This product is subsequently detected at the modified electrode, resulting in a signal that is proportional to the concentration of the analyte in the sample. More recent work has focused on the use of the NEEIA system for the detection of small molecules via competitive binding/immunoassays. Experiments were carried out in a manner identical to that for the noncompetitive assays. For both competitive and noncompetitive NEEIA formats, unbound enzyme remains in solution following completion of the binding reaction owing to the absence of a washing step. Therefore NEEIA, as other nonseparation techniques, requires a means to differentiate between this unbound enzyme and enzyme that is bound during the course of the binding reaction. This is achieved in the NEEIA system through the spatial resolution of bound enzyme from unbound enzyme. This spatial resolution arises when enzyme bound at the electrode surface is exposed to the diffusing substrate, which is converted to an electrochemical product by the immunobound enzyme, prior to encountering the unbound enzyme in the bulk sample phase (Fig. 6.1).

Enzyme immunoassays yield optimal results when a stable enzyme label with a relatively high turnover rate is employed. Enzyme labels used in NEEIA have the additional requirement that the product of the enzymatic reaction be electrochemically oxidized or reduced at a potential low enough to minimize interference by endogenous species in the sample (e.g., ascorbate, urate, catecholamines). Alkaline phosphatase (ALP) was initially chosen as the enzyme label for the NEEIA system

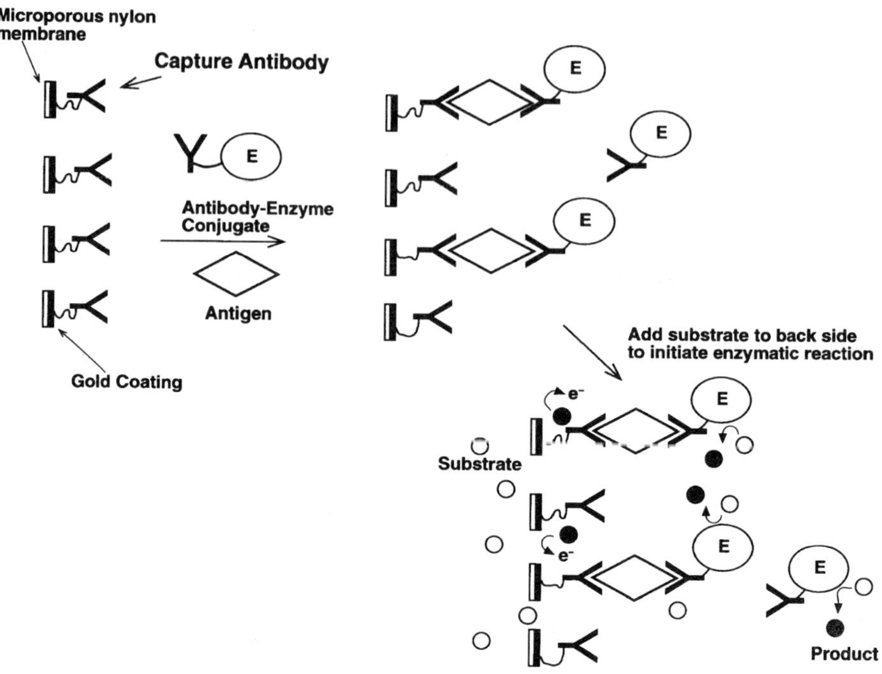

Figure 6.1. Schematic representation of a noncompetitive sandwich-type nonseparation electrochemical enzyme immunoassay (NEEIA).

owing to the low redox potential of *p*-aminophenol ($+190\,\mathrm{mV}$ vs. Ag/AgCl),[35] the product of enzymatic dephosphorylation of *p*-aminophenyl phosphate (pAPP) (See Fig. 6.2). Since the optimal pH for ALP activity is 10, the substrate solution is buffered to this pH. This results in the establishment of a pH as well as a substrate concentration gradient between the substrate solution on the back side and the sample solution (buffered to pH 7.4) on the front side of the microporous membrane. Surface-bound enzyme is exposed to a higher pH than unbound enzyme in the bulk of the sample, resulting in a higher turnover rate. It was experimentally determined[31] that ALP has 96% lower activity at pH 7.4 than at pH 10. Thus, the pH gradient combined with the spatial resolution between the bound and unbound enzyme adds to the sensitivity of the NEEIA system.

In addition to ALP, it is also possible to employ β-GAL as an enzyme label in NEEIA by using *p*-aminophenyl β-D-galactopyranoside (pAPG) as the substrate (Fig. 6.2). Since the optimum pH for β-GAL is 7.7, the pH gradient is not as steep as that for ALP; however, the substrate concentration gradient and high local concentration of β-GAL on the electrode still permit differentiation between bound and unbound enzyme. In this chapter we provide experimental details and a summary of analytical results relating to the use of the NEEIA system for detection of both large proteins, using a sandwich assay scheme, and low-molecular-weight analytes, using a competitive binding assay approach.

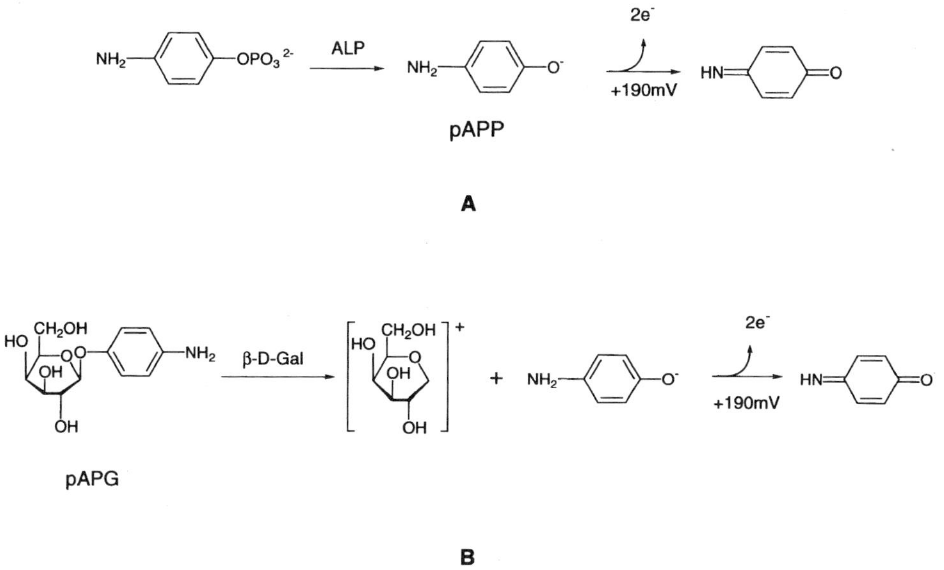

ALP

pAPP

$2e^-$

+190mV

A

β-D-Gal

pAPG

$2e^-$

+190mV

B

Figure 6.2. Enzymatic and subsequent electrochemical detection reactions for two enzymes that can be employed as labels in the NEEIA system: (A) Alkaline phosphatase (ALP); (B) β-D-galactosidase (β-GAL).

6.2. EXPERIMENTAL

6.2.1. Apparatus

Amperometric measurements were performed with a BAS CV-27 potentiostat in a three-electrode configuration, where the potential applied to the microporous gold electrode was $+190$ mV vs. Ag/AgCl reference and a Pt auxiliary. The diffusion cells employed were built in-house (Fig. 6.3). Data were recorded on a Fisher series 5000 chart recorder or a MacLab A/D converter using Chart software and a Macintosh computer. Gold-coating of the nylon membranes was carried out with a Denton Vacuum Desk II or a Plasma Sciences LVC-100 cold sputter-etch unit.

6.2.2. Reagents

Antihuman chorionic gonadotropin (anti-hCG) and anti-prostate specific antigen (anti-PSA) monoclonal capture antibodies, ALP–anti-hCG and ALP–anti-PSA monoclonal reporter antibody conjugates, and human PSA were generously provided by Hybritech (La Jolla, CA). Monoclonal anti-digoxin was purchased from Biodesign International (Kennebunk, ME). The digoxin–ALP conjugate was purchased from Diatech Research Biochemicals (Boston, MA). Avidin, biotin, biotin–ALP conjugate, hCG heat-shock bovine serum albumin, porcine gelatin, tris buffer salts,

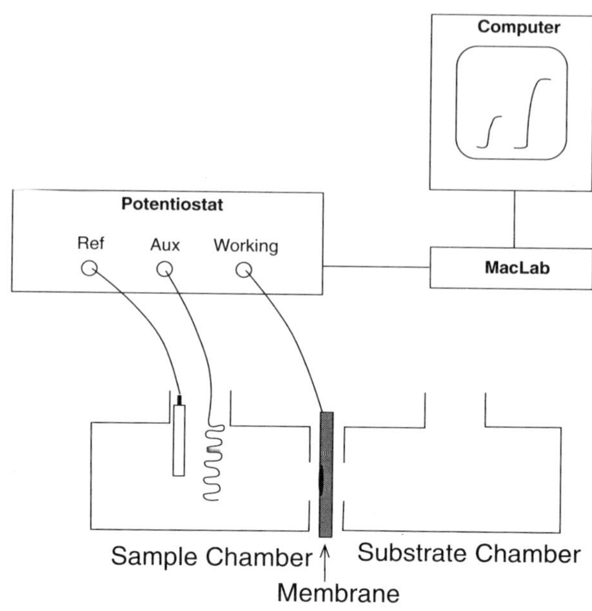

Figure 6.3.   Schematic representation of the experimental setup for conducting NEEIA assays.

thioctic acid and 1-ethyl-3,3-dimethylaminopropyl carbodiimide (EDAC) were purchased from Sigma (St. Louis, MO). A potato virus A enzyme-linked immunosorbent assay (ELISA) kit was obtained from Boehinger Mannheim Corp. (Indianapolis, IN). 4-Nitrophenyl phosphate was purchased from Pierce (Rockford, IL). Microporous neutral nylon membranes (0.2 μm pore size) were purchased from Gelman Science (Ann Arbor, MI) and Schleicher & Schuell (Keene, NH). The alkaline phosphatase substrate, pAPP, was synthesized by the reduction of *p*-nitrophenyl phosphate according to a previously reported procedure.[36] All other chemicals were of reagent grade or better purity.

6.2.3.   Preparation of Microporous Gold Electrodes and Immobilization of Binding Proteins

Single-use microporous gold electrodes were prepared as previously described (Fig. 6.4).[32] Briefly, gold disks 6 mm or 4 mm in diameter (approximately 60 nm thick) with 2-mm-wide outlet strips to the edge of the membrane (for electrical connection) were vapor-deposited onto 0.2-μm-pore-size nylon membranes using an appropriate mask. Electrodes were then placed into a solution of thioctic acid (1.5% w/v) in anhydrous ethanol for a minimum of 12 h with shaking. Following an ethanol washing step, the carboxylic acid moiety of the chemisorbed thioctic acid was then activated with EDAC in anhydrous acetonitrile for 5 h with shaking.

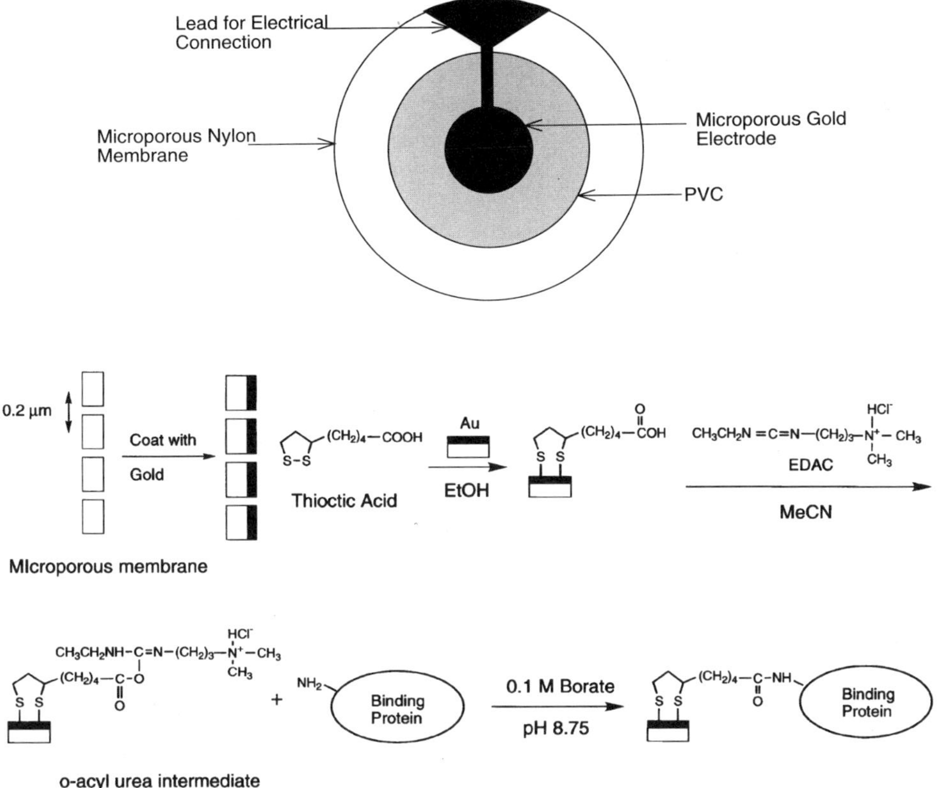

Figure 6.4. (A) pictorial representation of a microporous gold electrode used in NEEIA. (B) Reaction sequence for the covalent immobilization of binding proteins (antibodies) to the surface of a microporous gold electrodes.

Following an acetonitrile washing step, a layer of plasticized polymer [33% (w/w) poly(vinyl chloride), 67% (w/w) bis(2-ethylhexyl) sebacate] dissolved in tetrahydrofuran (1:6 w/v) was cast around the electrode. This insulates the gold lead and prevents wicking of the sample and substrate solutions through any portion of the membrane other than that covered by the electrode. The binding protein specific for the analyte of interest was then immobilized onto the electrode by placing $30\,\mu L$ or $10\,\mu L$ (for electrodes of 6-mm and 4-mm diameter, respectively) of the protein (typically $2\,mg/mL$) in 0.1 M sodium borate, pH 8.75, directly onto the electrode. The electrodes were then stored at 4°C for at least 24 h prior to use. Before use the electrodes were treated with 5% ethanolamine in 0.1 M borate buffer, pH 8.75, to block any unreacted O-acylurea intermediates remaining on the electrode surface. The electrodes were then rinsed in water or TBG and dried before being mounted within the diffusion cell for measurements.

6.2.4. Nonseparation Sandwich-Type Electrochemical Enzyme Immunoassays

Microporous gold immunoelectrodes (6 mm), prepared by depositing 30 μL of a 2-mg/mL solution of the capture antibody onto the electrode surface, were placed in the diffusion cell. A sample containing the analyte of interest and an ALP-labeled reporter antibody (recognizing a second epitope on the analyte of interest) in either a buffered solution or some complex matrix (i.e., serum or whole blood) was placed into the chamber of the diffusion cell that faced the gold electrode. The sample was then incubated with the electrode for a fixed time period with stirring. Following completion of the incubation period (30 min for hCG and PSA, 4 h for potato virus A), the reference and auxiliary electrodes were placed into the sample chamber and a potential of $+190$ mV was applied to the microporous gold electrode. Following decay of the charging current, the substrate solution (5 mg/mL pAPP in 1.0 M sodium carbonate, pH 10, 1 mM $MgCl_2$, and 0.1 mM $ZnCl_2$) was added to the empty chamber of the diffusion cell. The resulting steady-state current recorded after 1 min was directly proportional to the concentration of the analyte in the sample (Fig. 6.5).

6.2.5. Nonseparation Competitive Electrochemical Enzyme Binding/Immunoassays

Microporous gold electrodes (4 mm) were prepared by depositing 10 μL of a solution of the binding protein or antibody onto the electrode surface. The optimal mass of binding protein as well as the optimal concentration of the enzyme-labeled competitor were determined for each analyte by a limited checkerboard titration.

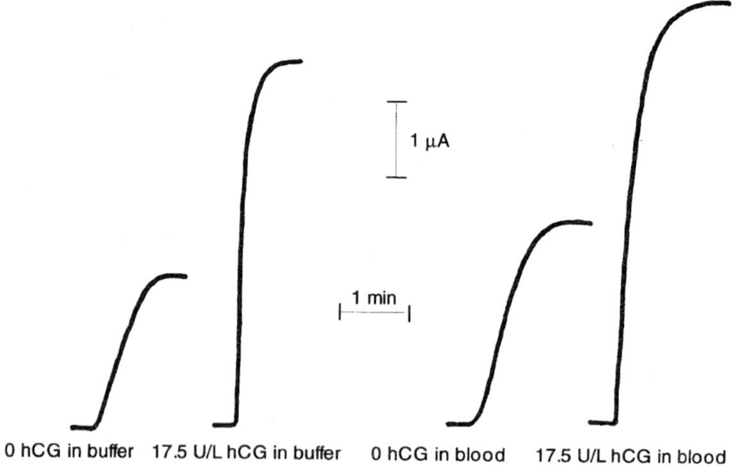

Figure 6.5. Typical dynamic responses observed for a noncompetitive sandwich-type NEEIA (response toward hCG in citrate buffer and whole blood depicted) (from Duan and Meyerhoff[31] with permission).

Following placement of the electrode into the diffusion cell, sample containing the analyte of interest was placed into the sample side of the diffusion cell and incubated for an optimized time period with stirring (30 min for biotin and 60 min for digoxin). Following completion of the incubation period, the reference and auxiliary electrodes were put into the sample chamber and a potential of $+190\,mV$ was applied to the microporous gold electrode. A solution of substrate was added to the substrate side of the diffusion cell and the steady-state current recorded (Fig. 6.6). Owing to the competitive nature of the assay the analytical signal is inversely proportional to the concentration of analyte in the sample.

6.3. RESULTS AND DISCUSSION

6.3.1. Noncompetitive Electrochemical Enzyme Immunoassay for Proteins and Microorganisms

Work in this area has resulted in assays for three different model analytes of interest, hCG, PSA, and potato virus A. Analytes could be detected in sample matrices ranging from buffer to undiluted serum and whole blood with relatively short analysis times, demonstrating the utility of NEEIA in the detection of macromolecular species.

The hormone hCG was detected by employing an anti-hCG monoclonal capture antibody and an ALP–anti-hCG conjugate. When hCG was present in the samples, it bound to the surface-immobilized capture antibody through one epitope and to the ALP-labeled conjugate through a second distinct epitope, resulting in an antibody:hCG:antibody–ALP "sandwich" immobilized onto the surface of the microporous gold electrode. Increases in surface-bound enzyme activity were thus directly proportional to the concentration of hCG in the samples. Detection of hCG was demonstrated in citrate buffer as well as in whole blood (Fig. 6.7).[31] It should be noted that for all concentrations of hCG, the steady-state current observed in whole blood was slightly elevated compared to the values obtained in buffer for the same concentration of analyte. This is probably due to the presence of endogenous ALP in the blood sample as well as to differences in viscosity and buffer capacity of blood compared to buffer. As a result, slightly higher detection limits were obtained for hCG in whole blood [2.5 U/L (using $S/N = 3$)], as compared to citrate buffer (0.8 U/L). The assay was linear in both cases from the detection limit up to approximately 40 U/L.

Assays for PSA were developed using a monoclonal anti-PSA capture antibody and an ALP–anti-PSA conjugate. Detection of PSA was demonstrated in three different sample matrices — tris buffered saline (TBS), whole blood, and undiluted female serum. Dose–response curves for PSA in both TBS and whole blood are shown in Fig. 6.8. The assays gave comparable linear ranges in both mediums (TBS: 0.1 to 15 μg/L; whole blood: 0.5 to 15 μg/L). Again, steady-state signals for PSA were elevated in whole blood as compared to TBS, for the same reasons described for hCG. PSA assays in whole blood also exhibited higher detection

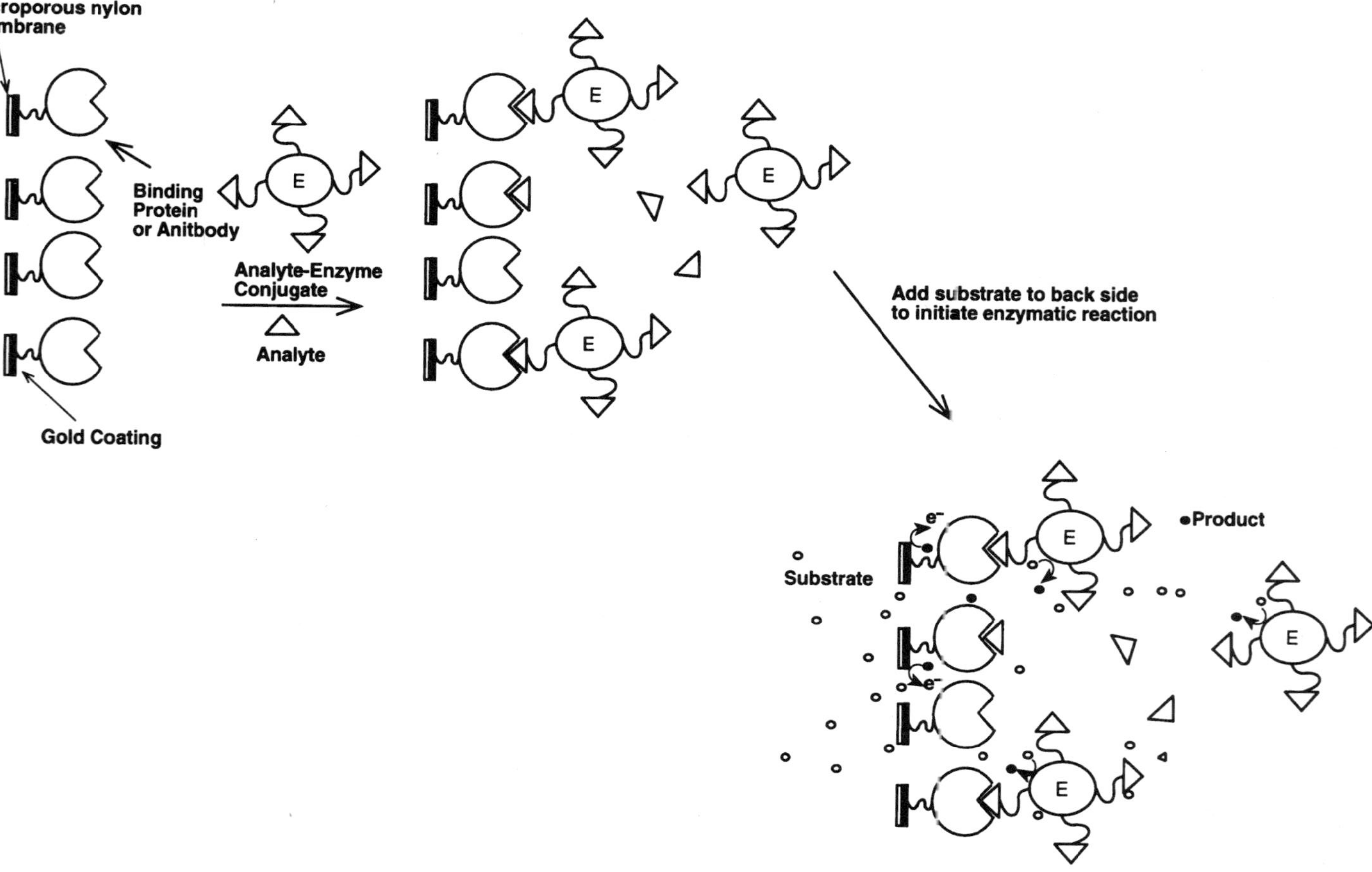

Figure 6.6. Schematic representation of a competitive binding assay using the NEEIA format.

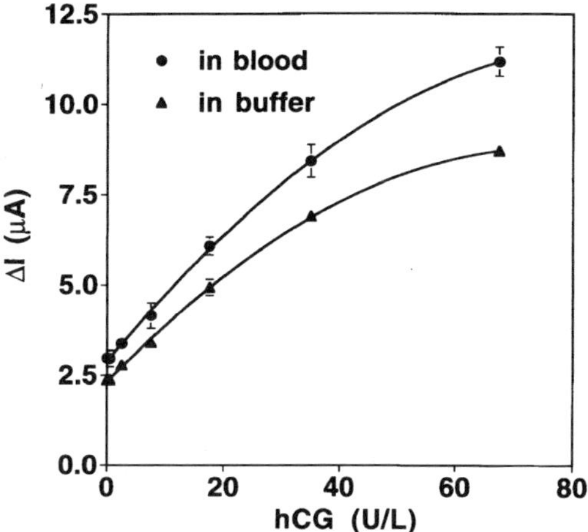

Figure 6.7. Dose–response curves for a noncompetitive sandwich NEEIA toward hCG in citrate buffer and whole blood (from Duan and Meyerhoff[31] with permission).

limits (0.37 μg/L PSA) as compared to TBS (0.09 μg/L PSA). Assays performed in undiluted female serum yielded results similar to those obtained in other matrices, providing a linear range from 0.3 to 15 μg/L PSA, and a detection limit of approximately 0.3 μg/L PSA. In addition, assays carried out in undiluted plasma were found to correlate well with measurements performed by an Abbott IMx analyzer (NEEIA = 0.719 IMx + 1.08; r = 0.896, $S_{y/x}$ = 2.02, N = 64).[33] This correlation is reasonable when one takes into account that the capture and reporter antibodies employed in each assay were not the same and might thus have recognized different epitopes on PSA. In addition, the reaction pH of each assay was

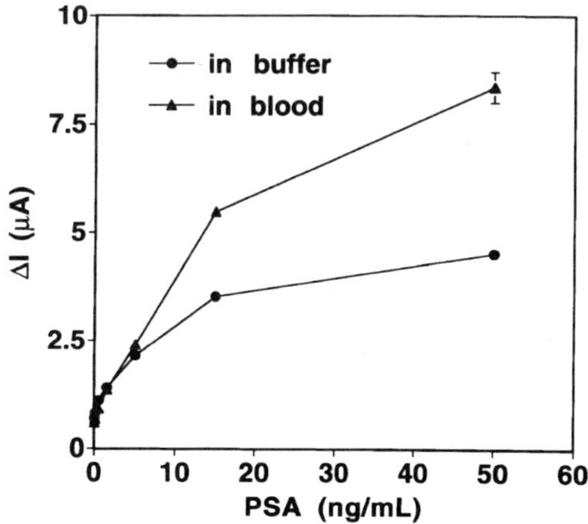

Figure 6.8. Dose–response curves for a noncompetitive sandwich NEEIA toward PSA in tris buffer and whole blood.

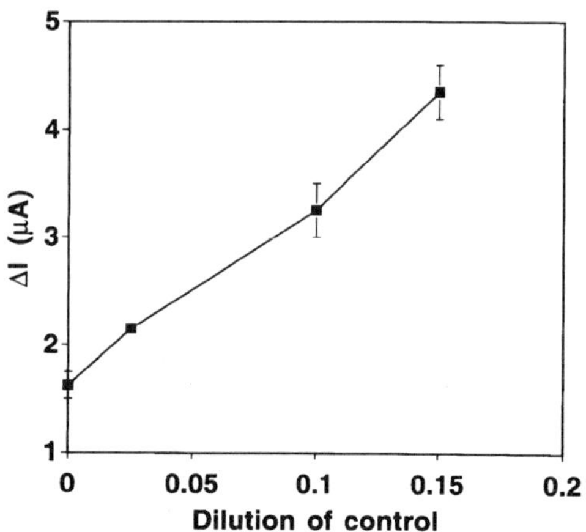

Figure 6.9. Dose–response curve for a noncompetitive sandwich NEEIA toward potato virus A in diluted potato plant leaf sap.

not the same, which may lead to different fractions of PSA being bound for each assay.

A NEEIA method for determination of potato virus A was also developed to provide a model system for detection of larger species, such as virus particles and/or bacteria. Detection of potato virus A was accomplished by diluting potato plant sap into TBS and incubating the sample for 4 h with a polyclonal anti-potato virus A capture antibody immobilized onto the microporous gold electrode. This was followed by addition of an ALP-labeled polyclonal anti-potato virus antibody and an additional 25 min incubation. Work involving potato virus A used a stock sample of unknown concentration; therefore, no detection limit or dynamic range could be determined. However, a definite increase in signal with increasing relative concentration of the virus was observed (Fig. 6.9).[37] These results clearly suggest that NEEIA is also applicable to the detection of intact microorganisms using the principle of a sandwich EIA.

6.3.2. Competitive Electrochemical Enzyme Binding Assay for Small Molecules

The ability to detect low-molecular-weight analytes by employing the NEEIA concept has been demonstrated with two small molecules—biotin and digoxin. Assays were carried out in a competitive format utilizing immobilized binding protein/antibody and ALP-labeled analyte competitors.

Competitive binding assays for biotin employed avidin immobilized on the surface of the microporous electrode and a biotin–ALP conjugate in solution. When biotin is present in the sample it competes with the biotin–ALP conjugate for the binding sites of the immobilized avidin. Owing to the competitive nature of the assay, lower concentrations of biotin in the sample result in larger fractions of the competitor being bound to the surface. The steady-state signal is thus inversely proportional to the concentration of biotin in the sample. By employing this format, biotin concentrations as low as 1 nM in aqueous buffered solutions could be detected (see Fig. 6.10).

Competitive immunoassays for the cardiac glycoside digoxin were carried out by employing an immobilized monoclonal antidigoxin antibody and a digoxin–ALP conjugate.[34] Competition between digoxin and the ALP conjugate in aqueous buffered solutions yielded detection limits as low as 100 pM (based on three times the standard deviation of the signal at zero dose) below the clinically relevant range of 600 pM to 2.4 nM (Fig. 6.11).[38] A linear dose–response is observed from 1 to 10 nM with good CV% within this range (6% at 3 nM). The detection limit in undiluted sheep serum was determined to be 1 nM, demonstrating the ability of NEEIA to assay small molecules in complex mediums.

Assays for each low-molecular-weight analyte were optimized in terms of the incubation time required to reach equilibrium for the binding reactions. The ratio of the signals obtained in the absence of analyte as well as in the presence of a saturating concentration of the analyte were plotted as a function of incubation time. In this manner it was determined that the biotin–avidin binding reaction had reached equilibrium after 30 min with stirring while the digoxin–anti-digoxin binding reaction required 60 min to reach equilibrium (data not shown). Enhancement in the speed of such assays could be achieved through the use of smaller sample volumes and redesign of the diffusion arrangement employed (i.e., improve the surface area to volume ratio).

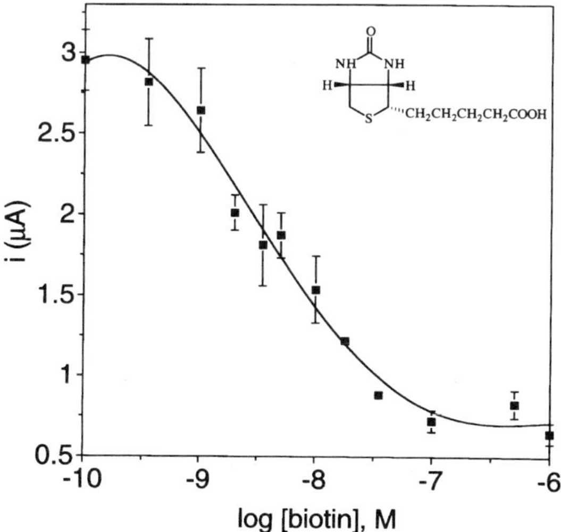

Figure 6.10. Dose–response curve for a competitive binding assay using the NEEIA format for the detection of biotin in tris buffer.

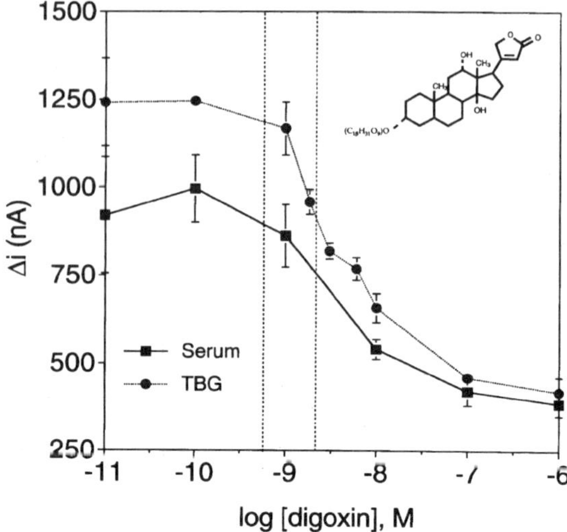

Figure 6.11. Typical dose–response curve for digoxin in tris buffer and undiluted sheep serum via a competitive NEEIA. The dashed lines define the clinically relevant range for digoxin in serum.

6.4. FUTURE DIRECTIONS

It has previously been demonstrated that by vacuum-depositing two gold electrodes onto one microporous nylon membrane (Fig. 6.12) and immobilizing the appropriate capture antibodies to each electrode, it is possible to detect two analytes (e.g., hCG and PSA) simultaneously in whole blood with the noncompetitive NEEIA format (Fig. 6.13).[33] This is possible by making the distance between the two

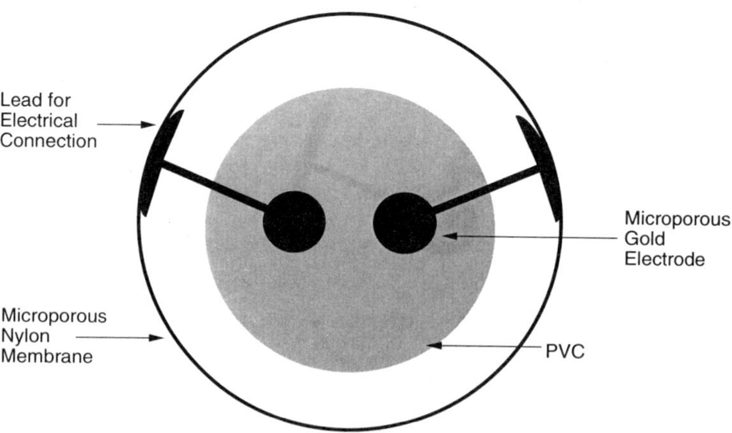

Figure 6.12. Schematic of a dual microporous gold electrode design for simultaneous multianalyte detection (from Meyerhoff et al.[33] with permission).

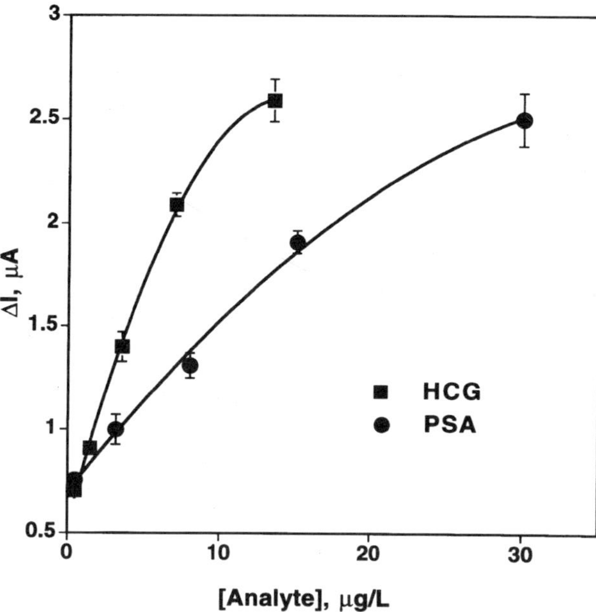

Figure 6.13. Dose–response curves for the simultaneous detection of hCG and PSA in whole blood using the NEEIA format (from Meyerhoff et al.[33] with permission).

electrodes large enough that the electroactive enzymatic product generated at one electrode cannot diffuse to the second electrode over the time course of the measurement. Despite the fact that both reporter antibodies are present simultaneously in the sample, the antibody specificity prevents cross-reactivity between anti-PSA and hCG, and anti-hCG and PSA. Based on these preliminary results, further investigation of simultaneous multianalyte testing is ongoing in our laboratory. One can envision numerous cases where the detection of several species in the same sample may be of interest, i.e., cardiac markers (CK-MB, troponin T) and tumor markers (prostatic acid phosphatase, CK-BB). It may also be possible to use this same dual-electrode configuration with one electrode serving as a control to subtract out signal due to nonspecifically bound protein as well as any interferences present in the sample (i.e., endogenous ALP). This could be accomplished by immobilizing some nonspecific protein onto the control electrode and the specific binding protein/antibody onto the primary electrode.

Efforts are also currently under way to demonstrate the use of enzyme labels other than ALP for NEEIA. Alternative enzymes include β-GAL, as noted in the introduction, as well as acetylcholine esterase and horseradish peroxidase (HRP). This area of investigation is important in demonstrating the general utility of the NEEIA system as well as providing alternative methods to overcome problems specific to sample matrices. This is especially apparent when measuring species in whole blood or serum samples where endogenous alkaline phosphatase activity can be significant.

Two fundamental aspects of the NEEIA system are also being investigated at this time. The first relates to determining the effect of the relative binding affinities of the immobilized binding protein or antibody on assay sensitivity (ED_{50}) for both the competitive and noncompetitive assay formats. Of course, one expects to lose sensitivity as the affinity constant between binding protein and analyte is reduced; however, the lack of a washing step in NEEIA adds a degree of complexity. As the unbound enzyme conjugate is not removed prior to addition of the conjugate, some signal will always be observed for this species. As lower affinity binding proteins are employed, lower fractions of enzyme conjugate are bound to the electrode. This results in a signal from specifically bound enzyme label that is increasingly indistinguishable from the signal resulting from unbound conjugate present in the detection zone adjacent to the surface of the electrode. In addition, lower-affinity species may require longer incubation periods before reaching equilibrium. By examining the effect of binding affinity on assay sensitivity, the conditions under which NEEIA can be performed will be more clearly defined.

The second major area of research on the NEEIA system involves characterizing the nature of the thioctic acid layer on the gold surface (i.e., monolayer, multilayer, polymer layer) and the amount of binding protein immobilized to the surface of the electrode via this layer. Preliminary studies using radiolabeled IgG indicate that under noncompetitive conditions, 95 pmol (yielding a surface coverage of 125 pmol/cm^2) of IgG is covalently immobilized to the electrode surface (based on the electroactive area of the electrode). The large surface coverage may be explained by several possibilities: (1) the electroactive area does not account for all of the gold surface area on the electrode, (i.e., some of the gold on the membrane is not in intimate contact with the bulk of the electrode and thus is not subject to electrochemical measurement); (2) protein aggregation may be occurring in solution leading to very high nonspecific binding; and (3) thioctic acid may form a polymerized layer on the electrode surface. The latter would yield an extended network of immobilized carboxylate sites to which a large amount of protein could be immobilized.

6.5. CONCLUSIONS

Although much work remains to be done before a full understanding of the surface structure of proteins immobilized on microporous gold electrodes is gained, it is clear that the NEEIA method offers a unique immunosensor format enabling separation-free immunoassays of both large and small molecules. The ability to conduct such assays in samples as complex as undiluted serum, plasma, or whole blood with detection limits rivaling conventional immunoassays suggests that the NEEIA method may be particularly attractive for developing future point-of-care test products. The possibility of coating multiple gold electrodes on a single membrane to conduct simultaneous assays for two or more species makes the NEEIA system also potentially useful for diagnostic panel testing.

At present, the main limitation of NEEIA is the rather long equilibrium times required to transport the reagents to the surface of the microporous gold electrode.

Enhanced mass transport could be facilitated by reducing the sample volume relative to the electrode surface area or by mechanically pulling the sample back and forth through the membrane prior to introduction of the substrate from the back side. Such practical aspects as well as the effect of other physical parameters (e.g., surface layer structure, electroactive area, and antibody surface density) on the sensitivity of the NEEIA method are currently being investigated in this laboratory.

ACKNOWLEDGMENTS: We gratefully acknowledge the National Science Foundation (Grant 9423063) for supporting these studies.

REFERENCES

1. Place JF, Sutherland RM, Dahne C. Opto-electronic immunosensors: a review of optical immunoassay at continuous surfaces. *Biosensors* 1985;1:321–353.
2. Di Gleria H, Hill HA, McNeil CJ. Homogeneous ferrocene-mediated amperometric immunoassay. *Anal Chem* 1986;58:1203–1205.
3. Bush DL, Rechnitz GA. Monoclonal antibody biosensor for antigen monitoring. *Anal Lett* 1987;20:1781–1790.
4. Bier FF, Stochlein W, Bocher M, et al. Use of a fibre optic immunosensor for the detection of pesticides. *Sens Act B* 1992;7:509–512.
5. Lu B, Lu C, Wei Y. A planar quartz waveguide immunosensor based on TIRF principle. *Anal Lett* 1992;25:1–10.
6. Miura N, Higobashi H, Sakai G, et al. Piezoelectric crystal immunosensor for sensitive detection of methamphetamine (stimulant drug) in human urine. *Sens Act B* 1993;13/14:188–191.
7. Wong RB, Anis N, Eldefrawi ME. Reusable fiber-optic-based immunosensor for rapid detection of imazethapyr herbicide. *Anal Chim Acta* 1993;279:141–147.
8. Souteyrand E, Martin JR, Martelet C. Direct detection of biomolecules by electrochemical impedance measurements. *Sens Act B* 1994;20:63–69.
9. Bier GG, Jockers R, Schmid RD. Integrated optical immunosensor for s-triazine determination: regeneration, calibration and limitations. *Analyst* 1994;119:437–441.
10. Bier FF, Schmid RD. Real time analysis of competitive binding using grating coupler immunosensors for pesticide detection. *Biosens Bioelectron* 1994;9:125–130.
11. Dubrovsky T, Vakula S, Nicolini C. Preparation and immobilization of Langmuir–Blodgett films of antibodies conjugated to enzymes for potentiometric sensor application. *Sens Act B* 1994;22:69–73.
12. Minunni M, Skladal P, Mascini M. A piezoelectric quartz crystal biosensor as a direct affinity sensor. *Anal Lett* 1994;27:1475–1487.
13. Deasy B, Dempsey E, Smyth MR, et al. Development of an antibody-based biosensor for determination of 7-hydroxycoumarin (umbilliferone) using horseradish peroxidase labeled anti-7-hydroxycoumarin antibody. *Anal Chim Acta* 1994;294:291–297.
14. Kalab T, Skladal P. A disposable amperometric immunosensor for 2,4-dichlorophenoxyacetic acid. *Anal Chim Acta* 1995;304:361–368.
15. Morgan CL, Newman DJ, Price CP. Immunosensors: technology and opportunities in laboratory medicine. *Clin Chem* 1996;42:193–209.
16. Rubenstein KE, Schneider RS, Ullman EF. Homogeneous enzyme immunoassay: new immunochemical technique. *Biochem Biophys Res Commun* 1972;47:846–851.
17. Ullman EF, Yoshida RA, Blakemore JI, et al. Mechanism of inhibition of malate dehydrogenase by thyroxine derivatives and reactivation by antibodies. *Biochim Biophys Acta* 1979;567:66–74.
18. Ullman EF, Maggio ET. Principles of homogeneous enzyme-immunoassay. In: Maggio ET, ed. *Enzyme Immunoassay*. Boca Raton: CRC Press, 1980:105–134.
19. Kabakoff DS, Greenwood HM. Homogeneous enzyme assay. In: Albert KGM, Price CP, eds. *Recent Advances in Clinical Biochemistry*. Edinburgh: Churchill Livingstone, 1981, pp 1–30.

20. Van Leite F, Galen RS. Determination of thyroxin by enzyme-immunoassay. In: Maggio ET, ed. *Enzyme Immunoassay*. Boca Raton: CRC Press, 1980, pp 135–153.
21. Gibbons I, Skold C, Rowley GL, et al. Homogeneous enzyme immunoassay for proteins employing β-galactosidase. *Anal Biochem* 1980;102:167–170.
22. Henderson DR, Firedman SB, Harris JB, et al. 'CEDIA', a new homogeneous immunoassay system. *Clin Chem* 1986;32:1637–1641.
23. Khanna PL, Worthy TE. A recombinant protein-based homogeneous immunoassay. *Am Clin Lab* 1989; October: 14–19.
24. Khanna PL, Dworschack RT, Mannin WB, et al. A new homogeneous enzyme immunoassay using recombinant enzyme fragments. *Clin Chim Acta* 1989;185:231–240.
25. Khana P. Homogeneous enzyme immunoassay. In: Price CP, Newman DJ, eds. *Principles and Practice of Immunoassay*. New York: Stockton Press, 1991, pp 326–364.
26. Badley RA, Drake RAL, Shanks IA, et al. Optical biosensors for immunoassays: the fluorescence capillary device. *Phil Trans R Soc London Ser* 1987;316:143–160.
27. Kronick MN, Little WA. A new fluorescent immunoassay. *Bull Amer Phys Soc* 1973;18:782.
28. Kronick MN, Little WA. A new immunoassay based on fluorescence excitation by internal reflection spectroscopy. *J Immunol Meth* 1975;8:235–242.
29. Sutherland RM, Dahne C, Place JF, et al. Optical detection of antibody–antigen reactions at a glass–liquid interface. *Clin Chem* 1984;30:1533–1538.
30. Parry RP, Love C, Robinson GA. Detection of rubella antibody using an optical immunosensor. *J Virol Meth* 1990;27:39–48.
31. Duan C, Meyerhoff ME. Separation-free sandwich enzyme immunoassays using microporous gold electrodes and self-assembled monolayer/immobilized capture antibodies. *Anal Chem* 1994;66:1369–1377.
32. Duan C, Meyerhoff ME. Immobilization of proteins on gold coated porous membranes via an activated self-assembled monolayer of thioctic acid. *Mikrochim Acta* 1995;117:195–206.
33. Meyerhoff ME, Duan C, Meusel M. Novel nonseparation sandwich-type electrochemical enzyme immunoassay system for detecting marker proteins in undiluted blood. *Clin Chem* 1995;41:1378–1384.
34. Ducey MW, Smith AM, Guo X, et al. Competitive nonseparation electrochemical enzyme binding/immunoassay (NEEIA) for small molecule detection. *Anal Chim Acta* 1997;357:5–12.
35. Tang H, Lunte C, Halsall H, et al. *p*-Aminophenyl phosphate: an improved substrate for electrochemical enzyme immunoassay. *Anal Chim Acta* 1988;214:187–195.
36. Christie I, Treloar P, Koochaki Z, et al. Simplified measurement of serum alkaline phosphatase utilizing electrochemical detection of 4-aminophenol. *Anal Chim Acta* 1992; 257:21–28.
37. Duan C. Development of a separation-free sandwich type electrochemical enzyme immunoassay for measuring proteins in undiluted whole blood. Ph.D. Dissertation. University of Michigan, 1995:199–205.
38. Chen I, Heminger LA. Digoxin and digitoxin. In: Pesce AJ, Kaplan LA eds. *Methods in Clinical Chemistry*. St. Louis: CV Mosby Co., 1987 pp 897–902.

7

Liposomes as Signal-Enhancement Agents in Immunodiagnostic Applications

Anup K. Singh, Joseph S. Schoeniger, and
Ruben G. Carbonell

7.1. INTRODUCTION

Liposomes are hollow spherical structures that consist of a phospholipid bilayer enclosing an aqueous core. They are self-assembled structures and form a stable suspension in water. Different methods of preparation result in liposomes with different structures and diameters from 25 nm to a few microns. Liposomes can have one (unilamellar) or multiple bilayers (multilamellar). The latter, which are formed spontaneously when a dried layer of lipids is hydrated, tend to be quite large in size and sediment over time. Ultrasonication[1] or extrusion[2] through controlled pore-size membranes converts multilamellar liposomes, into unilamellar liposomes, which are smaller and much more stable. We have prepared formulations that are stable for at least a year. Unilamellar liposomes can also be prepared by detergent dialysis[3] and reverse-phase evaporation[4] methods. A wide variety of synthetic and natural lipids that can be used to construct liposomes are available. This permits one to tailor them for specific purposes. For example, lipids having a functional group (e.g., an amine or carboxylic acid) as the head group can be used to covalently link molecules to the outer surface of a liposome. To obtain anionic liposomes, anionic lipids such as phosphatidic acid can be added to the lipid mixture used to prepare liposomes.

Liposomes have found a wide variety of applications in areas as varied as cheese-making and cosmetics.[5] In many of these applications they are used as storage and delivery vehicles, which is the feature that gives them their greatest usefulness — their capacity to carry a large number of small molecules. Figure 7.1

Anup K. Singh and Joseph S. Schoeniger • Sandia National Laboratories, Livermore, California 94551-0969. Ruben G. Carbonell • Department of Chemical Engineering, North Carolina State University, Raleigh, North Carolina 27695.

Biosensors and Their Applications, edited by Yang and Ngo, Kluwer Academic/Plenum Publishers, New York, 1999.

131

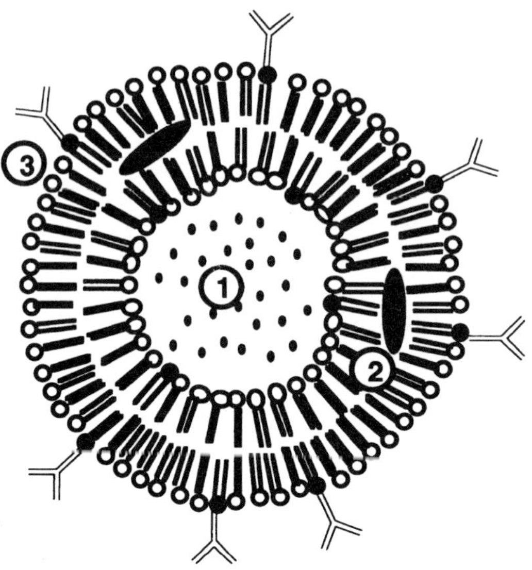

Figure 7.1. Sketch showing the cross section of a unilamellar liposome. Hydrophilic molecules can be entrapped in the aqueous core (region 1) or linked to the outer surface (region 3); hydrophobic molecules can be entrapped in the bilayer (region 2).

shows the various regions in a liposome where molecules can be trapped. Liposomes can carry both hydrophilic and hydrophobic payloads. Hydrophilic molecules can either be entrapped in the aqueous core or electrostatically or covalently conjugated to the outer surface. For example, doxorubicin can be entrapped in the core for delivery to cancerous cells, and DNA fragments can be electrostatically bound to the outer surface of a cationic liposome for potential use in gene delivery. Hydrophobic material can be entrapped in the interstitial space between the lipid layers. Another useful feature of liposomes is their similarity to biological cell membranes. Because of this property, they have been used to study transport of molecules across cell membranes, cell–cell interactions, and mechanisms of transfection.

In this chapter we discuss the utility of liposomes as a label in immunodiagnostic and biosensor applications. Specific recognition and binding of one biological molecule to another forms the fundamental basis of any biosensor or immunoassay. The prototypical example is the binding of an antibody to an antigen. Antigen–antibody binding is generally not accompanied by an easily detectable physical or chemical change. Therefore, one of the reagents is routinely conjugated with a label, translating a binding event into signal generation. Frequently used labels are fluorophores, enzymes, and radioisotopes. These are covalently conjugated to either an antibody or an antigen and typical stoichiometry of 1–2 enzyme molecules or 1–6 fluorophores or isotopes are obtained per antibody or antigen molecule.

Now let us consider a liposome as a label. A liposome of diameter 100 nm offers an internal volume of approximately 5.2×10^{-19} L and an outer surface area of 3.1×10^{-14} m^2. Starting with a concentration of 100 mM, it is possible to entrap

30,000 small labels such as carboxyfluorescein or calcein in the core. If we choose instead to incorporate lipids with labeled headgroups in the bilayer, up to 30,000 labels can be inserted into one liposome. For these label-loaded liposomes to be useful in an immunodiagnostic application, there is another requirement: they need to have a ligand or receptor conjugated to enable them to bind specifically to the complementary receptor or ligand. There are different methods by which these so-called "immunoliposomes" can be prepared: covalent linkage of antibodies or antigens to the outer surface or insertion of lipophilic receptors such as membrane proteins and glycolipids in the bilayer. There are two basic reasons that label-carrying immunoliposomes have a potential for improving enormously both the sensitivity and the signal strength in an immunodiagnostic application compared to a labeled antibody: liposomes carry a large number of labels and they are capable of multivalent binding thereby increasing the association constant significantly.

Liposomes have been used to enhance the signal in immunodiagnostic applications for almost two decades now.[6] Most of the earlier efforts employed antigen-sensitized liposomes in complement-based homogeneous immunoassays. In a typical assay of this type, antigen-sensitized liposomes, carrying encapsulated fluorophores or enzymes in the core, compete with free antigen in a sample to bind to a limited number of antibodies present in the solution. Guinea pig complement is added to the solution resulting in lysis of liposomes that have an antibody–antigen complex formed on the surface. Rupture of liposomes releases the entrapped labels, which can then be detected analytically. These assays tend to be fast and simple as separation of bound species from unbound is not required before signal measurement. Drawbacks are the instability and high cost of complement, the instability of liposomes in serum samples, and the nonspecific lysis of liposomes. Heterogeneous assays, where bound species are separated from unbound before measurement, circumvent most of these difficulties and have been implemented in a variety of formats. Examples include flow-injection immunoassays using liposomes for detection of theophylline,[7,8] a regenerable planar waveguide immunosensor to detect theophylline using theophylline-sensitized liposomes carrying fluorescein,[9] and an immunomigration assay for the herbicide alachlor using alachlor-sensitized liposomes.[10]

Most of the liposome-based diagnostic applications described to date achieve signal enhancement by encapsulating labels such as fluorophores or enzymes (Fig. 7.2). This approach is limited because in the case of large labels such as enzymes, only 5–10 can be entrapped in a liposome. In the case of smaller molecules such as fluorophores, a huge number can be entrapped but leakage owing to diffusion and nonspecific lysis results in the loss of labels.[11,12] We have suggested covalently attaching a large number of labels together with a few recognition molecules (e.g., antibody) to the lipid headgroups, before or after formation of liposomes.[13–15] We have successfully immobilized 100–200 enzyme molecules or up to 20,000 fluorophores per liposome using this approach. This not only provides a large number of labels per liposome, but also eliminates the problem of leakage of labels from the liposome core.

In the following sections we discuss the application of liposomes to enhance the sensitivity of three different type of assays: (1) an enzyme-linked immunosorbent assay, (2) a fluoroimmunoassay, and (3) a novel assay for bacterial toxins using glycolipids embedded in liposomes as recognition molecules.

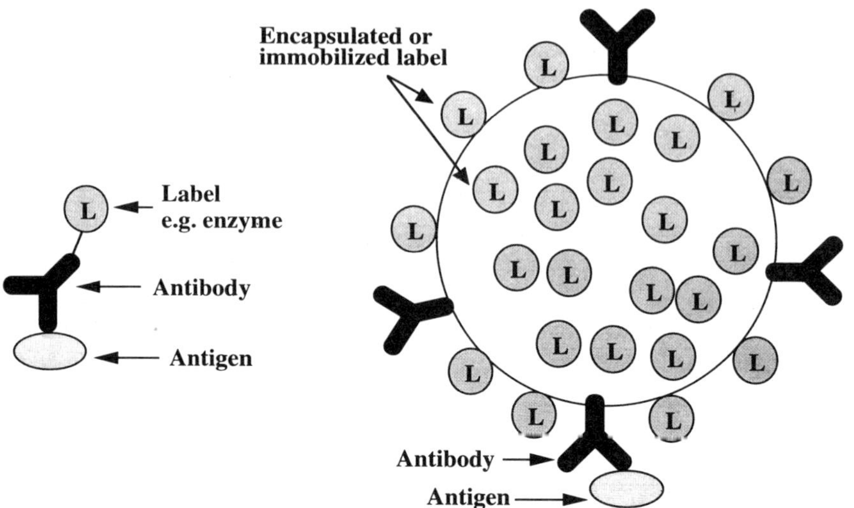

Figure 7.2. Amplification of signal by use of liposomes. Only a few (1–5) labels can be attached to an antibody molecule. Liposomes, on the other hand, can carry thousands of fluors or hundreds of enzymes together with a few antibodies attached to the outside surface.

7.2. AMPLIFICATION OF AN ENZYME IMMUNOASSAY USING LIPOSOMES

Enzyme-linked immunosorbent assay (ELISA) is one of the most commonly used immunodiagnostic techniques for detection of clinical analytes and other antigens. In a typical ELISA, one of the reagents (antibody or antigen) is adsorbed on a solid support such as a microtiter well and the analyte is detected using a secondary enzyme-labeled reagent. We present an example of the application of liposomes to increase the sensitivity of an ELISA where bifunctional liposomes, carrying both an enzyme and an antibody, are used as the enzyme-labeled reagent. The performance of liposome assay is compared to a conventional ELISA where an enzyme–antibody conjugate is used as the labeled reagent.

The bifunctional liposomes were prepared by covalently attaching an enzyme, horseradish peroxidase (HRP), and monoclonal anti-*d*-dimer immunoglobulin (IgG) to the outside of a liposomes. The reaction conditions were optimized to obtain 7–12 antibody molecules and 100–200 HRP molecules per liposome. The presence of immobilized proteins on the liposome surface also apparently increased liposome stability. To demonstrate the utility of these liposomes, we used them in a sandwich ELISA for diagnosis of thromboembolic disorders by assaying for *d*-dimer, the final and smallest proteolytic product in the degradation of cross-linked fibrin by plasmin. The assay results using liposomes led to a detection limit for *d*-dimer in human plasma an order of magnitude lower than what was achieved using the conventional enzyme–antibody conjugate assay.

7.2.1. Preparation of Enzyme- and Antibody-Bearing Liposomes

Small unilamellar liposomes were prepared with a mixture of distearoyl phosphatidyl choline (DSPC), cholesterol, and dimyristoyl phosphatidyl ethanolamine (DMPE) in molar ratio of 0.4:0.4:0.2, respectively, using probe sonication procedure.[1] HRP and monoclonal anti-d-dimer IgG were covalently linked to liposomes simultaneously using the periodate oxidation method.[16,17] The two glycoproteins, HRP, and the antibody were oxidized with sodium periodate to convert carbohydrate groups to aldehydes. These aldehydes reacted readily with the amines present on the liposome surface to form a Schiff-base intermediate. In the final step, unstable Schiff bases were converted to stable secondary amines by reduction with sodium cyanoborohydride. Liposomes with immobilized enzyme and antibody were separated from unreacted proteins by size exclusion chromatography.

7.2.2. Preparation of Enzyme–Antibody Conjugate

The periodate oxidation method was also used to conjugate the enzyme (HRP) to the monoclonal anti-d-dimer antibody. The carbohydrate moieties of HRP were oxidized to aldehydes and subsequently reacted to the lysines in the IgG molecule. The resulting Schiff-base intermediate was reduced to form a secondary amine bond between the two proteins.

7.2.3. Characterization of Liposomes with Immobilized HRP and Antibody

The immobilized HRP concentration was determined by measuring the absorbance of liposome solution at 403 nm and subtracting the contribution due to scattering by liposomes. The kinetic parameters of the enzyme were determined as described previously.[18] The amount of antibody conjugated to liposomes was determined by radiolabeling the antibody by reductive methylation[19] with ^{14}C-formaldehyde prior to conjugation to the liposome surface. The concentration of phospholipid in a liposome sample, and ultimately liposome concentration, was determined by a phosphate assay using the method of Chen et al.[20] Hydrodynamic radii of liposomes before and after immobilization were determined using quasi-elastic light scattering (QLS). Measurements were performed at a 90° scattering angle using a Coherent Innova 70-3 argon-ion laser with a Brookhaven BI-2030 AT correlator and goniometer.

7.2.4. Sandwich ELISA with Liposomes and Enzyme–Antibody Conjugate

The inner 60 wells of a 96-well microtiter plate were coated with monoclonal anti-d-dimer at a concentration of 40 μg/mL. The wells were washed with phosphate buffer saline (PBS) and blocked with bovine serum albumin (BSA) to saturate the unbound sites on the polystyrene surface. The stock solution of antigen, d-dimer, was serially diluted in PBS containing 1 wt% BSA. Dilutions of antigen samples or controls were applied to the coated wells. The plate was covered with a plate-sealer and incubated at 37°C for 1 h in a mechanical convection incubator. After the plate was washed four times with buffer, 100 μL/well of liposomes were added and the plate was incubated at 37°C for 1 h. The wells were washed six times with PBS to remove the unbound and nonspecifically bound liposomes. Then 100 μL/well of substrate [3,3′,5,5′ tetramethyl-benzidine-dihydrochloride (TMB) + urea hydrogen peroxide] were added and the plate was shaken at room temperature for 30 min. The reaction was stopped with 2 N sulfuric acid and the absorbance at 450 nm was read in an absorbance plate-reader. The control ELISA with HRP-antibody conjugate was performed in a similar manner except that 0.05 wt% tween-20 was added to the wash buffer.

7.2.5. Results and Discussion

Liposomes made with DSPC, DMPE, and cholesterol were very stable and no significant change in size was observed over a year. Characteristics of liposomes are presented in Table 7.1. The initial average diameter was found to be 60 ± 5 nm. After conjugating HRP and antibody to liposomes, the diameter increased to 112 ± 6 nm. An average of 10 IgG molecules and 103–204 HRP molecules were immobilized per liposome. The enzyme retained 60–80% of its activity upon immobilization.

The immunoassay protocol was also optimized to reduce nonspecific binding and maximize the specific signal. The variables considered were the concentration of coating antibody, blocking agent, diluents for liposomes and antigen samples, incubation duration, and incubation temperature. Coating with 40 μg/mL antibody for 16 h at 4°C was found to provide a saturated surface. BSA at a concentration of 1%(w/w) was found to be the optimum both as a blocking agent and as a diluent.

Table 7.1. Characterization of Liposomes

Application	ELISA for *d*-dimer	Fluoroimmunoassay for *d*-dimer	Fluoroimmunoassay for tetanus
Liposome composition	DSPC:DMPE:cholesterol (40:20:40)	DSPC:caproyl–PE:DPPE–FITC:cholesterol (40:5:15:40)	DSPC:DPPE–TRITC:GT1b:Cholesterol (40:15:5:40)
Initial liposome diameter	60 ± 5 nm	60–75 nm	120–135 nm
Liposome size after attaching receptor and marker	112 ± 6 nm	80–101 nm	—
Receptor	Monoclonal anti-*d*-dimer IgG	Monoclonal anti-*d*-dimer IgG	GT1b
Number of receptors/liposome	10	10–19	9000
Marker	HRP	Fluorescein–DPPE	Rhodamine–DPPE
Number of markers/liposome	103–204	10,000–20,000	27,000

It was found that 1 h at 37°C was the optimal condition for incubation of wells with both antigen and liposomes.

Figure 7.3 shows results of an ELISA performed with the two different conjugates — liposomes with HRP and anti-*d*-dimer immobilized on the surface, and HRP–anti-*d*-dimer conjugate. The curves have a sigmoidal shape and were fit by a four-parameter logistic model[21] of the form

$$S = \beta_2 + \frac{\beta_1 - \beta_2}{1 + ([A]/\beta_3)^{\beta_4}} \tag{1}$$

where S and $[A]$ are absorbance and antigen concentration, respectively. The model fits the experimental data very well ($R^2 = 0.9999$ and 0.9967 for liposome and HRP–antibody, respectively). For qualitative comparison of the two assays, absorbance values were normalized with respect to the parameter β_2 (the predicted maximum signal) and plotted on the same graph. It is evident from the Fig. 7.3 that liposomes performed significantly better (an order of magnitude more sensitive) than the conventional enzyme–antibody conjugate. For quantitative analysis we defined minimum detectable concentration (MDC) as the lowest concentration of analyte that results in an expected response (absorbance) two standard deviations higher than the mean response at zero concentration. With use of the logistic model, the MDC values were calculated to be 9 pM and 88 pM for liposomal and HRP–antibody assays, respectively (Table 7.2).

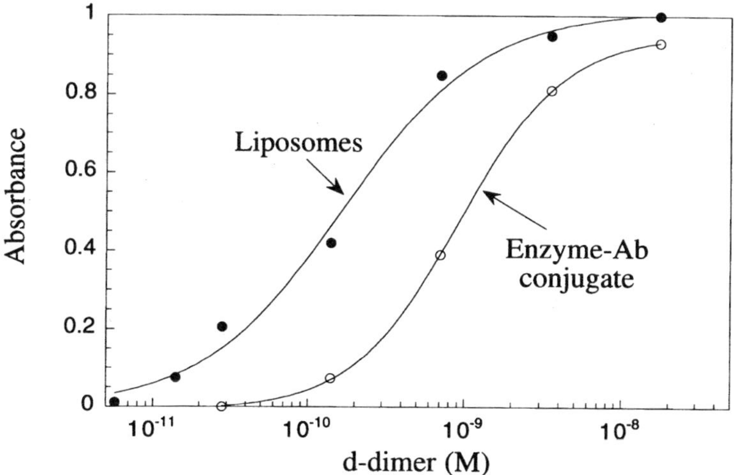

Figure 7.3. ELISA results for detection of *d*-dimer using liposomes and enzyme-labeled antibodies. Open and filled circles represent experimental values and solid lines represent the curve fits using the logistic model. Signal is normalized by dividing the absorbance values by the maximum absorbance obtainable (as predicted by the logistic model). Liposomes provide an order of magnitude lowering of detection limit. (Adapted from Singh et al.[14])

Table 7.2. Comparison of Various Immunoassays

Immunoassay	MDC (M)[a]	K_a (M^{-1})[b]
ELISA for *d*-dimer		
Liposomes	8.8×10^{-12}	4.2×10^9
Control (enzyme–Ab)	8.8×10^{-11}	1.1×10^9
Fluoroimmunoassay for *d*-dimer		
Liposomes	2.4×10^{-11}	3.98×10^8
Control (fluor–Ab)	3×10^{-9}	1.1×10^7
Fluoroimmunoassay for tetanus *C* fragment		
Liposomes	1.2×10^{-9}	1.2×10^{-8}

[a] Minimum detectable concentration—the lowest concentration of analyte that results in an expected response (fluorescence signal) that is two standard deviations higher than the response at zero concentration.
[b] Apparent association constant—inverse of parameter β_3.

The liposome immunoassay described here combines the higher sensitivity of a noncompetitive ELISA, which is not limited by the association constant of antibody to antigen and a large signal amplification achieved by linking hundreds of enzyme molecules to a liposome. In theory, liposomes should perform 100 to 200 times better than HRP–antibody conjugate as the number of enzyme molecules per liposome is 100–200 compared to 1–2 per HRP–antibody conjugate. However, liposomes are much larger (1100 Å) compared to the IgG molecule (100 Å) and hence binding of liposome to one IgG on the plate may render neighboring IgG molecules sterically unavailable for binding to other liposomes. This does not happen in the case of smaller HRP–antibody conjugates (150 Å). These two effects suggest that an order of magnitude improvement obtained by using liposomes appears reasonably close to the theoretical maximum enhancement possible.

7.3. AMPLIFICATION OF FLUOROIMMUNOASSAY USING LIPOSOMES

Fluorescent signal generation has been utilized to detect analyte binding in many biosensor applications and immunoassays. Fluorophores have significant advantages over chromophores and radiolabels. They can be detected at concentrations substantially lower than chromophores and do not pose the health and environmental hazards associated with radiolabels. Owing to their small size, unlike enzyme labels, they do not significantly alter the conformation of molecules to which they are linked. In this section, we discuss immunoassays that combine the sensitivity of fluorescence detection with the signal amplification provided by liposomes. Small unilamellar liposomes comprised of cholesterol and phospholipids, in which one of the lipids is labeled with fluorescein, were covalently functionalized with antibodies (monoclonal anti-*d*-dimer IgG). Each liposome contained thousands of fluorescein molecules and a few antibodies (~ 10). The immunoassay using liposomes was compared to a conventional immunoassay that uses fluor–antibody conjugate as the labeled reagent. Liposomes, by virtue of having thousands of fluorophores coupled to one liposome in contrast to one or a few reporter molecules in the conventional fluor–antibody conjugate, performed better on two counts: (1) they lowered the detection limit by two orders of magnitude, and (2) they provided an order of magnitude amplification in signal.

7.3.1. Preparation of Antibody-Bearing Fluorescent Liposomes

Liposomes were prepared with compositions of DSPC:cholesterol:DMPE:fluorescein–DPPE (40:40:5:15 mole percent) using sonication or extrusion. Lipids were dissolved in chloroform:methanol (9:1) and dried in a rotary evaporator to form a thin lipid film on the inside wall of a flask. The lipids were then hydrated in citrate buffer at 65°C to form multilamellar liposomes. These were converted to unilamellar liposomes either by sonicating with a sapphire-tipped probe or extruding through a 50-nm-pore-size polycarbonate membrane.

Monoclonal anti-*d*-dimer IgG was attached to preformed liposomes by the periodate oxidation method as described earlier. Liposome diameter, concentration, and number of antibody molecules immobilized were determined as described previously.

7.3.2. Preparation of Fluor–Antibody Conjugate

Isothiocyanate derivatives of fluorescein readily react with the amines in an antibody at an alkaline pH (>9) to form thiourea bonds that are moderately stable.[22] Monoclonal anti-*d*-dimer IgG was reacted with fluorescein isothiocyanate (FITC) in fluor/protein molar ratios varying from 6 to 40. The reaction was carried out for 2 h at room temperature at a pH of 9.2. The separation of excess FITC was done in a desalting column equilibrated with borate buffer at pH 8.5.

7.3.3. Fluoroimmunoassay with Liposomes and Fluorescein–Antibody Conjugate

Fluoroimmunoassay for *d*-dimer was performed in a microtiter plate coated with anti-*d*-dimer antibody. Wells were blocked with BSA. Serum samples containing *d*-dimer or buffer were added to the wells. After washing with PBS buffer, wells were exposed to liposomes or fluorescein-labeled antibody. Unbound and non-specifically bound reagents were removed by rinsing the wells with buffer. Wells were then incubated with $100\,\mu\text{L}$ of a nonionic surfactant, Triton X-100™ and fluorescence was measured using a fluorescence plate-reader (Cambridge Instruments).

7.3.4. Results and Discussion

Fluorescent liposomes prepared using fluorescein-labeled lipid, phosphatidylcholine, and cholesterol had an average hydrodynamic diameter of 67 nm and contained 10,000–20,000 fluorescein molecules (see Table 7.1). Fluorescein molecules in a liposome are in very close proximity to each other and undergo concentration-dependent quenching by forming nonfluorescent dimers. Hence, fluorescence from an intact liposome is 20-fold lower than the sum of fluorescence from individual fluorescein-labeled lipids. As a result, before measuring the fluorescence signal, liposomes are solubilized into micelles by a nonionic surfactant such as Triton X-100™. This leads to dilution of fluorescein molecules, thereby dequenching them.

To enable liposomes to bind to an antigen, 10–20 anti-d-dimer IgG molecules were conjugated to the outer surface by the periodate oxidation method. Although periodate coupling is a relatively slow and inefficient chemistry, the antigen-binding capacity of the conjugated antibodies is minimally affected as the coupling is done through carbohydrates present in the F_c portion of IgG. The results of fluoroimmunoassays performed with liposomes and fluorescein–antibody conjugate for detection of d-dimer are depicted in Fig. 7.4. The assay data were fitted by the logistic model as described above, and the results are tabulated in Table 7.2. The minimum detectable concentrations (MDCs) for d-dimer were 24 pM and 3 nM for liposome and fluorescein–antibody conjugate, respectively. Use of liposomes also amplified the signal 10–20 times for a given analyte concentration compared to fluorescein labeled antibody (data not shown). Hence, liposomes performed better than fluor–antibody conjugate both in terms of lowering the detection limit (by 125-fold), and amplifying the signal (by an order of magnitude). Liposomes also exhibited an approximately 40-fold higher association constant relative to the fluor–antibody conjugate (Table 7.2), implying that liposomes bind more strongly to antigen than fluorescein–antibody conjugates. This can be explained by multivalent binding of liposomes to the plate surface. Liposomes carry 10–20 bivalent antibodies on the surface and hence a single liposome can bind to multiple antigens simultaneously.

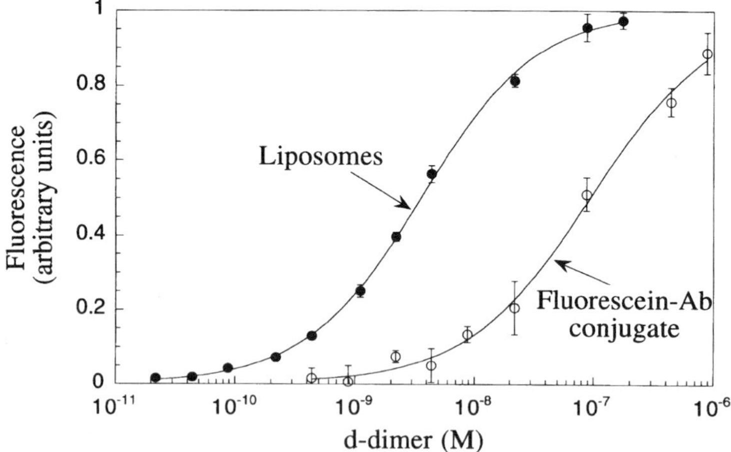

Figure 7.4. Fluoroimmunoassay results for detection of d-dimer using liposomes and fluorescein-labeled antibodies. Open and filled circles represent experimental values and solid lines represent the curve fits using the logistic model. Signal is normalized by dividing the fluorescence by the maximum fluorescence obtainable (as predicted by the logistic model). Liposomes yield a detection limit 125-fold lower than the fluor–antibody conjugate. (Adapted from Singh et al.[15])

This effect has also been reported by other researchers.[7,23] A detailed study of liposomal stability was not undertaken in this study but liposomes did not undergo any significant aggregation (less than 10% lipid loss) for up to 3 months when stored in solution at 4°C.

7.4. APPLICATION OF GANGLIOSIDE-BEARING LIPOSOMES AS SENSITIVE PROBES FOR POTENT NEUROTOXINS

Bacterial toxins and viruses are often specifically targeted to certain cell types that exhibit characteristic binding sites or specific receptors. The majority of receptors for bacterial toxins are carbohydrates, either in the form of glycoproteins or glycolipids. Neurotoxins such as botulinum and tetanus toxins are the most potent bacterial toxins identified to date. An essential step in intoxication by these neurotoxins involves binding of the toxin protein to specific receptors on the presynaptic neuronal membrane. Botulinum and tetanus toxins have been determined to bind selectively to gangliosides of the series G1b, which are present in neuronal cell membranes. Trisialoganglioside (GT1b) has been reported to have the highest affinity for tetanus toxin.[24–26] If we prepare liposomes containing GT1b, these liposomes essentially resemble neuronal cells and are immediately "attacked" by tetanus toxin. We can thus selectively probe for tetanus toxin using liposomes containing GT1b as one of the constituent lipids. To impart signal generation capability to these liposomes, we can incorporate thousands of fluorescent-labeled lipids in the bilayer.

The bifunctional liposomes, containing both a label (rhodamine) and a receptor (GT1b) in the bilayer, were used in a heterogeneous sandwich immunoassay for recombinant tetanus fragment C. Fragment C is a 50 kD C-terminal domain of the tetanus toxin heavy chain that is responsible for recognition and binding to neuronal membranes mediated by ganglioside GT1b. Researchers have shown that fragment C displays almost all the binding ability of intact toxin to a neuron.[25] We used the recombinant fragment C to avoid the exposure hazard associated with the use of intact toxin. As low as 50 ng/mL (1.2 nM) of the analyte could be detected using liposomes in an immunoassay. Hence, these bifunctional liposomes serve not only as receptors for neurotoxins, but also as vehicles to amplify the signal. Furthermore, as these liposomes do not contain any proteinaceous component, they are potentially less prone to deactivation upon storage than labeled antibodies.

7.4.1. Preparation and Characterization of GT1b Liposomes

Liposomes were prepared from a mixture of DSPC:cholesterol:DPPE–TRITC:GT1b in a mole ratio of 40:40:15:5 by extrusion using a syringe extruder (Avestin, Inc., Vancouver, Canada). GT1b was obtained from the Sigma Chemical Company (St. Louis, MO) and DPPE-TRITC from Molecular Probes, Inc. (Eugene, OR). Recombinant tetanus toxin fragment C and a monoclonal antibody against it were supplied by Boehringer Manheim Corp. (Indianapolis, IN). All the other chemicals used have been described in preceding sections. Liposome diameters were measured by quasi-elastic light scattering. The number of fluorescent lipids incorporated in a liposome were determined by measuring absorbance at 550 nm after lysing liposomes with Triton X-100™. When the size of liposomes, the number of fluorescent lipids, and their initial mole fraction in the lipid mixture were known, it was possible to estimate the concentration of liposomes and the number of receptors (GT1b) per liposome.

7.4.2. Fluoroimmunoassay with GT1b Liposomes

Monoclonal antitetanus toxin fragment C was physically adsorbed in the wells of a microtiter plate. Remaining hydrophobic sites on the polystyrene surface were blocked with BSA. Wells were exposed to dilutions of recombinant tetanus toxin fragment C or buffer. After incubation at 37°C for 1 h, wells were washed with phosphate buffer. Liposome solution was dispensed into the wells and the microtiter plate was incubated for 1 h at 37°C. Wells were washed again and a detergent, Triton X-100™, was added. Fluorescence was read in a fluorescence plate-reader.

7.4.3. Results and Discussion

We successfully prepared liposomes carrying approximately 9000 GT1b and 27000 rhodamine–DPPE (see Table 7.1). Liposome stability was monitored by measuring liposome size and the lipid concentration and no significant change was observed over 3 months. Most of the liposomes did precipitate out after 1 year of storage. Quasi-elastic light scattering yielded a mean diameter of 130 nm for liposomes. As discussed above, rhodamine molecules in a liposome are very close to each other (separation $< 10\,\text{Å}$) and hence undergo concentration–dependent quenching. Triton X-100 was used to disrupt liposomes and dilute rhodamine-labeled lipids. The fluorescence increased by 33-fold after liposome disruption.

Figure 7.5 shows the result of a fluoroimmunoassay performed with liposomes for detection of tetanus toxin fragment C. The results are also shown in Table 7.2. We found the MDC to be 1.2 nM of tetanus fragment C in a sample. The apparent association constant of liposome binding to the surface was $1.2 \times 10^8\,\text{M}^{-1}$, which is quite close to what one expects from an antigen–antibody complexing. This strong affinity of liposomes for toxin may be due to multivalent attachment of one liposome to several toxin molecules.

7.5. CONCLUSIONS

In the ever-advancing field of biosensors and immunoassays, there is always a need for methods to amplify the signal resulting from binding of one biological molecule to another. There is also a desire for flexibility in the types of ligands (i.e., not just antibodies) used and the chemical characteristics of the interacting surfaces. Liposomes offer one such solution owing to some of their unique features: (a) they provide a large internal volume and outer surface area where molecules can be

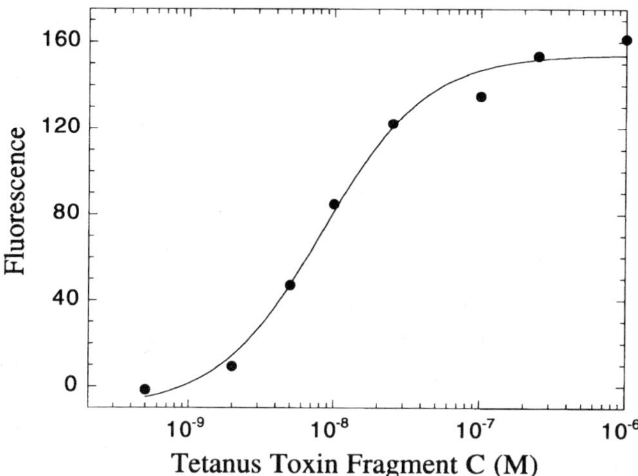

Figure 7.5.   Fluoroimmunoassay results for detection of tetanus toxin fragment C using GT1b-bearing fluorescent liposomes. GT1b is a glycolipid that binds specifically to tetanus toxin.

entrapped or attached; (b) their bilayers provide a very flexible, cell-membrane-like environment, where biological molecules can maintain their native conformation; (c) they form stable suspension in water and; (d) they can be tailor-made to offer sites for covalent chemistry or decrease nonspecific binding. The most common method for using liposomes in an immunoassay application has been to attach receptor molecules such as antibodies or antigens on the surface and entrap labels inside. A number of researchers have used this method successfully to amplify the signal and decrease the limit of detection. A few drawbacks associated with the entrapment technique are leakage of entrapped material over time and loss of labels due to nonspecific lysis of liposomes upon exposure to serum. We successfully circumvented these problems by preparing liposomes with labels such as enzymes and fluorophores covalently conjugated together with receptors such as antibodies or glycolipids. Liposomes with entrapped material also need to be maintained in aqueous solution, making long-term storage expensive. Liposomes with covalently linked molecules, on the other hand, can be lyophilized and resuspended without any loss of immobilized material.

Liposome assays, both enzyme- and fluorescence-based, performed better than their nonliposomal counterparts (see Table 7.2). In the ELISA for *d*-dimer, liposomes led to an order of magnitude lowering of detection limit compared to an assay performed with antibody–enzyme conjugate. In the fluoroimmunoassay for *d*-dimer, liposomes performed even better and their use resulted in lowering the detection limit by more than two order of magnitudes compared to fluor–antibody conjugate. In both types of assays, use of liposomes also amplified the signal by approximately an order of magnitude.

We also presented results from assays for detection of tetanus toxin where we used liposomes that contained glycolipids as receptors to recognize the toxin.

Glycolipids have been known as receptors for a variety of biological molecules for a long time but have not been much used in the development of biosensors and immunoassays. One major reason is the low solubility of glycolipids in water, prohibiting their use as a reagent. Another reason is that being small molecules, they do not offer many sites for covalent attachment without disrupting their activity. This makes it hard to conjugate them to a biosensor surface or to labels such as enzymes or fluorophores. An alternative is to use self-assembled structures such as liposomes, where the receptor glycoproteins are incorporated together with other lipids in a bilayer. This arrangement obviates any need for covalent conjugation and also presents glycolipids in an aqueous environment. Liposomes with GT1b incorporated in the bilayer performed well in a sandwich immunoassay to detect tetanus toxin. Many bacterial toxins and viruses bind to glycolipids quite specifically allowing liposomes containing suitable glycolipids to be used as sensitive probes for a variety of analytes of significance in clinical diagnostics, food science and food-quality monitoring, national defense, and basic research applications.

REFERENCES

1. Szoka F, Papahadjopoulos D. Comparative properties and methods of preparation of lipid vesicles (liposomes). *Ann Rev Biophys Bioeng* 1980;9:467–508.
2. Mayer LD, Hope MJ, Cullis PR. Vesicles of variable sizes produced by a rapid extrusion procedure. *Biochim Biophys Acta* 1986;858:161–168.
3. Zumbuehl O, Weder HG. Liposomes of contollable size in the range of 40 to 180 nm by defined dialysis of lipid/detergent micelles. *Biochim Biophys Acta* 1981;640:252–262.
4. Szoka F, Papahadjopoulos D. Procedure for preparation of liposomes with large internal aqueous space and high capture by reverse-phase evaporation. *Proc Natl Acad Sci USA* 1978;75(9):4194–4198.
5. Lasic D. Liposomes. *Ame Sc* 1992;80:20–31.
6. Singh AK, Carbonell RG. Liposomes in immunodiagnostics. In: Lasic DD, Barenholz Y, eds. *Handbook of Nonmedical Applications of Liposomes, Vol. 4.* Boca Raton: CRC Press, 1995, pp 209–228.
7. Locascio-Brown L, Plant AL. Horvath V, et al. Liposome flow injection immunoassay: implications for sensitivity, dynamic range, and antibody regeneration. *Anal Chem* 1990;62:2587–2593.
8. Yap WT, Lacasio-Brown L, Plant AL, et al. Liposome flow injection immunoassay: model calculations of competitive immunoreactants involving univalent and multivalent ligands. *Anal Chem* 1991;63:2007–2011.
9. Choquette SJ, Locascio-Brown L, Durst RA. Planar waveguide immunosensor with fluorescent liposome amplification. *Anal Chem* 1992;64:55–60.
10. Durst RA, Seibert STA, Reeves SG. Immunosensor for extra-lab measurements based on liposome amplification and capillary migration. *Biosens Bioelectron* 1993;8:xiii–xv.
11. Fiechtner M, Wong M, Bieniarz C, et al. Hydrophilic fluorescein derivatives: useful reagents for liposomes immunolytic assays. *Anal Biochem* 1989;180:140–146.
12. Katoh S, Sohma Y, Mori Y, et al. Homogeneous immunoassay of polyclonal antibodies by use of antigen-coupled liposomes. *Biotech Bioeng* 1993;41:862–867.
13. Jones M A, Singh AK, Kilpatrick PK, et al. Preparation and characterization of ligand-modified labelled liposomes for solid phase immunoassays. *J Lip Res* 1993;3:793–804.
14. Singh AK, Kilpatrick PK, Carbonell RG. Non-competitive immunoassays using bifunctional unilamellar vesicles (or liposomes). *Biotech Prog* 1995;11:333–341.
15. Singh AK, Kilpatrick PK, Carbonell RG. Application of antibody and fluorophore-derivatized liposomes to heterogeneous immunoassays for *d*-dimer. *Biotech Prog* 1996;12:272–280.

16. Nakane PK, Kawaoi A. Peroxidase-labeled antibody: a new method of conjugation. *J Histochem Cytochem* 1974;22:1084–1091.
17. Heath TD, Robertson D, Birbeck MSC, et al. Covalent attachment of horseradish peroxidase to the outer surface of liposomes. *Biochim Biophys Acta* 1980;599:42–62.
18. Jones MA, Kilpatrick PK, Carbonell RG. Preparation and characterization of bifunctional unilamellar vesicles for immunosorbent assays. *Biotech Prog* 1993;9:242–258.
19. Jentoft N, Dearborn DG. Protein labeling by reductive alkylation. In: *Methods in Enzymology, Vol. 19*, 1983:570–579.
20. Chen PS, Toribara TY, Warner H. Microdetermination of phosphorus. *Anal Chem* 1956;28:1756–1758.
21. O'Connell MA, Belanger BA, Haaland PD. Calibration and assay development using the four-parameter logistic model. *Chemometrics and Intelligent Laboratory Systems* 1993;20:97–114.
22. Goding JW. Conjugation of antibodies with fluorochromes: modifications to the standard methods. *J Immunol Meth* 1976;13:215–226.
23. Jones MA, Kilpatrick PK, Carbonell RG. Competitive immunosorbent assays for biotin using bifunctional unilamellar vesicles. *Biotech Prog* 1994;10:174–186.
24. Gustafsson B, Whitmore E, Tiru M. Neutralization of tetanus toxin by human monoclonal antibodies directed against tetanus toxin fragment C. *Hybridoma* 1993;12(6):699–708.
25. Schengrund CL, DasGupta BR, Ringler NJ. Binding of botulinum and tetanus neurotoxins to ganglioside GT1b and derivatives thereof. *J Neurochem* 1991;57(3):1024–1032.
26. Calappi E, Masserini M, Schiavo G, et al. Lipid interaction of tetanus neurotoxin *FEBS* 1992, 309(2):107–110.

8

Recent Development in Polymer Membrane-Based Potentiometric Polyion Sensors

Bin Fu, Mark E. Meyerhoff, and Victor C. Yang

8.1. INTRODUCTION

Many biologically significant macromolecules such as DNA, RNA, highly charged proteins (e.g., protamine), polysaccharides (e.g., heparin) and other species (e.g., polyphosphate) are polyions.[1-4] Quantitative measurement of such polyionic species in complex sample matrices (e.g., whole blood) is important and challenging. For example, heparin, a highly sulfated glycosaminoglycan, is the anticoagulant of choice for numerous clinical procedures, especially those involving extracorporeal blood circulation such as open heart surgery and kidney dialysis.[4,5] It is critical to monitor and adjust the blood heparin levels since an excess of heparin can lead to potentially fatal bleeding complications.[5,6] Currently, blood heparin levels are estimated indirectly by measuring clotting times such as, e.g., activated clotting time (ACT) or activated partial thromboplastin time (APTT).[5,6] However, as clotting times can be influenced by a wide range of additional physiological processes,[5,6] direct measurement of heparin via a sensor may provide a more precise as well as a complementary tool to aid clinicians in monitoring heparin therapy. Development of simple, convenient sensors for other biologically important polyionic species such as DNA, RNA, and polyphosphate is also of considerable interest. For example, a chemical sensing device that can accurately determine the concentration of free, uncomplexed DNA in the solution phase could provide a convenient means of following DNA binding reactions via simple titration of DNA with other binding species.

Bin Fu • Diagnostic Division, Bayer Corporation, Tarrytown, New York 10591.   Mark E. Meyerhoff • Department of Chemistry, University of Michigan, Ann Arbor, Michigan 48109.   Victor C. Yang • College of Pharmacy, University of Michigan, Ann Arbor, Michigan 48109.

Biosensors and Their Applications, edited by Yang and Ngo, Kluwer Academic/Plenum Publishers, New York, 1999.

147

In the search for a device that can detect an ionic species in a complex sample matrix, the ion-selective electrode (ISE) may be a good candidate. ISE-based techniques have several advantages over other available analytical methods for routine quantitative analysis of ions in clinical or other complex samples.[7,8] Since they were first introduced in the 1970s, polymer membrane-based ISEs have become the most successful sensing tools for the measurement of biomedically important ionic species. For example, modern clinical analyzers equipped with such electrodes are utilized routinely for detecting levels of Ca^{2+}, K^+, Na^+, Cl^-, etc. in undiluted whole blood samples.[8,9] However, polymer membrane-based ISEs have yet to be applied for the detection of polyionic species. First, it is difficult to identify the appropriate membrane chemistry that would allow for the effective extraction of the analyte polyion into the membrane phase. Second, even if such a membrane chemistry were available, according to classic ISE theory, the prospects of devising a polyion potentiometric sensor would seem improbable. That is, in accordance with the Nernst equation, the response slope of a potentiometric sensor ($59.2/z_i$ at 25°C) is inversely proportional to the charge of the analyte ion (z_i).[7] Therefore, the high charge of a polyion (e.g., -70 for heparin) should result in a device that exhibits an impractically low slope (<1 mV/decade for heparin). Such ideal Nernstian behavior would not provide sufficient sensitivity for practical analytical purposes.

In this chapter, we review the recent development of polymer membrane-based potentiometric sensors for multiply charged polyions, including polyanionic heparin, DNA, RNA, polyphosphate, and polycationic protamine (an antidote used during extracorporeal procedures to reverse heparin's anticoagulant activity).[10-12] Specifically, it is shown that poorly plasticized PVC membranes doped with appropriate lipophilic ion-exchangers exhibit significant and reproducible EMF responses to sub-μM concentrations of polyionic heparin and protamin.[10-12] The unique underlying principles for the potentiometric polyion sensors are described.[13,14] The utility of polyion sensors has been demonstrated in both practical and fundamental bioanalytical applications. The limitations of this type of sensors are also discussed.

8.2. DEVELOPMENT OF POLYMER MEMBRANE-BASED POLYION SENSORS

8.2.1. Extraction Chemistry

As noted above, in order to develop a membrane-based potentiometric sensor for a polyion, it is necessary to identify a membrane chemistry that can effectively extract the analyte polyion into the membrane phase. It has been reported in the literature that various water-soluble quaternary ammonium salts (QAS) form tight complexes with heparin and have been used effectively for its purification.[15] To this end, extensive research by Ma *et al* focused on the commercially available lipophilic QAS that can be doped into a conventional, plasticized poly (vinyl chloride) (PVC) membrane.[10,11] It has been found that the precise structure of the lipophilic ion-exchanger used within the membrane phase has a profound influence on the magnitude of the EMF responses. For example, when tridodecylmethylammonium chloride (TDMAC) is doped into a plasticized PVC membrane, the resulting electrode yields the largest potentiometric response to heparin.[11] Other QAS structures that have different numbers of long, lipophilic alkyl chains exhibit much less sensitive responses.[11]

Further thermodynamic studies revealed that a much tighter ion pair is formed in the membrane phase when TDMAC is incorporated into the membrane.[14] The findings have been fully explained with a cooperative interaction model, as shown in Fig. 8.1. It is clear that the maximum electrostatic interaction between the polyanionic heparin and the positively charged QAS site in the membrane can only be achieved when a small methyl group is present in the QAS to avoid steric hindrance.

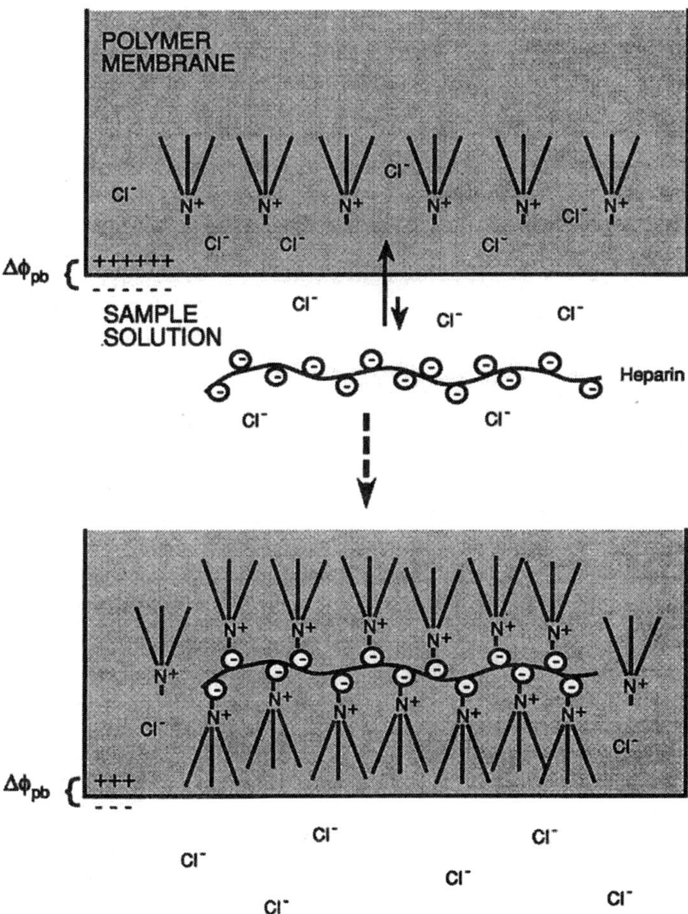

Figure 8.1. Schematic representation of how the heparin molecule may interact with the lipophilic quaternary ammonium sites and be extracted into the organic membrane phase to yield a large potentiometric response.

Indeed, when tetradodecylammonium chloride (TDAC) is used, the resulting membrane exhibits only minimal EMF response toward heparin.[10,11] On the other hand, more lipophilic alkyl groups would stabilize the ion pair formed between heparin and QAS in a lipophilic membrane environment. For example, membranes incorporated with didodecyldimethyl- or dodecyltrimethylammonium chloride exhibit less sensitive EMF responses toward heparin.[11] Toward this end, tridodecylmethylammonium chloride was selected to be doped into a plasticized PVC membrane for the selective detection of heparin and other polyanionic species.

It should be noted that various lipophilic quaternary ammonium structures have previously been incorporated into plasticized PVC membranes for the detection of small anions.[7,8] However, all QAS structures show similar responses in this type of application, where the observed EMF response depends only on the lipophilicity of the small ions (the so-called Hofmeister series) and not on the QAS structures.[7,8] Since the QAS structures in these membranes serve only as the ion-exchange sites, there is no specific interaction between the QAS and the small ions.

8.2.2. Response Slope

As discussed previously, even with an appropriate membrane-phase extraction chemistry, the corresponding electrode should not yield an analytically significant response under thermodynamic conditions.[7] Indeed, a dioctyl sebacate (DOS) plasticized PVC membrane doped with TDMAC exhibits an equilibrium response slope of less than 1 mV/decade toward heparin (see Fig. 8.2).[13] It should be pointed out, however, that the equilibrium response to a solution containing a low concentration (sub-μM) of heparin can be obtained only after the electrode is equilibrated in the solution for a long period of time (>20 hr). Interestingly, when the electrode was in contact with the solution for a much shorter period of time (around 5 min), a reproducible, sensitive response was observed for low concentrations of heparin. Considering the low heparin concentration and the slow diffusion of macromolecular heparin, it appears that the amount of heparin extracted into the membrane does not fully replace the counterchloride ion initially present at the interface of the membrane during the short measuring period. Therefore, the apparent steady-state response must be due to a nonequilibrium process, where the amount of heparin extracted into the membrane phase equals the amount of heparin that diffuses into the bulk of the membrane.[13] This apparent nonequilibrium steady-state response has been modeled and a mathematical expression has been derived for the EMF response[13]:

$$\Delta \text{EMF} = \frac{RT}{F} \ln \left(1 - \frac{z}{[R^+]_m} \frac{D_a \delta_m}{D_m \delta_a} c_{\text{Hep}} \right)$$

where ΔEMF is the potential change in going from a sample containing chloride only to the same sample after the addition of heparin; D_a and D_m represent the heparin diffusion coefficients in the aqueous and membrane phases, respectively; δ_a

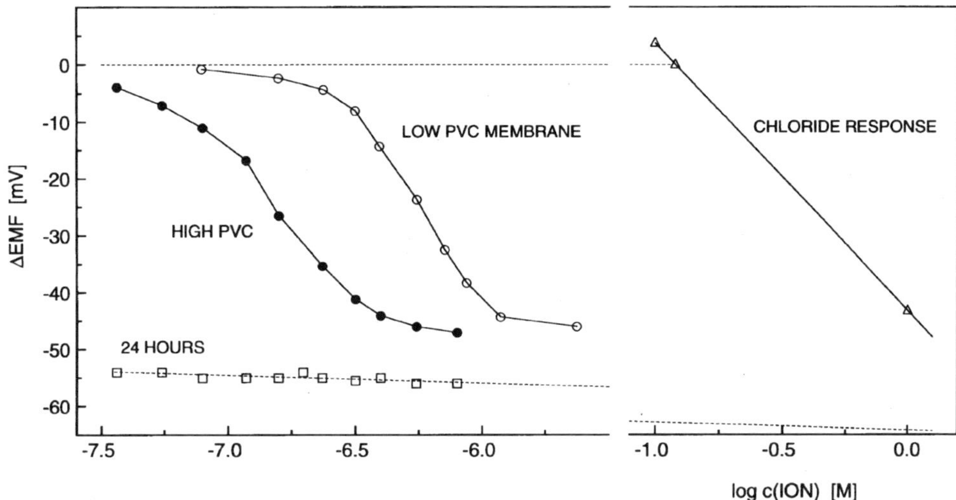

Figure 8.2. Typical potentiometric heparin responses of polymer membranes doped with TDMAC in 0.12 M NaCl background solution: steady-state heparin response with 66 wt% (●) and 32.5 wt% (○) PVC membranes; and equilibrium heparin response (□).

and δ_m denote the diffusion layer thickness in aqueous and membrane phases, respectively; c_{Hep} and $[R^+]$ are the heparin concentration in the sample and the TDMA^+ concentration in the membrane, respectively.

Based on this equation, it is clear that the electrode's steady-state response toward heparin (or other polyanions) can be manipulated by controlling the diffusion processes in the aqueous and the membrane phases. That is, in order to obtain a sensitive response toward low concentrations of heparin, the diffusion of heparin in the membrane should be slowed down while its diffusion in the aqueous phase should be accelerated. Indeed, when the fluidity of the membrane is reduced by decreasing the plasticizer (DOS) content from 66 to 33 wt %, the dose–response curve shifts to a lower concentration range (see Fig. 8.2).[13] Another option for increasing the mass transport efficiency in the aqueous phase is to control the geometric configurations of the electrode.[13] Toward this end, a tubular and a regular planar electrode was fabricated using the same membrane composition. As shown in Fig. 8.3, the tubular electrode exhibits a more sensitive response toward low concentrations of heparin than a regular planar electrode. This observation is not surprising considering that cylindrical diffusion is more efficient than planar diffusion.[16] It should also be clear from this model that the aqueous solution should be stirred constantly in order to achieve a better detection limit.[13] Further, unless otherwise indicated, a tubular membrane electrode that contains 62.5 wt% PVC, 33 wt% DOS, and 1.5 wt% TDMAC was chosen for all the subsequent practical applications for measuring polyanions, including heparin.

The same principle has been applied for devising a polycationic potentiometric sensor for protamine and other polypeptides.[12,17] For the protamine sensor, a

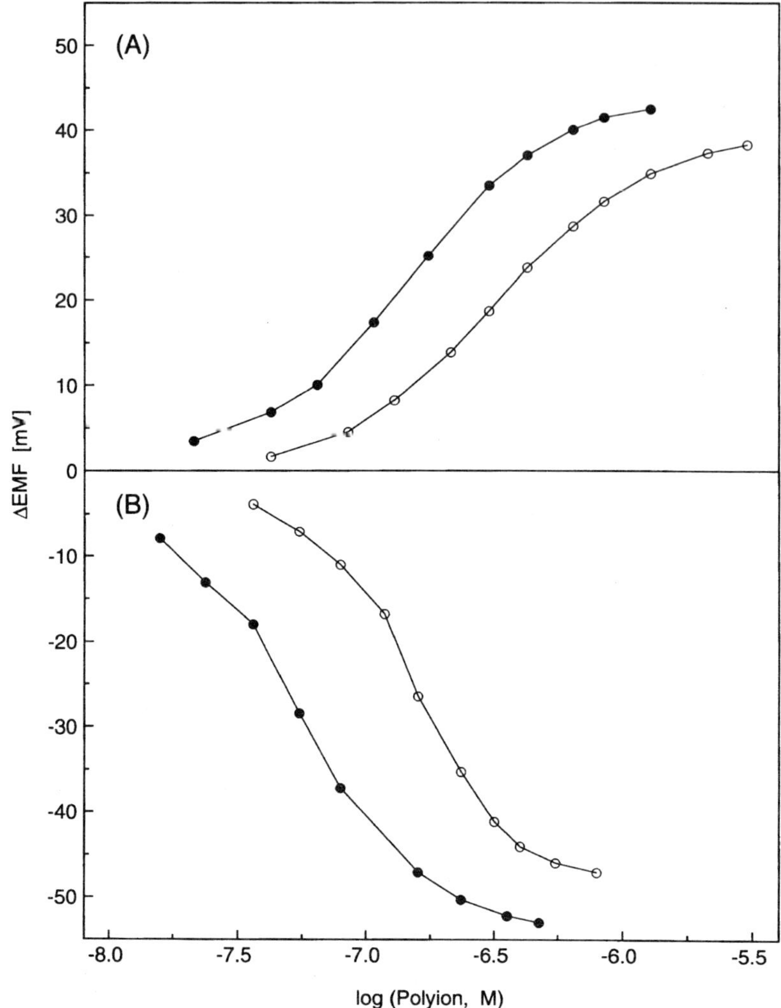

Figure 8.3. Effect of electrode geometry on the potentiometric response of protamine sensor [(A) 66 wt%
PVC, 33 wt% DOS, and 1 wt% KT*p*ClPB] and heparin sensor [(B) 66 wt% PVC, 32.5 wt% DOS, and
1.5 wt% TDMAC] in 0.12 M NaCl background solutions: (●) tubular electrode; (○) planar electrode.

lipophilic cation-exchanger, tetra(*p*-chlorophenyl)borate (T*p*ClPB), was doped within an *o*-nitrophenyl octyl ether (*o*-NPOE) plasticized PVC membrane.[12] Protamine
concentration as low as 10^{-7} M can be detected in the presence of 0.12 M NaCl.[12]
The response behavior of the protamine sensor is similar to that of the heparin
sensor, e.g., a sensitive EMF response to low levels of protamine can be observed
only with a poorly plasticized membrane, and a tubular electrode shows a lower
detection limit than a planar electrode (see Fig. 8.3).[13] Further studies demonstrated
that a negatively charged dinonylnaphthalene sulfonate (DNNS) salt (Ca^{2+}), when

doped in a DOS plasticized PVC membrane, exhibited an even more sensitive EMF response toward protamine and other synthetic polypeptides.[17] Toward this end, membrane electrodes prepared with both TpClPB and DNNS have been used in various applications for detecting polycations, including protamine and other synthetic polypeptides (see below).[12,17]

It should be emphasized that an analytically useful response can only be obtained with a "clean" electrode over a short period of time for low concentrations of polyionic species. After the electrode has been in contact with an analyte-containing solution, it must be washed with a highly concentrated salt solution (e.g., 2 M NaCl) to dissociate the ion pair at the membrane interface prior to the actual polyion measurement. As noted above, in order to obtain a significant response the interaction between the analyte polyion and the ion-exchanger inside the membrane should be strong. Apparently, a compromise must be found between sensitivity and reversibility. In addition, the sensitivity of the polyion sensor depends on the diffusion and extraction of the polyion into the membrane phase; thus, the electrode actually detects the concentration rather than the activity of the polyion, which makes it different from typical ISEs. Moreover, owing to the extraction of the polyion from the sample, the actual concentration of the polyion in the sample phase is perturbed. In order to minimize the perturbation either a small electrode or a relatively large volume of sample should be used.

8.3. APPLICATIONS OF POLYION SENSORS

8.3.1. Measuring the Blood Heparin Levels

The potential number of applications of polymer membrane-based polyion sensors is enormous. As noted previously, the current methods for detecting heparin in blood samples are based on measurement of clotting time.[4,6] In a preliminary study, we have shown that the optimized heparin sensor exhibited a significant EMF response toward heparin in whole blood from patients undergoing open heart surgery.[11] However, the absolute potential change cannot be used to determine the blood heparin concentration accurately, since the direct EMF values can also be influenced by slight variations in the levels of other anionic species (e.g., Cl^- and HCO_3^-) in the blood.[11] Subsequently, the heparin sensor was used in potentiometric titrations for quantitative measurement of heparin.[18,19] In these experiments, known amounts of protamine were mixed with aliquots of a blood sample and the EMF responses were measured by the heparin sensor with a bare Ag/AgCl wire as a pseudoreference electrode.[18] A typical potentiometric protamine titration curve is shown in Fig. 8.4. As can be seen, when a blood sample with heparin is titrated with protamine, the potential becomes less negative and plateaus after the heparin is fully neutralized by protamine. By extrapolating the titration data, the blood heparin level can be determined accurately. In contrast, the blood sample without heparin does not exhibit any electrode potential change as protamine is added.

The accuracy of the method has been tested during open heart surgery, with data obtained by the electrochemical method correlating well with current pro-

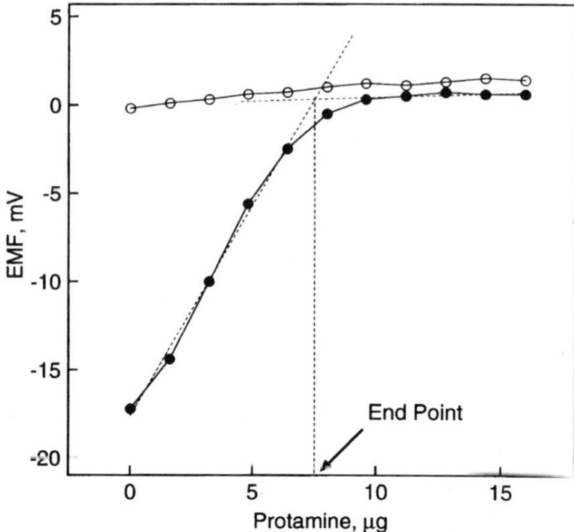

Figure 8.4. Potentiometric titration of (●) 3.3 U/mL; and (○) 0 U/mL of heparin in whole blood samples with protamine as followed by a tubular heparin sensor (membrane composition: 66 wt% PVC, 32.5 wt% DOS, and 1.5 wt% TDMAC).

tamine titration assay performed on a Hepcon HMS system.[18,19] The potentiometric titration method also appears to yield results that correlate better with a standard chromogenic assay than the commercial titration instruments in which clotting times are used to indicate the titration end-point.[18] It should be noted, however, that owing to the reversibility limitation of the heparin sensor described above, potentiometric response from samples should be measured going from the highest to the lowest levels of protamine added if the same electrode is to be used for all the aliquots. Otherwise, multiple electrodes with identical heparin response characteristics should be used for samples with different amounts of added protamine.

A more direct potentiometric titration was recently achieved using a newly formulated protamine membrane (containing DNNS, DOS, and PVC) electrode to indicate the titration end-point.[20] In such direct titration, a pair of electrodes (protamine sensor with a reference electrode) was used for the EMF measurement, and incremental amounts of protamine added to the blood sample were measured directly.[20] Initially, the potentiometric response from the protamine sensor remains constant as all the added protamine binds to the heparin. As more protamine is added, the free protamine in the sample results in a gradual increase of EMF response from the protamine sensor. The end-point is determined by the breakpoint of potential measured by the protamine sensor.[20] This direct titration has certain advantages over indirect titration, where the heparin sensor was used to indicate the end-point, e.g., there is no need to divide the blood sample into a number of aliquots. A correlation study has shown that the direct titration method correlates well with the commercial Hepcon HMS method and the chromogenic factor Xa assay. The direct titration method can be used for measuring heparin levels in whole blood from patients undergoing open heart surgery. In contrast, plasma samples had to be used for the chromogenic assay.[20]

8.3.2. Probing Binding Reactions

Beyond direct biomedical applications of measuring blood heparin levels, the polyion sensors also appear useful for more fundamental studies. For example, both the polyanion and the polycation sensors have been used successfully in potentiometric titration to quantify the solution-phase binding reactions of macromolecules with a host of other macromolecules.[12,21] Such thermodynamic information regarding the binding reaction between macromolecules is of considerable interest and is currently being investigated through various expensive and sophisticated techniques, such as solid-phase radiolabeled binding assays and homogeneous fluorescent polarization methods.[22–24] The simple potentiometric titration method using the polyion sensor as the indicator electrode provides a very attractive alternative. For example, potentiometric titrations using the heparin sensor as the indicator electrode have been successfully applied for probing the binding reactions between heparin and a host of other polycationic species, including protamine.[21] The stoichiometries of the binding reactions between heparin and other polycations can be obtained directly from the titration curves. Furthermore, by using prior potentiometric calibration data for heparin to determine the free concentrations of heparin at each titration point, the binding constants between heparin and other polycations can be estimated by recasting the data in the Scatchard plots. The binding constants between heparin and a number of other species have been determined and the data correlate well with the values reported in the literature.[21] Similarly, the electrode has also been shown to be useful for probing the binding reactions between DNA, RNA, and a number of other macromolecules in a relatively low concentration of buffer solution.[25] The protamine responsive electrode has also been used successfully to measure the binding constants between protamine and other polyanionic species.[12]

8.3.3. Detecting Protease Activities

Another exciting new application of polyion sensors relies on the observation that the appropriately formulated membranes exhibit little or no EMF response to small fragments resulting from digestion of larger parent polyion structures by proteases or other enzymes.[12,17,26] The rate of hydrolysis of larger polyions to smaller multivalent species can be monitored conveniently by measuring the rate of the potential change (dE/dt). Thus, it is possible to use the polyion sensors to monitor important enzyme activities that cleave larger polyionic structures into smaller fragments. For example, it was shown that the potentiometric response of an newly formulated membrane electrode (containing TpClPB, o-NPOE, and PVC) to low concentrations of protamine can be used to measure the activity of protease trypsin, which cleaves protamine specifically at arginine residues into smaller polycationic structures (Fig. 8.5).[12] As shown in the case of trypsin, this rate is directly related to the activity of the protease added (see inset in Fig. 8.5). Similarly, a polyanion responsive membrane electrode (containing TDMAC, DOS, and PVC) was applied for measuring the enzymatic activity of phosphatase, which can cleave the polymeric chain of polyphosphate and result in the potential change measured by the polyanion sensor.[26] In this assay, a 5-mer polyphosphate was employed as the substrate for phosphatase and a linear response (dE/dt vs. enzyme activity) was observed in the acid phosphatase activity range of 5 to 300 mU/mL.[26]

The concept of monitoring protease activities with polyion-sensitive membrane electrodes can be further extended to use specifically designed, multiply charged polypeptides for the detection of other significant protease including thrombolytic agents.[17] It is possible to design polypeptides that can serve as very selective substrates for given proteases and, at the same time, exhibit even more favorable extraction into polymer membranes doped with cation exchangers (and thus yield greater potentiometric response in the presence of physiological cations). To this end, a series of novel polypeptide structures have been designed where di- or triarginine sequences are linked to each other by one or more other amino acids recognized specifically by the peptidase to be determined. A polycation-sensitive membrane electrode (containing DNNS, DOS, and PVC) was applied for monitoring the enzymatic activities of renin and chymotrypsin in undiluted plasma samples.[17] The selectivity can be controlled by the sequence of the amino acids selected, and the sensitivity of the method can be controlled by varying the overall lipophilicity of the polypeptides, where the higher the lipophilicity the easier it is to extract the polypeptides into the membrane.[17] It is expected that this approach will enable the development of simple electrode-based assays for detecting thrombolytic drugs and other important protease activities directly in small discrete samples of undiluted

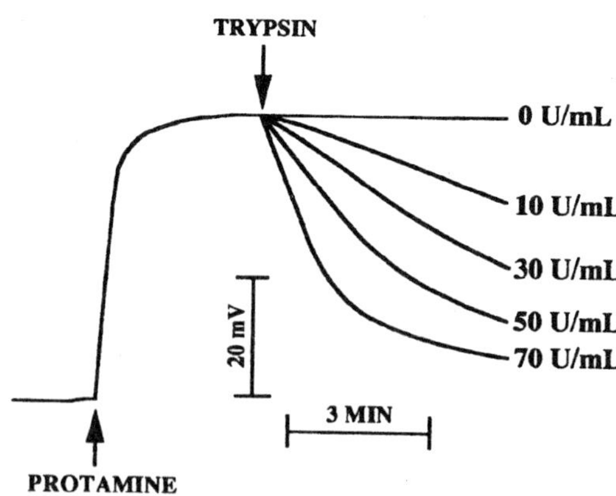

Figure 8.5. Response profile of the polycation-sensitive membrane electrode (66 wt% PVC, 33 wt% DOS, and 1 wt% KTpClPB) to protamine (20 mg/mL) and then subsequent digestion of the added protamine with different concentrations of trypsin.

blood. This measurement is not possible using existing protease assays based on smaller synthetic substrates that are hydrolyzed by proteolytic enzymes to yield colored products (detected spectrophotometrically in diluted plasma only).

8.4. CONCLUSIONS

The use of membrane electrodes to detect polyions represents a new and exciting research direction for electroanalytical chemistry. The immediate applications of this technology in the biomedical arena are obvious from initial studies with heparin/protamine summarized above. In addition, preliminary studies with DNA and other polyphosphates indicate that appropriately formulated membranes also exhibit a significant nonequilibrium response toward these species (at sub-μg/mL levels), suggesting the possibility of even more diverse applications in the future. Extensions of this measurement technology to other areas in which polyions are used (e.g., the food and pharmaceutical industries) should also be possible.

REFERENCES

1. Steiner R, Miller DBS. The nucleic acids. In: Veis A, ed. *Biological Polyelectrolytes*. New York: Marcel Dekker Inc, 1970, pp:65–130.
2. Moroson H, Rotman M. Biomedical applications of polycations. In: Rembaum A, Selegny E. eds. *Polyelectrolytes and Their Applications*. Dordrecht: Reidel Publishing Co, 1975, pp. 187–195.
3. Newman DJ, Olabiran Y, Price CP. Bioaffinity agents for sensing systems. In: Kress-Rogers E, ed. *Handbook of Biosensors and Electronic Noses — Medicine, Food, and the Environment*. Boca Raton: CRC Press Inc, 1997, pp 59–89.
4. Jaques LB. Heparins: anionic polyelectrolyte drugs. *Pharmacol Rev* 1979;31:99–166.
5. Metz S, Horrow JC. Protamine and newer heparin antagonists. *Pharmacol Physiol Anesth Prac* 1994;1:1–15.
6. Hirsh J. Heparin induced bleeding. *Nouv Rev Fr Hematol* 1984;26:261–266.
7. Morf WE. *The Principles of Ion-Selective Electrodes and of Membrane Transport*. New York: Elsevier Science Publishing Co, 1981, pp 1–25.
8. Yim H, Kibbey C, Ma SC, et al. New polymer membrane based ion, gas and bioselective sensors. *Biosens Bioelectron* 1993;5:719–724.
9. Oesch U, Ammann D, Simon W. Ion-selective membrane electrodes for clinical use. *Clin Chem* 1986;32:1448–1459.
10. Ma SC, Yang VC, Meyerhoff ME. Heparin-responsive electrochemical sensor — a preliminary study. *Anal Chem* 1992;64:694–697.
11. Ma SC, Yang VC, Fu B, et al. Electrochemical sensor for heparin: further characterization and bioanalytical applications. *Anal Chem* 1993;65:2078–2084.
12. Yun JH, Yang VC, Meyerhoff ME. Protamine-sensitive polymer membrane electrode: characterization and bioanalytical applications. *Anal Biochem* 1995;224:212–220.
13. Fu B, Bakker E, Yun JH, et al. Response mechanism of polymer membrane-based potentiometric polyion sensors. *Anal Chem* 1994;66:2250–2259.
14. Fu B, Bakker E, Yang VC, et al. Extraction thermodynamics of polyanions into plasticized polymer membranes doped with lipophilic ion-exchangers: a potentiometric study. *Macromolecules* 1995;28:5834–5840.
15. Hurst RE, Jennings GC, Lorincz AE. Partition techniques for isolation and fractionation of urinary glycosaminoglycans. *Anal Biochem* 1977;79:502–512.

16. Crank J, Park GS. Methods of measurement, In: Diffusion in polymers, Crank J, Park GS, ed. London: Academic Press Inc, 1968, pp 1–37.
17. Han IS, Ramamurthy N, Yun JH, et al. Selective monitoring of peptidase activities with synthetic polypeptide substrates and polyion-sensitive electrode detection. *FASEB J* 1996;10:1621–1626.
18. Meyerhoff ME, Yang VC, Wahr JA, et al. Potentiometric polyion sensors: a new measurement technology for monitoring blood heparin levels during open heart surgery. *Clin Chem* 1995;41:1355–1356.
19. Wahr JA, Yun JH, Yang VC, et al. A new method of measuring heparin levels in whole blood by protamine titration utilizing a heparin-responsive electrochemical sensor. *J Cardiothor Vasc Anesth* 1996;10:447–450.
20. Ramamurthy N, Baliga N, Wahr JA, et al. Improved protamine-sensitive membrane electrode for monitoring heparin concentrations in whole blood via protamine titration. *Clin Chem* 1998;44:606–613.
21. Yun JH, Ma SC, Yang VC, et al. Direct potentiometric membrane-electrode measurements of heparin-binding to macromolecules. *Electroanalysis* 1993;5:719–724.
22. Johnson NP, Hoeschele JD, Rahn RO. Kinetic analysis of the in vitro binding of radioactive cis- and trans-dichlorodiammineplatinum (II) to DNA. *Chem Biol Interact* 1980;30:151–169.
23. Revzin A. *The Biology of Nonspecific DNA-Protein Interactions*. Boca Raton, FL: CRC Press, 1990.
24. Millar DP, Ho KM, Aroney MJ. Modification of DNA dynamics by platinum drug binding: a time-dependent fluorescence depolarization study of the interaction of cis- and trans-diamminedichloroplatinum (II) with DNA. *Biochemistry* 1988;27:8599–8606.
25. Fu B, Meyerhoff ME. Unpublished results.
26. Esson JM, Meyerhoff ME. Polyanion-sensitive polymeric membrane electrodes for polyphosphates, DNA, RNA and acidic polypeptides. *Pittcon'97 Book of Abstracts* 1997:1245.

Piezoelectric Immunosensors: Theory and Applications

C. K. O'Sullivan and G. G. Guilbault

9.1. INTRODUCTION

A biosensor can be defined as a device incorporating biological material connected to or integrated within a transducer. The specificity and sensitivity is complemented by the transducer, which measures and computes the signal electronically. Many possible biological elements can be combined with various transducers to construct the biosensor.[1,2] Some of the more important biological elements and transducers that have been used are listed in Table 9.1. The purpose of this chapter is to provide a survey of the use of antigens as coatings in quartz crystal microbalances. This use was first proposed by Shons et al.,[3] and since this original publication, considerable research has been devoted to the development of the piezoelectric immunosensor.

9.2. QUARTZ CRYSTAL MICROBALANCE—THEORY

The theoretical foundation for the use of piezoelectricity was pioneered by Raleigh in 1885,[4] but the first thorough investigation was done by Jacques and Pierre Curie in 1880.[5] A quartz crystal microbalance (QCM) consists of a thin quartz disk with electrodes plated on it (Fig. 9.1). As the QCM is piezoelectric, an oscillating electric field applied across the device induces an acoustic wave that propagates through the crystal and meets minimum impedance when the thickness of the device is a multiple of a half-wavelength of the acoustic wave. A QCM is a shear mode device in which the acoustic wave propogates in a direction perpendicular to the crystal surface. Deposition of thin film on the crystal surface decreases the frequency in proportion to the mass of the film.

C. K. O'Sullivan and G. G. Guilbault • Laboratory of Sensor Development, Department of Chemistry, University College Cork, Cork, Ireland.

Biosensors and Their Applications, edited by Yang and Ngo, Kluwer Academic/Plenum Publishers, New York, 1999.

Table 9.1.   Important Biological Elements and
Transducers Incorporated in Biosensors

Biological elements	Transducers
Enzymes	Electrochemical
Receptors	Potentiometric
Antibodies	Amperometric
Whole cells	Conductimetric
Tissues	Optical
	Luminescence
	UV/Visible
	Fiber optic
	Colorimetric
	Piezoelectric
	Acoustical (SAW)

A resonant oscillation is achieved by including the crystal in an oscillation circuit where the electric and the mechanical oscillations are close to the fundamental frequency of the crystal, which depends upon the thickness of the wafer, its chemical structure, its shape, and its mass.

Several factors can influence the oscillation frequency, such as the thickness, the density, and the shear modulus of the quartz, which are constant, and the physical properties of the adjacent mediums (density or viscosity of air or liquid). As shown by Sauerbrey,[6] changes in the resonant frequency are simply related to the mass accumulated on the crystal by the following:

$$\Delta f = -2\Delta m n f_0^2 / \eta_q \rho_q \tag{1}$$

where η_q and η_q are the density and viscosity of the quartz (2.648 g/cm^3 and

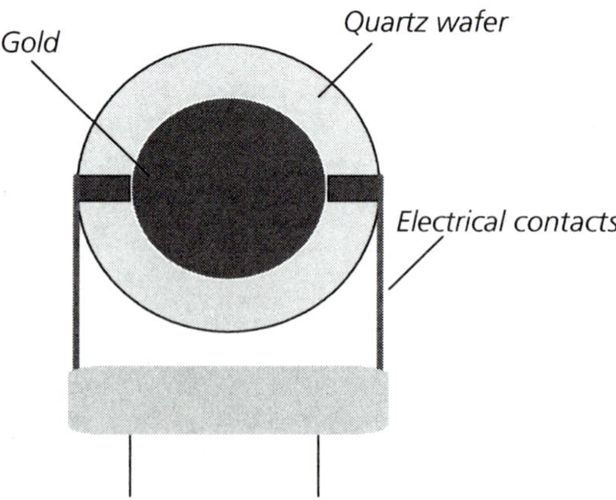

Figure 9.1.   The quartz crystal microbalance.

2.947×10^{-11} g/cm s, respectively), n is the overtone number, f_0 is the basic oscillator frequency of the quartz, and Δm is the material adsorbed on the surface per unit area. For an AT-cut crystal $\Delta f = -2.26 \times 10^{-6} f_0^2 \Delta m$.

When a crystal is dipped into a solution, the oscillating frequency depends on the solvent used. The question as to which factors determine the frequency is important for understanding the mechanism of oscillation of a crystal in solution and for its development as a sensor in solution.[7-9]

When an overlayer is thick, the relationship between the f and Δm is no longer linear and corrections are necessary. The coupling of the crystal surface to a liquid drastically changes the frequency when a quartz crystal oscillates in contact with it, a shear motion on the surface generates motion in the liquid near the interface. The oscillation surface generates plane-laminar flow in the liquid, which causes a decrease in the frequency proportional to $(\rho\eta)^{1/2}$, where ρ and η are the liquid density and the viscosity:

$$\Delta f = f_0^{3/2}(\rho\eta/\pi\eta_q\rho_q)^{1/2} \tag{2}$$

The penetration depth of this sheer wave depends on $(\pi f_0 \eta_q \rho_q)^{-1/2}$. For water at 20°C this is 2500 units for a 5-MHz crystal. Kurosawa et al.[8] stressed that Δf in solution is a linear function of $(\eta\rho)^{1/2}$ except for salts and high-polymer solutions. They, in fact, tested the linearity of the dependence of the frequency decrease on $(\eta\rho)^{1/2}$ using many solvents selected on the basis of differences in their viscosity, density, and electrical conductivity. For a 10-MHz crystal with one face exposed to dilute aqueous solutions near room temperature Δf is ca. 2 kHz. Crystal surface roughness may increase the magnitude of the liquid-phase-induced frequency change by several kilohertz.[9]

Two types of oscillator sensors are available: surface acoustic wave (SAW) and piezoelectric (PZ) crystal. SAW devices can operate either on the Raleigh wave propagation principle at solid thin-film boundaries or as a bulkwave piezoelectic device.[1,2,10,11] The latter follows the Sauerbrey equation, i.e., small thin films can be treated as equivalent mass changes on the crystal. SAW crystals oscillate at several hundred megahertz, instead of 9–14 MHz for the PZ. Theoretically greater sensitivity should result with the SAW, since the frequency term in the Sauerbrey equation is squared. Hence, a frequency of $(100)^2 = 10{,}000$ as compared to $(10)^2 = 100$ in the PZ detector. This two-decade increase in sensitivity is, however, diminished by the requirement of the SAW detector for a very thin coating film.

The first analytical application of the PZ detector was described by King.[12] PZs also served as detectors in gas chromatography.[13] Most of the early applications were limited to measurements in the gas phase and these have been reviewed.[13-16] Subsequent technological advances, especially in circuit design, that facilitate oscillation in the liquid phase have led to the rapidly growing field of piezoimmunosensors.[1,2,17,18]

The PZ immunosensor can use single- or multistep binding to the crystal surface and direct or indirect measurement of analyte. A single step measures the binding of one component to the modified crystal surface. Multistep methods rely on sequential binding of two or more components. Direct measurement depends on interaction of the analyte itself with the modified crystal surface (i.e., IgG determination using immobilized anti-IgG antibodies). The resonating frequency decreases with increasing amount of analyte. Indirect measurements rely on the interaction of the analyte with other components free in solution. In the competitive immunoassay, the antigen is immobilized on the crystal surface and the analyte present in solution competes for the binding sites of the antibody with the antigen immobilized. The observed frequency shift is inversely related to the analyte concentration.

In the displacement assay, the antigen is immobilized on the crystal surface. The addition of the relevant antibody results in the formation of the immunocomplex with an increase in mass and a corresponding frequency decrease. The antigen present in the added sample has a higher affinity for the antibody site bound to the antigen on the crystal surface and causes the displacement of the antibody with a resulting increase in frequency. The displacement of the antibody from the surface is proportionally related to the concentration of the analyte in the sample.

In performing an immunoassay, two crystals are used — a reference and an indicator. The ratio of the resonant frequencies of the waves generated in the reference and indicator crystals, when immersed in a blank solution, is determined. The crystals are, in turn, exposed to the test sample containing the corresponding antigen or antibody. After an immunochemical reaction, the change in the resonant frequency of the two crystals is measured, and a new resonant frequency ratio is determined.

The kinetics of the formation of the immunocomplex can be followed in real time. The reaction between the immobilized compound and the molecule in solution can often be assumed to follow pseudo-first-order kinetics. The reaction:

$$Ag + Ab \leftrightarrow AgAb \tag{3}$$

can be described by the kinetic association and dissociations k_a and k_b, respectively:

$$d[AgAb]/dt = k_a[Ag][Ab] - k_b[AgAb] \tag{4}$$

Considering the Sauerbrey equation and using f_m as the frequency change after, e.g., a complete saturation of the surface of the crystal with antibodies, one finds that the concentration of free antigen is proportional to $f_m - f$ and that the concentration of the complex AgAb is proportional to f. Then,

$$df/dt = k_a c(f_m - f) - k_b f \tag{5}$$

where c is the concentration of the free antibody A, held constant in a continuously flowing solution.

Antigens as PZ crystal coatings were first demonstrated by Shons et al.,[3] who attached a layer of antigen onto the surface of a 9-MHz crystal and used this for direct measurement of the amount of specific antibody in a liquid sample.

Microgravimetric immunoassays for the determination of antigens in solution were disclosed in a patent by Olivera and Silver.[19] Rice[20] patented a method for determining both the type of antibody subclass and the concentration of antibody present. In a follow-up patent[21], he described a sandwich assay that can be used to determine both the amount of a particular class or species of antibody and also the amount of total antibody–antigen present in the antibody layer.

Both PZ quartz crystals and SAW devices operating as bulk wave devices were coated with either antigens or antibodies for the assay of the corresponding antibody or antigen.[10] A method for immunogravimetric immunoassay was reported by Bastiaans[11] using ST-cut piezoelectric SAW devices in a solution phase. PZ immunosensors have found widespread application in the fields of environmental, food, gas, and clinical analysis.

9.3. QUARTZ CRYSTAL MICROBALANCE—APPLICATIONS

9.3.1. Clinical Analysis

Various infectious agents have been detected using the QCM. A study of five Herpes viruses was reported by König and Grätzel.[22] Herpes simplex types 1 and 2, Varicella-zoster virus, Cytomegalovirus, and Epstein–Barr were detected. Antiviral antibodies were immobilized on the crystal and were able to detect from 5×10^4 to 1×10^9 cells for all the viruses. The sensor was reusable 18 times without any detectable loss of activity when the regeneration process used synthetic peptides to elute the bound antigen (Fig. 9.2).

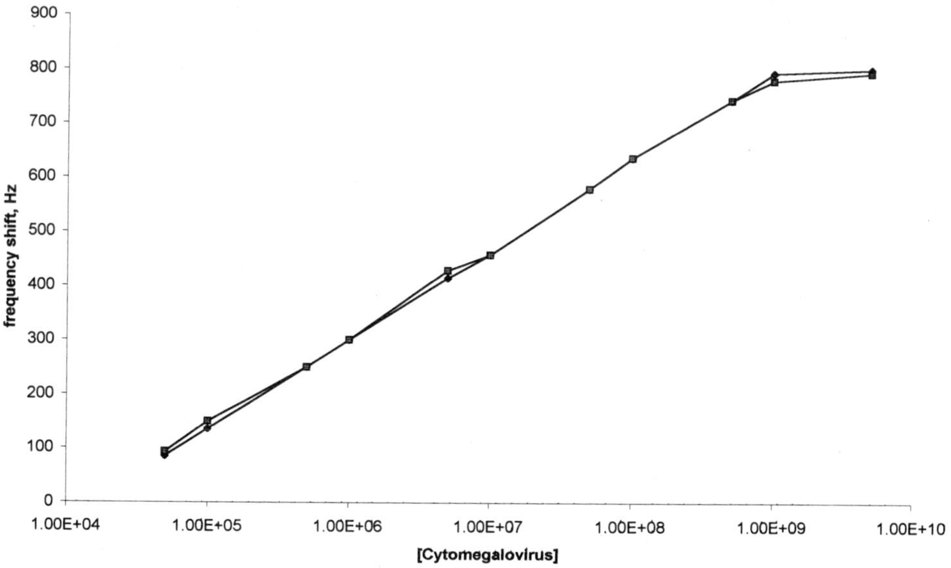

Figure 9.2. Frequency changes with increasing numbers of pure cytomegalovirus (◆) and cytomegalovirus in human species (■). Each point represents the average of five experiments.

Rotavirus, which causes acute diarrhea has been detected in stool specimens. Other viruses and bacteria are also found in stool — *Adenovirus, Salmonella, Shigella, Camphylobacter, E. coli,* and *Yersinia*. A QCM-based sensor for the detection of these microorganism was reported by König and Grätzel.[23] The linear range was 1×10^6 to 1×10^{10} cells for the *Rotavirus* and *Adenovirus* and from 1×10^6 to 1×10^8 cells for the *Camphylobacter* and the enterobacteria *Salmonella, Shigella, E. coli,* and *Yersinia* (Fig. 9.3). The long-term stability and reusability of a PZ immunosensor for the detection of purified human erythrocytes in whole human blood was studied by the same team[24]. The antibody-coated gold electrodes provided a long-term stability of 10 weeks stored either at room temperature or at 4°C. The crystal was reused 12 times without a decrease in activity. A reusable PZ quartz crystal immunosensor for the detection of hepatitis A and B viruses has also been reported.[25]

Subsequent to this the number of purified human B-lymphocytes in blood samples was determined by immobilizing an anti-B-cell antibody layer onto a 10-MHz-AT cut crystal 5×10^3 to 5.6×10^5 cells; resulted in a linear frequency change and a long-term stability of 6 weeks. The coated crystal can be used eight times without detectable loss of activity.[22]

Kosslinger et al.[26,27] used a model system of the HIV virus to demonstrate the feasibility of detecting viruses using the QCM as this pathogenic agent is well characterized on the molecular level and the established screening tests are highly selective and sensitive. Antibodies specific for the HIV were adsorbed on the crystal surface and 11 specimens from different patients were tested; there was good correlation between the response from the ELISA and the QCM tests. The overall

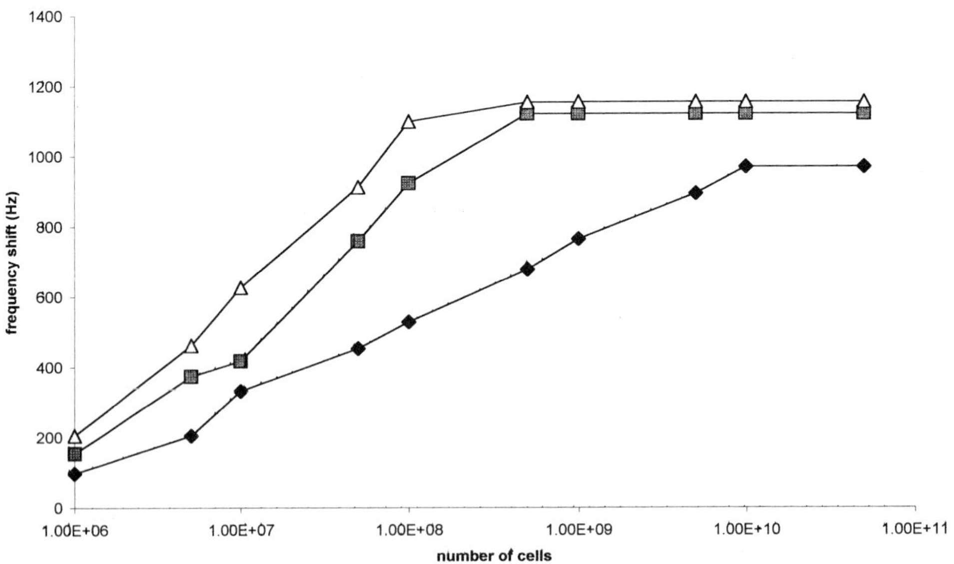

Figure 9.3. Frequency change with increasing antigen concentration of Rotavirus (♦), Campylobacter (■) and Shigella (△). Each point represents the average of ten experiments.

time demand for testing one serum with QCM in a flow system is 10 min, whereas for ELISA requires 2 h. The specificity of the system was also tested and no response was found with a nonspecific antibody.

Attili and Suleiman[28] developed a PZ immunosensor for the detection of cortisol. Cortisol antibody was layered onto the gold electrodes of a 10-MHz PZ crystal and was used successfully for the detection of cortisol in standard solutions in the range 36–3628 ppb. The detection limit with an immunosensor consisting of a PZ crystal coated with a water-insoluble polymer to which an antibody was covalently bound using thyroxine antibody on a 10-MHz crystal was 5×10^{-12} g of thyroxine.

Determination of immunoglobulins was reported by Muramatsu et al.[29] who detected IgG in the range 10^{-2} to 10^{-6} mg/mL under flow conditions. Immunoglobulin M (IgM) was determined using the specific binding between IgM and protamine. The IgM concentration up to a level of 10 ng/mL could be detected without interference of IgG.[30] Human chorionic gonadotropin (h-CG) and adenosine 5′-phosphosulfate (APS) reductase using an immunoassay amplified by an enzymatic reaction were determined by QCM[31]; 5 ng/mL of APS reductase could be detected.

An antibody-based PZ sensor has been applied to the detection of *M. tuberculosis* antigens in the liquid and vapor phase. A simultaneous immunoassay technique has been developed for the determination of dual analytes immunoglobulin M (IgM) and C-reactive protein (CRP) by constructing a PZ-array system made up of five PZ quartz crystals. Each crystal was immobilized with an antibody mixture that had a fixed ratio of anti-IgM to anti-CRP antibodies but worked under different detection conditions.[32]

9.3.2. Environmental Analysis

Immobilization of proteins (i.e., enzymes and antibodies) for the detection of atmospheric pollutants has proven successful in a large number of applications. The first use of protein as a coating for a direct assay of a gaseous compound was described by Guilbault in 1983,[33] thus making this method an attractive alternative to some of the conventional techniques then in use. Formaldehyde in the concentration range 1–100 ppb was assayed in the gas phase using formaldehyde dehydrogenase and its cofactor layered onto a 9-MHz crystal. No interference from other aldehydes, alcohols, or substances was reported.

Parathion antibodies were immobilized on a PZ crystal for the specific detection of this pesticide at parts per billion levels. In the presence of air containing traces of moisture the immunological reaction proceeds well in the gas phase, making a new generation of detectors possible.[34] Subsequently, the team coated crystals with cholinesterase and with antibodies to organophosphorus nerve agents for the specific detection of these compounds in air at part per billion levels (Fig. 9.4).[35]

Biosensor devices for analysis of drugs was reported by Guilbault and Schmid.[36] One of the systems investigated involved the use of PZ crystals coated with specific antibodies coupled with an FIA format that allowed the

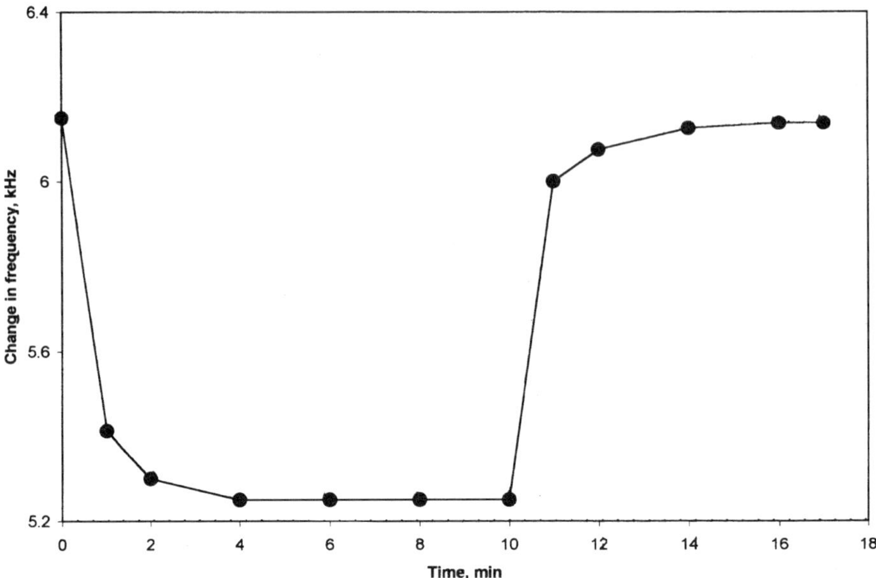

Figure 9.4.   Typical PZ crystal response to parathion-saturated carrier gas, 35 ppb at 30°C.

quasi-continuous detection of selected compounds in the parts per million range.

In 1992, Guilbault et al. reported a PZ sensor for the herbicide atrazine. Polyclonal antibodies to atrazine were layered onto the gold electrode of 10-MHz PZ crystals that were precoated with protein A. Determination from 0.03 to 100 ppb of atrazine can be made with a relative standard deviation of about $\pm 8\%$. The sensor is reversible, being reusable for up to nine assays.[37]

Minnuni et al.[38] presented an indirect assay for atrazine based on immobilization of an atrazine derivative on the crystal surface and the use of a competitive assay. Antiatrazine antibodies are incubated in the presence of different concentrations of the herbicide. The number of antibodies bound to the surface is inversely related to the atrazine concentration in the solution. Steegborn and Skladal[39] silanized the gold electrodes of the PZ quartz crystal and activated them using glutaraldehyde. The bioaffinity ligand atrazine was linked through albumin as a spacer molecule. The modified PZ crystal was placed in a flow cell and all measurements were performed directly in flowing solution. Repeated use of the crystals was achieved using a 5-min flow of sodium hydroxide for regeneration.

Red tide is a phenomenon caused by small algae or phytoplankton whereupon seawater appears red. Some kinds of red tides cause mass deaths of cultured fish. In Japan, the *Chattenella* species is one of the dominant representatives among many kinds of flagellates recognized in red tides. An immunosensor comprising a gold PZ crystal coated with monoclonal antibody MR − 21 showed high specific reactivity toward the cytoplasmic membrane of *Chattenella marina*. The immunosensor was able to directly detect 10^2 to 10^6 cells/mL of *C. marina* directly in seawater.[40]

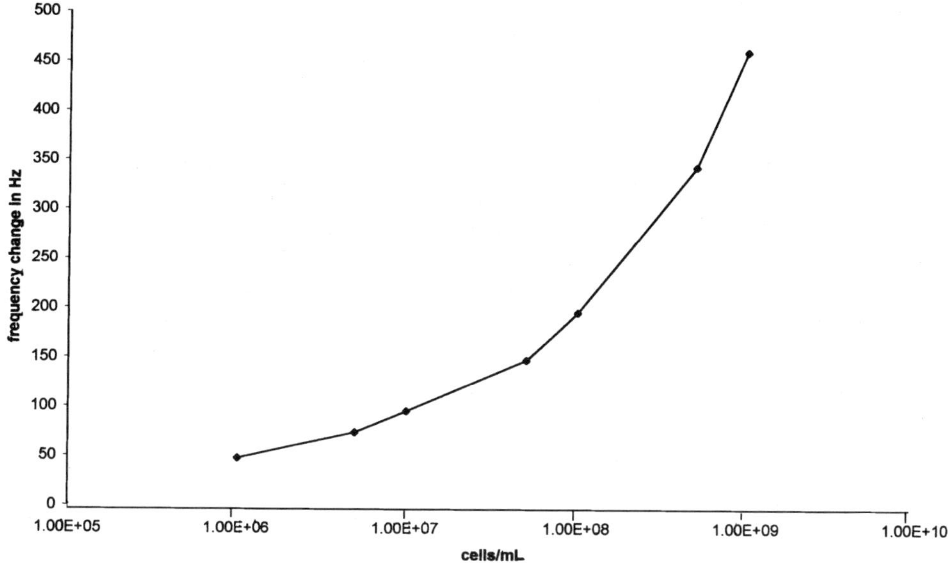

Figure 9.5.   Response of a 5 mg mL antienterobacterial antibody coated crystal to *E. coli.*

Carter et al.[41] proposed a QCM method for the assay of Vibrio cholerae 0139 using 10-MHz crystals coated with rabbit anti-Vibrio cholerae antibody specific for the 0139 Cholera serotype, but not for the more common 01(Ogawa) serotype. The coated crystals were exposed to 4×10^3 to 4×10^6 total cells/crystal of either the 0139 (specific) or Ogawa (nonspecific) serotypes. As few as 4×10^3 cells over background could be detected.

Carter et al.[42] also developed a system involving PZ detection of ricin. Ricin is many times more lethal than hydrogen cyanide and only small amounts would be needed to, e.g., contaminate a water reservoir or dust a large area were it to be used as a biological warfare agent. Detection levels as low as 0.5 μg ricin were achieved.

A rapid and reliable method for screening drinking water has been developed based on a PZ crystal immunosensor using antibodies against the enterobacterial common antigen.[43] When an antienterobacterial antibody layer via protein A immobilization is applied onto a 10-MHz crystal, a response is observed for 10^6 to 10^9 cells/mL of *E. coli* K12, and for various other antigens of the Enterobacteriaceae (Fig. 9.5). It was not possible to remove the bound antigen from the antibody and the only mode for crystal reuse was to remove both antigen and antibody.

9.3.3. Food Analysis

The use of PZ immunoassay techniques for assay of microbial contamination in food samples has been reviewed.[2,44–46] Whereas immunoassay techniques can satisfy the requirements for food quality control, a QCM biosensor may find applications in semicontinuous/continuous analysis of foods, especially with the use of instrumentation that contains a flow cell and microprocessor control.

The first immunogravimetric microbial assay was reported by Muramatsu et al.[44] A PZ crystal coated with anti-*Candida albicans* antibody was used for the determination of *C. albicans* concentrations in the range 1×10^6 to 5×10^8 cells/mL. The sensor showed no response to another yeast species, *Saccharomyces cerevisiae*, and frequency shifts owing to nonspecific adsorption were not significant.

A PZ immunosensor has been developed for the detection of *Salmonella typhimurium*.[45] The antibody to *Salmonella* was immobilized on the crystal in various ways. The best result was obtained when the crystal was precoated with a thin layer of polyethyleneimine.[46] The response of the coated crystal for *S. typhimurium* in a microbial suspension was in the range of 10^5 to 10^9 cells/mL (Fig. 9.6). The time required for a complete interaction between the crystal and the cells appeared to depend upon the cell concentration of the analyzed sample. The antibody-bound crystal lost no activity over 4 days at 4°C and could be reused for up to eight consecutive assays.

A separated electrode PZ crystal sensor was used to determine *E. coli*.[47] This method is based on the fact that the resonant frequency shifts with the culture time where *E. coli* is inoculated, and the frequency detection time (FDT) is linearly related to the initial number of *E. coli* in the range 10^1 to 10^6 cells/mL. *E. coli* determination in pure culture was tested by the FDT method and compared with the most probable number technique and standard plate count method. A correlation coeffi-

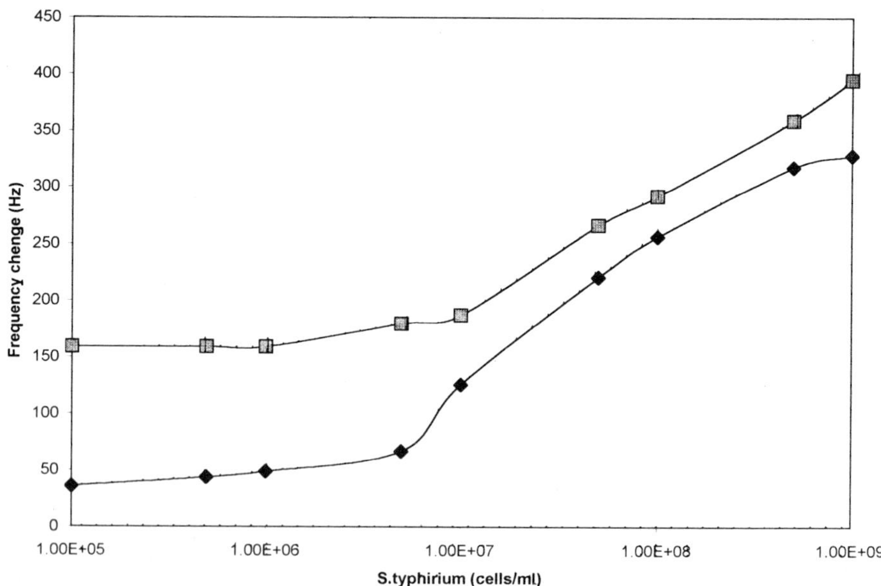

Figure 9.6. Relationship between *S. typhimurium* concentration and the resonant frequency of the PZ coated with anti-*Salmonella* antibody: (◆) 0.5 h of incubation of *Salmonella* with the crystal; (■) 5.0 h of incubation of *Salmonella* with the crystal.

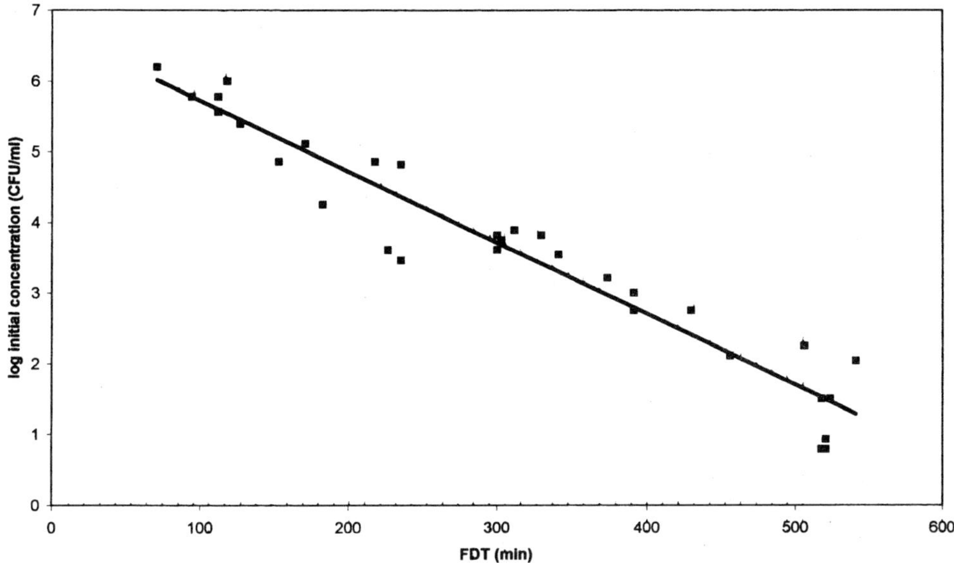

Figure 9.7.   FDT vs. *E. coli* concentration calibration graph. Frozen coliforms were used.

cient of 0.96 was obtained between the FDT and the decadic logarithm of initial concentration of bacteria. The proposed PZ method is much more rapid and sensitive for determining microorganisms than the traditional techniques (Fig. 9.7). The QCM can also be employed to monitor the growth of *Pseudomonas cepacia* biofilms.[48]

A novel PZ immunosensor has been developed for the detection of *Listeria monocytogenes*.[49,50] The *Listeria* was immobilized on the gold surface using different immobilization procedures. With the use of a liquid flow cell, it was possible to monitor the antigen–antibody binding directly without the use of a label, such as an enzyme or a radioactive compound (Fig. 9.8). The calibration curve was prepared using a displacement assay with a response range from 2.5×10^5 to 2.5×10^7 cells/crystal. The assay was also performed in milk that was injected with the nonspecific antigen *Serratia* or *Listeria*. The method is as sensitive as ELISA and results can be obtained in less than 15 min.

Karube[51] reported on the use of a PZ sensor for determining microbial concentrations and for monitoring the viscosity of a fermentation broth and has also described the use of a bioelectronic artificial nose for on-site monitoring to assess the reaction patterns of solid-state fermentations for control of microclimate at the sampling site. The nose consisted of fractionated olfactory cell proteins of bullfrog (as a receptor membrane) coated on a PZ quartz crystal connected to an oscillator and a data recorder with a PC. Five fractions of the olfactory cell proteins gave five probes. The response of the nose was analyzed by the PC either as profiles to show the kinds of odors or a scale factors to show their intensity.

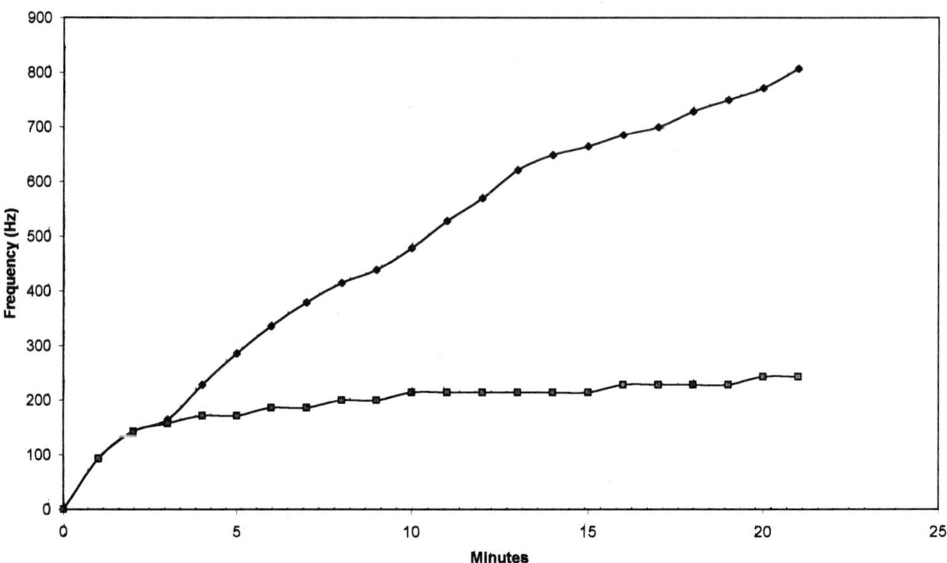

Figure 9.8.   Frequency change resulting from the addition of either nonspecific antigen or specific antigen on Protein-G-coated crystals. The slope differences became significant at approximately 5 min into the assay.

Table 9.2.   Representative Commercially Available Quartz Crystal Microbalances

Company	Model	Resolution	LOD
EG&G Princeton Applied Research, N.J.	QCA-917 Quartz Crystal Analyzer	0.1 Hz	1 ng/cm^2
Elchema, N.Y.	EQCN-900 Electrochemical Quartz Crystal Microbalance	0.01 Hz	0.05 ng
Maxtek, Cal.	PM-700 Series	0.03 Hz	0.7 ng/cm^2
QCM Research, Cal.	Mark Series Cryogenic/Thermoelectric QCM	0.1 Hz	4.42 ng/cm^2
Universal Sensors, La.	PZ-1000 Immunobiosensor/PZ-105 Gas Phase PZ detector	1 Hz	1 ng

9.4. QUARTZ CRYSTAL MICROBALANCE—COMMERCIAL SOURCES

Generally, commercial systems are designed to reliably measure mass changes up to approximately 100 μg, whereas the minimum detectable mass change is typically approximately 1 ng/cm^2. The simplicity of the device means that it is relatively inexpensive. AT-cut quartz crystals are used as QCMs because they have a low temperature coefficient at room temperature, and so the resonance frequency changes only minimally with temperature in that region. They can be rough or smooth but optically polished crystals are recommended for fluids because liquid can get caught in the crevices of rough crystals, which sometimes causes spurious frequency changes. When deciding on which QCM to use, one must know whether the applications will be in the gas or liquid phase. The fixtures for the gas- and liquid-phase systems are different; if a gas-phase system is used in the liquid phase, the signal will attenuate too much, and the fixture will most likely short-circuit. Liquid applications present a challenge because multiple effects occur simultaneously. The mass effects that cause a frequency shift in the gas phase engender a similar response in the liquid phase, but viscosity changes produce frequency shifts and a signal attentuation, requiring the use of an appropriate algorithm [previously described in Eq. (2)]. Table 9.2 lists representative companies selling QCMs.

9.5. QUARTZ CRYSTAL MICROBALANCE—CONCLUSIONS AND FUTURE DIRECTIONS

The QCM has found a wide range of applications in areas of food, environmental, and clinical analysis since it was first developed, owing to its inherent ability to monitor analytes in real time. Piezoimmunosensors using antibody–antigen coated crystals offer the greatest potential, with possible applications in food, environmental, and clinical analysis.

Several issues still need to be addressed, however, to produce commercially viable devices able to compete with the success of surface plasmon resonance (SPR), the technique used in the highly prosperous Pharmacia BIAcore instrument. A reproducible immobilization of the biological material on the crystal surface must be achieved as well as a method to overcome problems owing to nonspecific binding of proteins to the antibody test surface. The reusability of the PZ crystal remains an important issue that is currently the focus of much interest from a plethora of research groups. The availability of precoated gold crystals, which greatly reduces assay time and the requirement for skilled operators, is now viable with preactivated crystals available from Universal Sensors, Inc.

If, indecd, the QCM can achieve the same results as SPR, it will offer many advantages over the BIAcore. The basic instrumentation required is far less comprehensive and thus markedly less expensive. This coupled with a predicted reduction in the cost of crystals and an increase in the number of times the crystal can be regenerated promises that the QCM(PZ) will supplement existing techniques in immunochemical research and technology in a most favorable manner.

REFERENCES

1. Guilbault GG, Luong JH. Gas phase biosensors. *J Biotech* 1988;9:1–10.
2. Guilbault GG, Luong JH. Biosensors: current status and future possibilities. *Sel Elec Rev* 1989;11:3–16.
3. Shons A, Dorman F, Najarian J. Piezoelectric quartz immunosensor. *J Biomed Mater Res* 1972;6:565–575.
4. Raleigh L. Quartz piezoelectric (1885). In: Pacy DJ *Vacuum* 1960;9:261–270.
5. Curie J, Curie P. An oscillating quartz crystal mass detector. *Rendu* 1880;91:294–297.
6. Sauerbrey GZ. Use of quartz vibration for weighing thin films on a microbalance. *J Physik* 1959;155:206–212.
7. Kanasawa KK, Gordon JG. A liquid phase piezoelectric detector. *Anal Chem* 1985;57:1771–1775.
8. Kurosawa S, Tawara E, Kamo N., et al. A novel piezoelectric system. *Anal Chim Acta* 1993;230:230–240.
9. Bruckenstein S, Shay M. Dual quartz microbalance oscillator circuit. *Anal Chem* 1994;66:1847–1855.
10. Roederer JE, Bastiaans GJ. Highly sensitive SAW immunosensors. *Anal Chem* 1983;55:2333–2338.
11. Bastiaans GJ. A surface acoustic wave device for measurement in liquids. U.S. Patent 4,735,906, 1988.
12. King WH. Piezoelectric sorption detector. *Anal Chem* 1964;36:1735–1741.
13. Guilbault GG, Jordan J. Analytical uses of piezoelectric crystal. *CRC* 1988;19:1–28.
14. Alder JF, Callum J. Piezoelectric crystals for mass and chemical measurements. *Analyst* 1983;108:1169–1189.
15. Guilbault GG, Ngeh-Ngwainbi J. Use of protein coatings on piezoelectric crystals for assay of gaseous pollutants. *Biotech* 1988;2:17–22.
16. Guilbault GG, Ngeh-Ngwainbi J. In: Guilbault GG, Mascini M, eds. *Analytical Uses of Immobilised Biological Compounds for Detection, Medical and Industrial Uses.* NATO Advanced Science Institute Series, Reidel Publishing Co 1988 Ch. 9.
17. Luong JH, Guilbault GG. Analytical application of piezoelectric crystal biosensors. In LU Blum and P Coulet, eds. *Biosensors: Principles and Applications* New York: Marcel Dekker, 1991, pp 107–138.
18. Suleiman AA, Guilbault GG, Piezoelectric immunosensors and their applications. *Anal Lett* 1991;24:1283–1292.
19. Oliveira RJ, Silver SF. U.S. Patent 4,242,096, 1980.
20. Rice TK. U.S. Patent 4,236,893, 1980.
21. Rice TK. U.S. Patent 4,314,821, 1982.
22. König B, Grätzel M. A novel immunosensor for Herpes virus. *Anal Chem* 1994;66:341–348.
23. König B, Grätzel M Detection of viruses and bacteria with piezoelectric immunosensors. *Anal Lett* 1993;26:1567–1575.
24. König B, Grätzel M. Development of a piezoelectric immunosensor for the detection of human erythrocytes. *Anal Chim Acta* 1993;276:329.
25. König B, Grätzel M. Long term stability and improved reusability of a piezoelectric immunosensor for human erythrocytes. *Anal Chim Acta* 1993;280:37–42.
26. Kösslinger C, Drost S, Aberl F, et al. A quartz crystal microbalance for measurements in liquids. *Biosens Bioelectron* 1992;7:397–410.
27. Kösslinger C, Drost S, Aberl F, et al. Quartz crystal microbalance for immunosensing. *J Anal Chem* 1994;349:349–357.
28. Attili BS, Suleiman AA. Piezoelectric immunosensor for the detection of cortisol. *Anal Lett* 1995;28:2149–2159.
29. Muramatsu H, Dicks JM, Tamiya E, et al. A piezoelectric crystal biosensor modified with protein A for determination of immunoglobulins. *Anal. Chem* 1987;59:2760–2763.
30. Raman Suri C, Raje M, Gyran Mishra C. *Biosens Bioelectron* 1994;9:325.
31. Ebersole RC, Miller JA, Moran JR, et al. pH Sensors: AT cut resonators with polymer films. *J Am Chem Soc* 1990;122:2553–2562.
32. Chu X, Jiang JH, Shen GL, et al. *Anal Chim Acta* 1996;336:185–193.
33. Guilbault GG. Determination of formaldehyde with an enzyme coated piezoelectric crystal. *Anal Chem* 1983;55:1682–1684.

34. Ngeh-Ngwainbi J, Foley PH, Kuan SS, et al. Parathion antibodies on piezoelectric crystals. *J. Am. Chem. Soc* 1986;108:5444–5448.

35. Ngeh-Ngwainbi J, Suleiman A, Guilbault GG. Piezoelectric crystal biosensors. *Biosens Bioelectron* 1990;5:13–26.

36. Guilbault GG, Schmid RD. Biosensors for the determination of drug substances. In: Turner APF, ed. *Advances in Biosensors*, London: Jai Press, 1991, vol I, pp 264–289.

37. Guilbault GG, Hock B, Schmid R. A piezoelectric immunosensor for atrazine in drinking water. *Biosens Bioelectron* 1992;7:411–419.

38. Minunni M, Guilbault GG, Hock B. The quartz crystal microbalance as a biosensor. *Anal Lett* 1995;28:749–764.

39. Steegborn C, Construction and characterization of the direct piezoelectric immunosensor for atrazine operating in solution. *Biosens Bioelectron* 1997;12:19–25.

40. Nakanishi K, Karube I, Hiroshi S, et al. Detection of red tide causing plankton chattonella using a PZ immunosensor. *Anal Chim Acta* 1996;325:73–80.

41. Carter RM, Mekalanos JJ, Jacobs MB, et al. Quartz crystal microbalance detection of Vibrio Cholerae 0139 serotype. *J Immunol Meth* 1996: 187:121–125.

42. Carter RM, Jacobs MB, Lubrano GJ, et al. Piezoelectric detection of ricin and affinity purified goat anti-ricin antibody. *Anal Lett* 1995;28:1379–1386.

43. Plomer M, Guilbault GG, Hock B Development of a piezoelectric immunosensor for detection of enterobacteria. *Enzyme Microb Technol* 1992;14:230–235.

44. Muramatsu H, Watanabe Y, Hikuma M, et al. A piezoelectric crystal biosensor system for detection of *Escherichia coli*. *Anal. Lett* 1989;22:2155–2166.

45. Prusak-Sochaczewski E, Luong JH, Guilbault GG. Development of a piezoelectric immunosensor for the detection of *Salmonella typhinurium Enzyme Microb Technol* 1990;12:173–177.

46. Prusak-Sochaczewski E, Luong JH. A new approach to the development of a reusable piezoelectric crystal biosensor. *Anal Lett* 1990;23:401–410.

47. He F, Geng Q, Zhu W, et al. Rapid detection of *E. coli* using a separated electrode piezoelectric crystal sensor. *Anal Chim Acta* 1994;289:313–319.

48. Nivens NE, Chalmers JQ, Anderson TA, et al. Long term on-line monitoring of microbial biofilms using a quartz crystal. *Anal Chem* 1993;65:65–73.

49. Minunni M, Mascini M, Carter RM, et al. A quartz crystal microbalance displacement assay for Listeria monocytogenes. *Anal Chim Acta* 1996;325:169–174.

50. Jacobs MB, Carter RM, Lubrano GJ, et al. A piezoelectric biosensors for *Listeria monocytogenes*. *Am Lab* 1995;27:11–20.

51. Karube I. Detection of odorant using an array of piezoelectric crystal and neural network pattern recognition. *Biochem. Eng. Stuttgart* 1991;1:7–12.

10

Surface Photovoltage-Based Biosensor

*Yuji Murakami, Eiichi Tamiya, Hidekazu Uchida,
Teruaki Katsube*

10.1. INTRODUCTION

The surface photovoltage (SPV) technique was first applied to a chemical sensor by
Hafeman et al. in 1988.[1] We applied the technique to immunosensors in 1990[2] and
also investigated possible applications to various kinds of chemical sensors, including ion sensors,[3] gas sensors,[4] biosensors,[5] and image sensors.[6] These studies
demonstrated various advantages of an SPV sensor over other chemical sensor: (1)
the fabrication process is simple; (2) by multiplexing different light sources in
different locations, the device can be a multisensor [light-addressable potentiometric
sensors (LAPS)] without additional process complexity; and (3) encapsulation is
easier and less critical. SPV also allows for flexibility in the signal processing method
as it is based on ac measurements. For example, a differential measurement
technique that we proposed made high-sensitive measurements possible by canceling
out the common noise and drift component.[3] These advantages also suggested that
the SPV technique might be used to develop a new generation of chemical sensors,
such as a highly integrated chemical and image sensor.

This chapter outlines possibilities for new kinds of biosensors, including
integrated structures and discusses the sensing mechanism and immobilization
techniques of biosensitive material using SPV.

10.2. MEASUREMENT PRINCIPLE

Figure 10.1 shows the cross section of a sensor structure and measurement
system. The sensor structure comprises a reference electrode/sample solution/ion-

Yuji Murakami and Eiichi Tamiya ● School of Materials Science, Japan Advanced Institute of Science
and Technology, Hokuriku, Tatsunokuchi, Ishikawa 923-12, Japan.   Hidekazu Uchida and Teruaki
Katsube ● Department of Information and Computer Science, Faculty of Engineering, Saitama
University, Urawa, Saitama 338, Japan.

Biosensors and Their Applications, edited by Yang and Ngo, Kluwer Academic/Plenum Publishers, New
York, 1999.

175

sensitive membrane/insulator/semiconductor. The semiconductor surface potential was measured by an ac photovoltage technique using the semiconductor in the depleted condition. An alternately modulated photon beam was irradiated onto the silicon surface and the photovoltage in the surface depletion layer was picked up through a capacitive coupling circuit and measured by a lock-in amplifier.[5]

10.3. ENZYME SENSOR

Multifunctional enzyme sensors were developed on a single semiconductor chip using a new type of photo-cross-linkable polymer (a copolymer of dimethylacrylamide and cinnamoyloxyethylmethacrylate),[7] which is water soluble and cross-linkable by UV irradiation at room temperature, as shown in Fig. 10.2. Upon UV irradiation, the polymers form a membrane that is insoluble in water and if the enzyme molecules are mixed into the polymers before the irradiation they are entrapped in the polymer networks. The immobilized-enzyme membrane was also fixed to the silicon surface by a photoreactive polymer, poly-(meta azide styrene), through

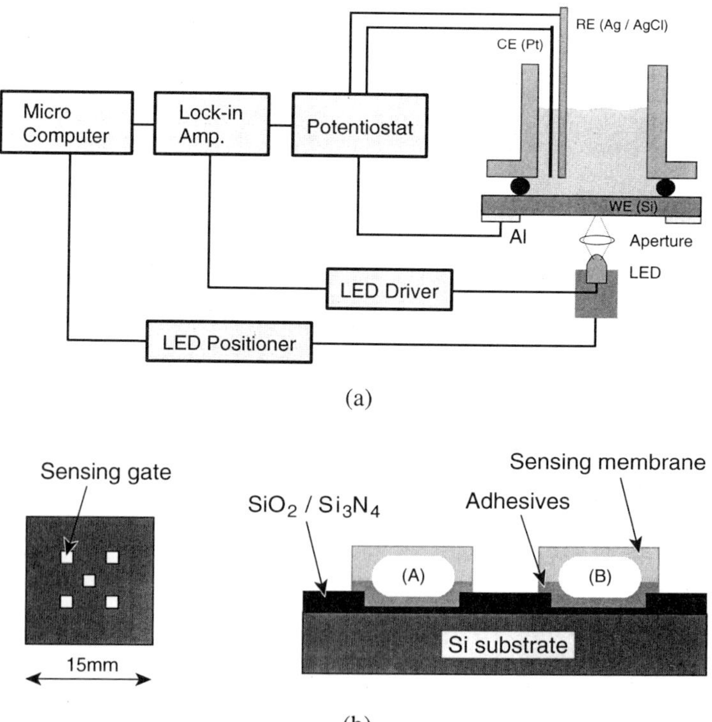

Figure 10.1. Illustration of the SPV sensing system and sensor structure.

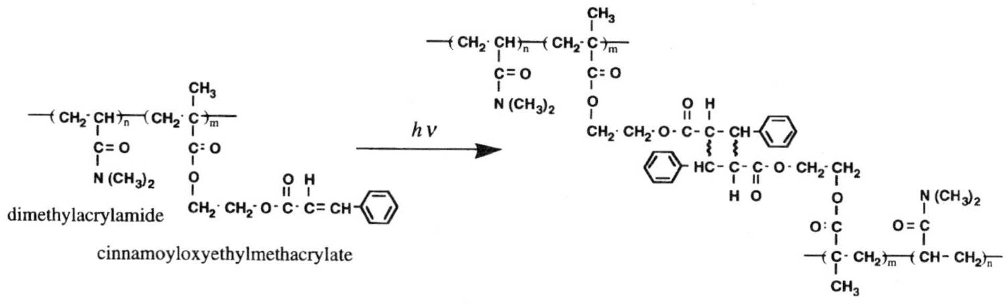

Figure 10.2. Photo-cross-linkable polymer.

covalent bonding. The phenyl azido group of poly-(meta azide styrene), which is inserted between the substrate and photo-cross-linkable copolymer, is converted to highly reactive nitrene upon UV irradiation and the generated nitrene reacts to form covalent bonding to both the substrate and immobilized-enzyme membrane.

The enzyme patterning procedure is as follows: The sample electrode was n-type silicon of $1-5\,\Omega\,cm$ resistivity with SiO_2 and Si_3N_4 as insulators. The sample area was $1.5 \times 1.5\,cm$. First, 1 wt.% poly-(meta azide styrene) dissolved in chloroform was spread on the sensor surface and dried at room temperature for 5 min in the dark. The surface was then overcoated with $60\,\mu L$ phosphate buffer solution (10 mM, pH 6.9) containing 10 mg glucose oxidase (GOD) (Sigma, 151 U/mg) and 6 mg water-soluble photo-cross-linkable copolymer (dimethylacrylamide and 3.5 mole% cinnamoyloxyethylmethacrylate) and left overnight in the dark. The coated film was then irradiated through a photomask for 10 min with UV light (>290 nm) from a 100-W halogen lamp, whereupon the irradiated polymer film ($200 \times 200\,\mu m$) became a water-insoluble membrane. Next, the peripheral nonirradiated polymer was washed out by deionized water. The enzyme (GOD) was fixed in the cross-linked photopolymer and was simultaneously immobilized on the sensor substrate by covalent bonding to the phenyl azide group of poly-(meta azide styrene). On the continuous region adjacent to the immobilized GOD films, urease (163 U/mg, Toyobo Co. Ltd) and enzyme-free film were immobilized using the procedure described above. The structure of the fabricated sample is shown in Fig. 10.1b.

Figure 10.3 shows responses for the urea and glucose sensors measured in a 1-mM phosphate buffer solution. The added enzyme concentration was 50 mg/dL for both the measurements. The rather low sensitivity of the glucose as compared to the urea sensor is consistent with previous results measured by ISFET(ion-sensitive field effect transistor) biosensor fabricated in our laboratory.[8] Sensitivity curves were obtained by plotting the output values measured 5 min after the substances were added, as shown in Fig. 10.4. The detectable concentration ranges for urea and glucose were 0.5–250 and 5–250 mg/dL, respectively. It should be noted that lower concentrations of urea and glucose were detectable in the present system than was possible with the ISFET biosensor mentioned above.[8] This result suggests that low noise characteristics may be obtainable with the SPV system.

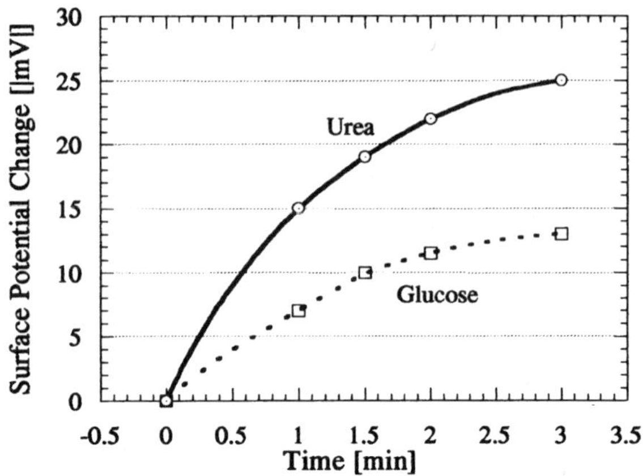

Figure 10.3.   Response transients of the sensor for urea (50 mg/dL) and glucose (50 mg/dL).

10.4.   SURFACE PHOTOVOLTAGE IMMUNOSENSOR

10.4.1.   A Highly Sensitive Immunosensor

The sensing principle of the conventional SPV is based on the detection of a surface-potential change across the depletion layer caused by a surface-generated current. It is generally considered that the output signal of SPV is proportional to this photopotential change. Thus it is impossible to measure charge-free substances adsorbed on the electrode surface with conventional SPV. However, we note here that the output current contains information about the impedance of the surface

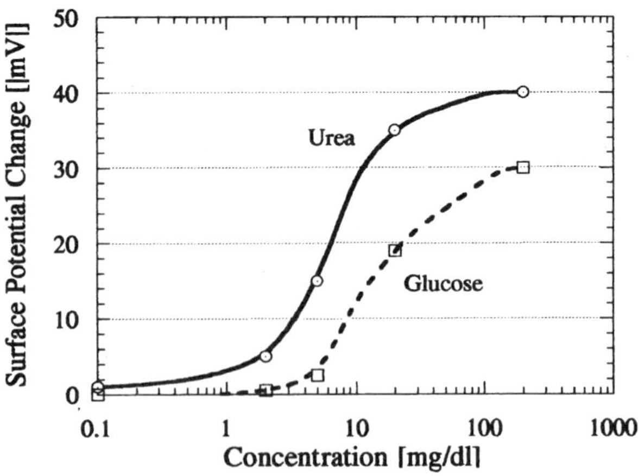

Figure 10.4.   Sensor response for various concentrations of (a) urea and (b) glucose. The response was measured 5 min after injection.

insulating layer of a semiconductor electrode as well as the surface-potential change. If the surface insulator is covered by adsorbed substances, there is an increase in the equivalent thickness of the insulator, leading to a change in the electrode impedance. We have proposed a new concept for measuring the change in impedance of insulating film caused by the adsorbed substances.

If the impedance of an adsorbed substance is assumed to be a capacitance C_s, the equivalent circuit of the electrode can be drawn as shown in Fig. 10.5, where C_d and r_d are the capacitance and resistance of the depletion layer of the semiconductor and r is the circuit resistance. Output current caused by the photogenerated current i_{ph} can be expressed as

$$i_{out} = \frac{i_{ph}}{1 + r/r_d + C_d/C + j(\omega r C_d - 1/\omega r_d C)} \tag{1}$$

where C is the series connection of the capacitance of the oxide and the adsorbed substance, i.e.,

$$C = \frac{C_s C_{ox}}{C_s + C_{ox}} \tag{2}$$

For a simplified model, C_s is assumed to be

$$C_s = \frac{\varepsilon_s S}{d_s} \tag{3}$$

where ε_s is the dielectric constant, S the area of the light beam, and d_s the equivalent thickness of the adsorbed substance. Thus the thickness of the adsorbed substance can be determined by the SPV measurement even if there is no detectable change in the surface photovoltage.

However, this system still requires an improvement in sensitivity as C_s is usually much higher than $C_{ox}(C_s \gg C_{ox})$. In order to achieve greater sensitivity, we proposed two techniques: differential measurement and frequency-sensitive measurement.

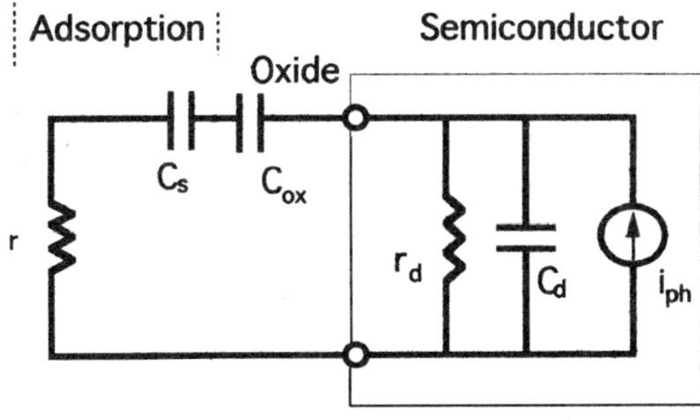

Figure 10.5.   Electrical equivalent circuit for SPV.

Differential measurement makes it possible to enhance the sensitivity for the detection of incremental change of output current by canceling the common noise and drift component, which is accomplished by taking the differential signal between the sensing and the reference regions. The procedure is illustrated by the time sequence shown in Fig. 10.6. Two light beams (LED_1 and LED_2), each with a 50% duty cycle, were used alternately to irradiate the sample surface. The photocapacitive currents generated in the sensing region I_{ph_1} (phase t_1) and the reference region I_{ph_2} (phase t_2) were fed into the lock-in amplifier. The output of the lock-in amplifier is the time integral of the product of the total photocurrent (I_{ph}, Fig. 10.6e) and the reference signal (Fig. 10.6f).

If reference region is designed so as to have no adsorption, the differential current is given by

$$\Delta i_{out} = i_{ph} \frac{(t - t_0) + j(s - s_0)}{(t + js)(t_0 + js_0)} \tag{1}$$

where

$$s = \omega r C_d - \frac{1}{\omega r_d C} \qquad s_0 = \omega r C_d - \frac{1}{\omega r_d C_{ox}}$$

$$t = 1 + \frac{r}{r_d} + \frac{C_d}{C} \qquad t_0 = 1 + \frac{r}{r_d} + \frac{C_d}{C_{ox}}$$

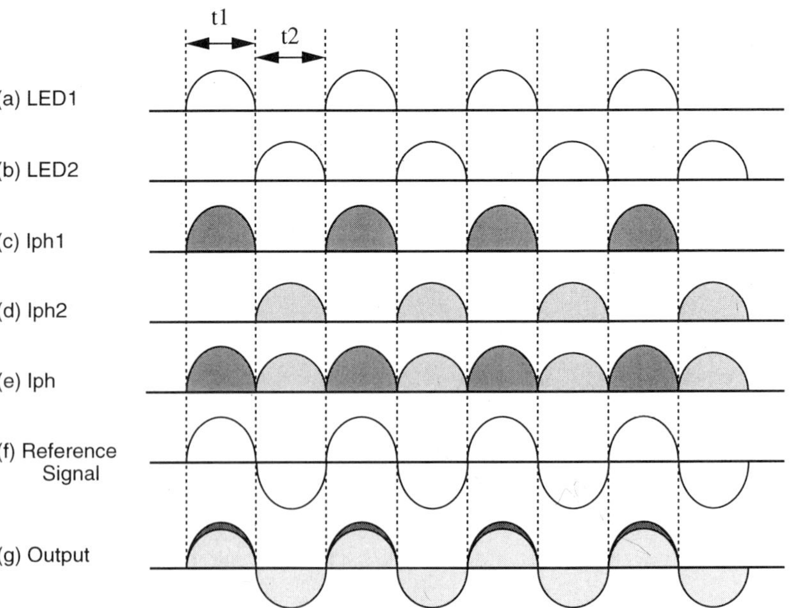

Figure 10.6.   Time sequence of differential measurement of SPV sensor.

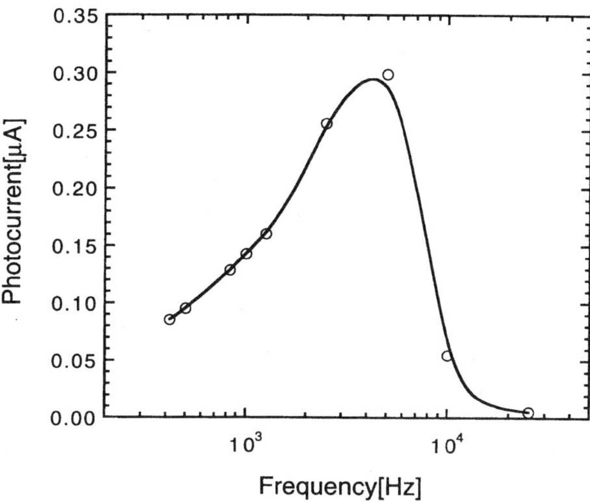

Figure 10.7.   Frequency dependence of photocurrent.

Further improvement in sensitivity is achieved by taking into account the frequency dependence of the output signal. Figure 10.7 shows the frequency dependence of the photogenerated current. It should be noted that the circuit works as a band pass filter, as would be expected from the electroequivalent circuit. Cutoff in the high-frequency region results from the impedance of the depletion layer in a semiconductor and a decrease in the low-frequency region is the result of capacitance due to the field oxide and adsorbed substance. Thus a higher level of sensitivity is reached by adjusting the frequency to maximum output, i.e.,

$$\omega_0 = \frac{1}{\sqrt{rr_d C_d C}} \tag{5}$$

The output current then becomes

$$\Delta i = i_{\text{ph}} \frac{1}{j\omega C_s} \frac{1 - j\omega r_d C_d}{r_d(1 + r/r_d + C_d/C_{\text{ox}})^2} \tag{6}$$

So that the thickness of the adsorbed substance is directly proportional to the output current, which indicates a more sensitive measurement.

10.4.2.   Sample Preparation and Measurement

Figure 10.8 shows a cross-sectional view of the electrode. The homogeneously oriented antibody (monoclonal antihuman IgM) was immobilized on the silicon surface with protein A. An *n*-type silicon wafer of $2-3\,\Omega\,\text{cm}$ was thermally oxidized to the thickness of 100 nm and the surface was pretreated with 2% (3-aminopropyl)

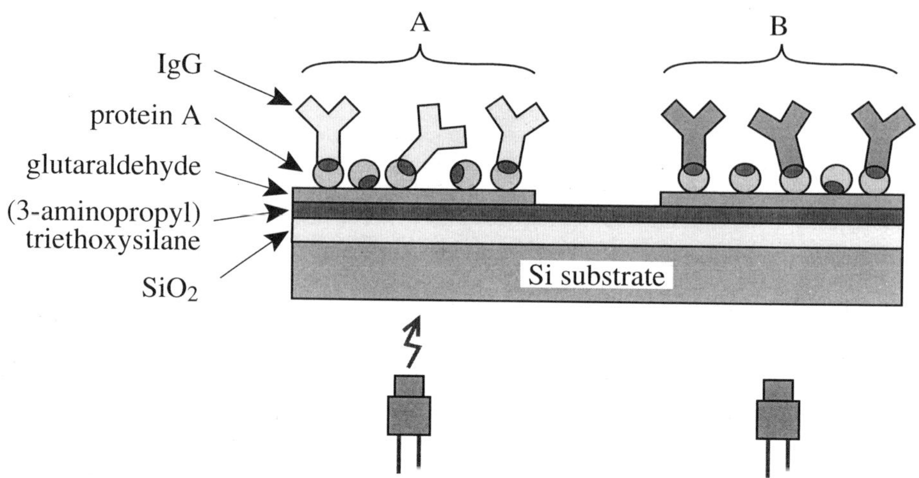

Figure 10.8.   Schematic illustration of differential SPV immunosensor.

triethoxysilane on which a glutaraldehyde membrane was coated from a 1% aqueous solution for 1 h at room temperature. Then, protein A (UCB) and antihuman IgM (mouse IgG2b/κ YAMASA M-01) were overcoated. As the reference of the differential measurement, antihuman IgG was immobilized at different positions as shown in Fig. 10.8 by sensor B. Two ac laser beams of 980 nm wavelength were focused on different areas of the semiconductor surface (the sensing area and the reference area). The two signals were differentially superimposed by a lock-in amplifier.

Figure 10.9 shows surface-potential change measured by a differential photocurrent signal for each process of electrode surface modification, i.e., glutaraldehyde (GLU), protein A (PA), and IgG (IGG). Clear photocurrent change was observed for different surface modifications.

Figure 10.10 shows the time response of the differential SPV signal after 10 μg/mL of human IgM antigen was added. The solid line is the signal of anti-IgM-modified electrode immobilized on protein A. The dotted/dashed line is the response of the anti-IgM electrode modified on glutaraldehyde. It should be noted that the use of protein A enhances the sensitivity. No response was seen by IgG-modified electrode, as is shown by the dotted line.

10.5. MICROBIAL BIOLOGICAL OXYGEN DEMAND SENSOR

10.5.1. Surface Photovoltage-Based Microbial Biological Oxygen Demand Sensor

Metabolism of microorganisms or cells produces acidic substances such as carbonate ion and organic acid. Microbial metabolism is affected by many factors in a cell's environment. Changes in the biological, chemical, and physical environment

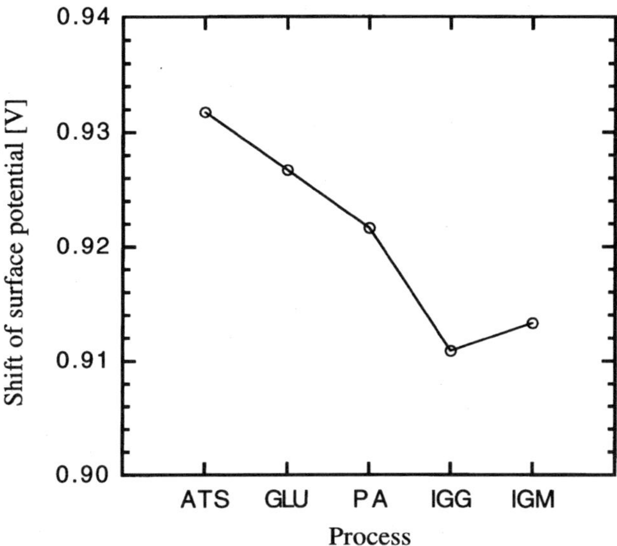

Figure 10.9. Surface potential shift for various surface treatments [ATS: (3-aminopropyl)triethoxysilane, GLU: glutaraldehyde, PA: protein A, IGG: antihuman IgM, IGM: human IgM].

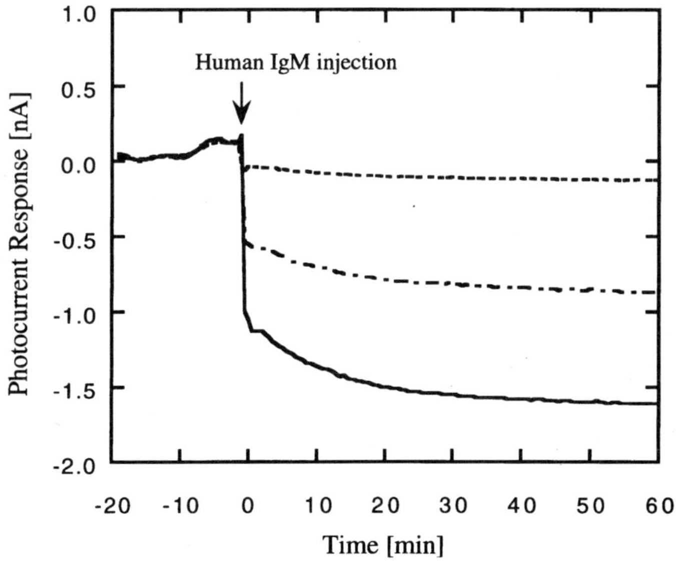

Figure 10.10. Time response of photocurrent signal of antihuman IgM reaction.

of a cell are reflected in the production of acidic compound. Thus, when microorganisms or cells are immobilized on the sensor side of the SPV device, the system is expected to sense the biological information of the analyte solution. The biochemical oxygen demand (BOD) sensor is the most widely used microbial sensors.[9] BOD evaluates the organic pollution in wastewater and is a typical parameter of the pollution. The 5-day BOD test has been a standard pollution-monitoring tool since 1936.[10] In practice, apart from the high level of skill it demands the BOD test requires a 5-day incubation period at 20°C and simpler and faster BOD techniques are needed for pollution control

Many methods have been developed recently for BOD measurements including the use of a sensor that consists of microbial cells immobilized on an oxygen electrode, which measures the current decrease resulting from a decrease in dissolved oxygen.[11–19] Several other methods have also been reported.[20–22] Both the 5-day method and the oxygen-electrode method use oxygen consumption as a parameter of the activity of the immobilized microorganism. The other parameters of metabolism are potentially useful for constructing a microbial sensor combined with an appropriate transducer.

However, no microbial sensor has used SPV. Therefore, we applied the SPV device to the fabrication of a novel BOD sensor. In this section, we will discuss the fabrication and characterization of this novel BOD sensor, which combines microorganisms with SPV. *Trichosporon cutaneum*, the designated microorganism for use in BOD sensor by the Japan Industrial Standard (JIS),[9] was employed as the immobilized microorganism. The SPV-based BOD sensor was fabricated and characterized compared with BOD_5 and BOD_S.

Measurements were carried out with a flow-cell system consisting of the SPV device (Shindengen, Japan) and silicone sheets as shown in Fig. 10.11. A peristaltic pump was used to pump carrier and sample solution from a carrier or a sample reservoir to the flow cell. The flow cell was connected to an SPV controller (SE1030, Technologue, Japan) and a computer (9801ns/t, NEC, Japan).

10.5.2.   Immobilization Method of T. cutaneum on the Device

Some microbe immobilization methods were evaluated with the response to 1000 mg/L of glucose in the buffer. First, *T. cutaneum* IFO10466(AJ4816) was cultured in GP medium containing 20 g glucose, 5 g polypepton, 2 g yeast extract, 1 g KH_2PO_4, 0.5 g $MgSO_4$ per liter at 37°C for 36 h. The cells were collected by centrifugation (3000 rpm, TMA-6, TOMY). Phosphate buffer and glycerol as cryoprotectants were added to the cells to reach 20% glycerol and 28 g/L wet cell weight. A 2-mL aliquot of the suspension was gradually frozen at -25°C, and stored -80°C.

In the first method, the microbe was suspended in the test solution and flowed as a suspension. The response was small and unstable and the suspension sometimes choked up the tube, so the method was unsuitable for the purpose. In a second approach, the microbe was immobilized in alginate gel. *T. cutaneum* was mixed with 4% sodium alginate, and the mixture was dropped into 0.05 M of calcium chloride

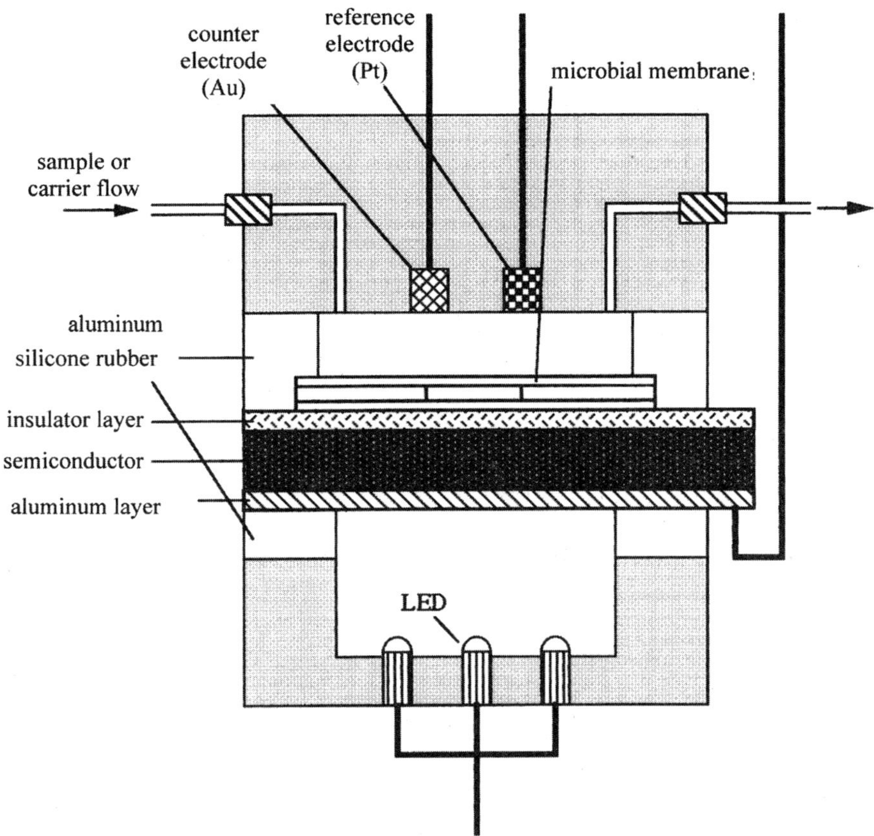

Figure 10.11. Schematic diagram of a flow cell. LEDs were located at back side against the sample flow side. Platinum and gold electrodes were located on the inside wall of the flow cell, and used as a reference and a counter electrode, respectively. Two jigs bound a silicon chip and a microbial membrane with silicone rubber sheets as a spacer of flow cell.

solution at 37°C through a needle (4 mm in diameter) and allowed to equilibrate overnight at 4°C. Several ratios of the microbial wet weight were tested, and the 50 wt.%, the highest one tested, gave the maximum response. However, this second method was unsuitable for repeated use owing to destruction of the gel. Calcium ion was added to the buffer to retard destruction, but it reacted with the phosphate in the buffer. Finally, the microbe was immobilized using an acetylcellulose membrane. The forzen cell stock was melted in a water bath at 37°C. The cell was washed three times with phosphate buffer using centrifugation. A double-sided adhesive tape having a 4-mm hole was put on an acetylcellulose membrane filter (DISMIC-25cs, pore size 0.8 μm, Advantec). One mg of cells was filtered through the membrane. A similar membrane was overlaid on the original one, positioned to cover the silicon device (Fig. 10.12). Several kinds of the immobilized samples were examined, as shown in Fig. 10.13. There was a stronger response with a microbe sample larger

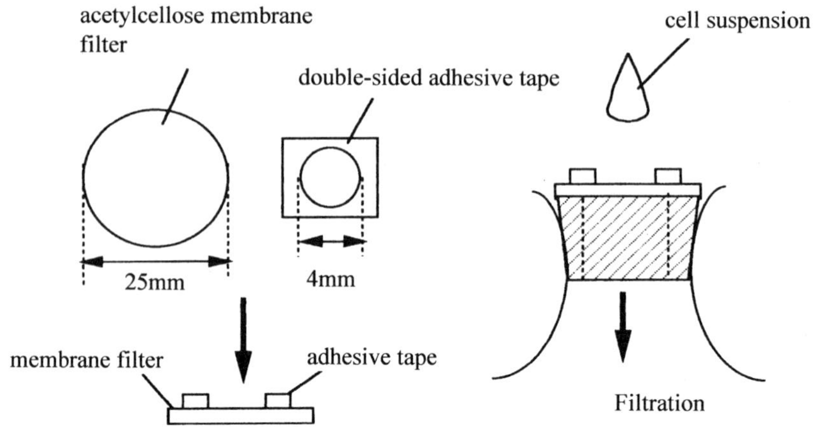

(1) preparation of a membrane filter   (2) Suction filtration of microorganism

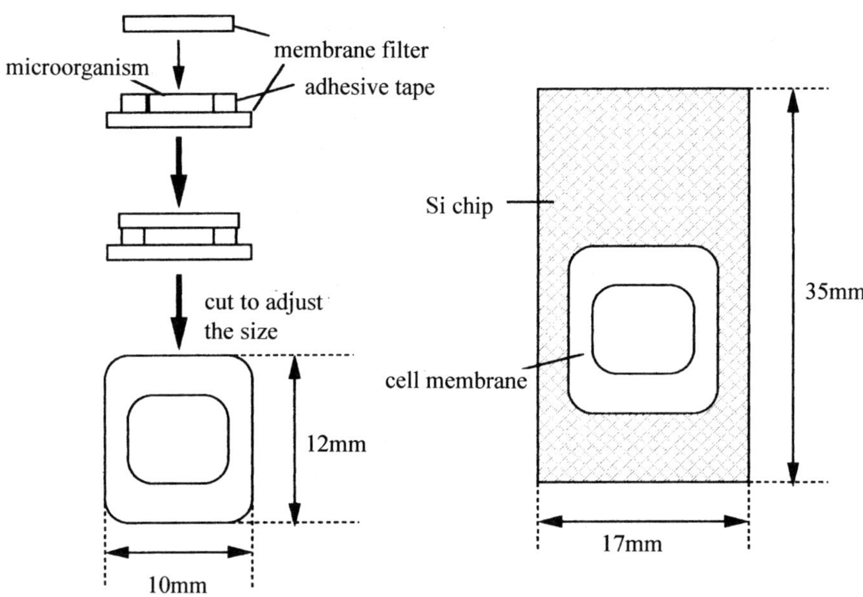

(3) Bind between membranes   (4) Fixation of the membrane on a device

Figure 10.12.   Immobilization procedure of microbial cells between membrane filters.

than 0.5 mg than with the alginate gel with 50 wt.%, while the microbe sample of more than 1 mg matched the alginate gel response. Following these results we immobilized 1.0 mg of the microbe for further testing. Both sides of the microbial membrane were composed of acetylcellulose so it was easy to handle and store. Figure 10.14 shows the long-term stability of the sensor system with a microbial

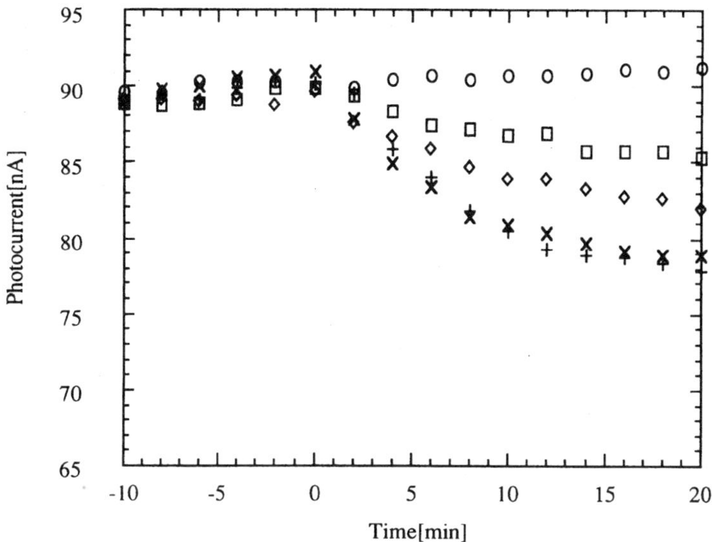

Figure 10.13.   Dependence of the response curve on the amount of immobilized microorganism. After stabilization of the response with the flow of 1 mM phosphate buffer (pH 7) containing 0.15 M NaCl at 250 μL/min, the carrier solution was changed to a sample solution containing 1.0 g/L glucose in the buffer. The time when the solution was changed is zero on the horizontal axis. The amount of immobilized *T. cutaneum* was 0.2 ($\square$), 0.5 ($\diamond$), 1.0($\times$), 2.0(+) mg in weight. The control sample did not contain either microorganism or glucose ($\bigcirc$).

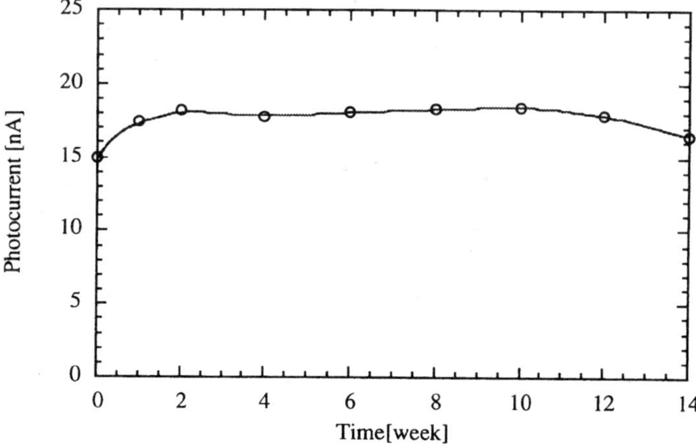

Figure 10.14.   Long-term stability of a microbial membrane. A membrane was used once in two weeks to measure 200 mgO$_2$/L of BOD standard solution in the 1 mM phosphate buffer (pH 7) at 250 μL/min. The membrane was stored in phosphate buffer saline at 4°C.

membrane. A microbial membrane was stored in the 1 mM phosphate buffer (pH 7) containing 0.15 M of NaCl at 4°C, and used once in two weeks. The membrane can be used for more than 14 weeks.

10.5.3.   Optimization of the System

Biosensors that measure pH shift are highly dependent on the capability of the buffer. High buffering capacity inhibits the pH shift at the sensor surface and reduces the sensitivity, while a low buffering capacity renders the response unstable. Figure 10.15 shows the dependence of the response on the concentration of the buffer — lower concentration resulted in a stronger response, while even the lowest concentration (1.0 mM) showed enough stability.

Figure 10.16 shows the dependence of the response on the flow rate. Generally speaking, slower flow makes the response stronger, but it takes a longer time for the system to become stable. The response time for a flow rate of 50 μL/min, about 70 min, was too long, compared with the response time of BOD_S, i.e., 10–20 min. The flow rate of 250 μL/min was chosen to make the response time about 25 min. We employed a continuous sample flow to obtain a steady-state response, and the method requires more sample than the injection technology. The quantity of sample required for a measurement, however, is less than 10 mL, and usually the test

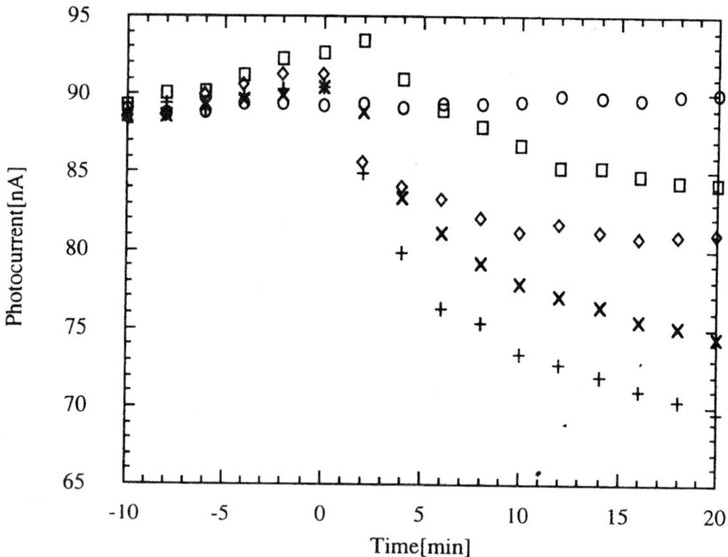

Figure 10.15.   Dependence of the response on the concentration of the buffer. After stabilization of the response with the flow of 1 mM phosphate buffer (pH 7) containing 0.15 M NaCl at 250 μL/min, the carrier solution was changed to a sample solution containing 1 g/L glucose in various concentrations of the phosphate buffer. *T. cutaneum* was 1.0 mg. The time when the solution was changed is zero on the horizontal axis. The concentrations of the buffer phosphate were 1.0(+), 2.0(×), 5.0(◇), 10(□) mM. The control sample was 1.0 mM of phosphate buffer without glucose in the sample circle (○).

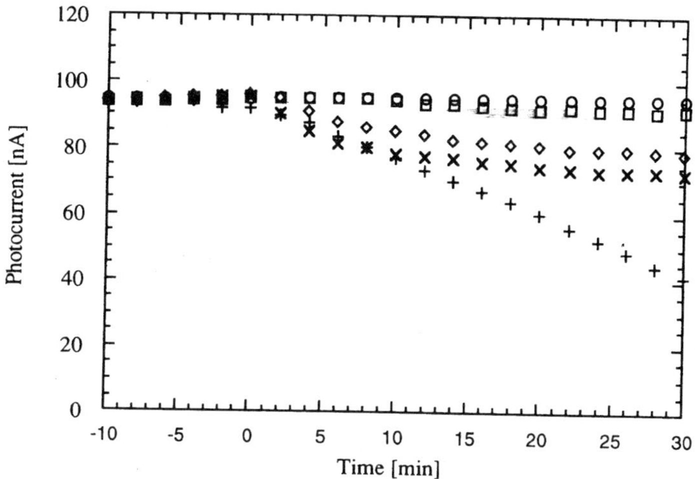

Figure 10.16. Dependence of the response on flow rate. After stabilization of the response with the flow of 1 mM phosphate buffer (pH 7) containing 0.15 M NaCl at various flow rates, the carrier solution was changed to a sample solution containing 1 g/L glucose in the buffer. *T. cutaneum* was 1.0 mg. The time when the solution was changed is zero on the horizontal axis. The flow rate was 50($+$), 250($\times$), 300($\diamond$), 500($\square$) μL/min. The control experiment was at 250 μL/min without glucose in the sample ($\bigcirc$).

solution is diluted before use. Thus, the amount of test solution required is still small enough for environmental monitoring.

Various concentrations of BOD solution (0–500 ppm) were applied to the system to obtain a calibration curve as shown in Fig. 10.17. A BOD standard solution containing glucose (150 mg/L) and glutamic acid (150 mg/L) was employed as a wastewater model according to the Japan Industrial Standard (JIS). The BOD response increased with increasing BOD concentration of standard solution in a linear relationship between the decrease in photocurrent (BOD response) and the BOD concentration from 0 to 100 ppm ($r = 0.988$).

10.5.4. Comparison with BOD_S and BOD_5

The BOD value for various kinds of organic substances was also measured by the SPV-based sensor. Table 10.1 summarizes the substrate specificity of the SPV method, comparing it with reported BOD_5 and BOD_S data for similar compounds.[13] For organic substances, including lactose, soluble starch, and lactic acid, the SPV-based sensor and the sensor based on the oxygen electrode show lower BOD values as compared with the BOD_5. Because these two methods employ short-term metabolism of the same kind of microorganism, the compounds that are difficult to degrade would have smaller BOD values than BOD_5.

The SPV-based sensor, however, gives lower values for some compounds, such as acetic acid, citric acid, and ethanol. The most likely explanation is that the system responds primarily to the acidic compounds produced by the microbe in its early

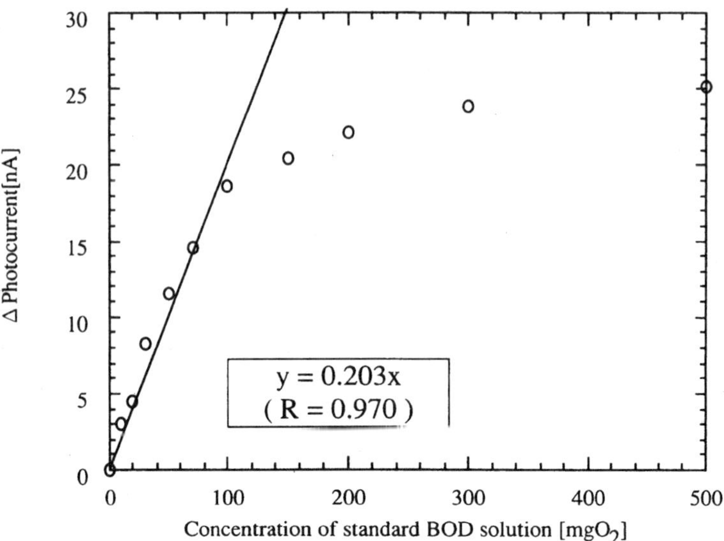

Figure 10.17. Calibration curve of the SPV-based BOD sensor for the standard BOD solution. A phosphate buffer solution of 1 mM, pH 7 was pumped at 250 μL/min. The amount of immobilized cell was 1.0 mg.

Table 10.1. Comparison of BOD Values of Various Organic Samples
(mg O$_2$/mg substrate)[a]

Substrate	SPV	Microbial electrode[b]	5-day method[b]
Glucose	0.66	0.72	0.50–0.78
Fructose	0.73	0.54	0.71
Sucrose	0.45	0.36	0.49–0.76
Lactose	0.04	0.06	0.45–0.72
Soluble starch	0.07	0.07	0.22–0.71
Glycine	0.36	0.45	0.52–0.55
Glutamic acid	0.40	0.70	0.64
Histidine	0.34	0.35	0.55
Acetic acid	0.39	1.77	0.34–0.88
Citric acid	0.18	0.72	0.63–0.88
Lactic acid	0.14	0.17	0.40
Ethanol	0.49	2.90	0.93–1.67
Glycerol	0.44	0.51	0.62–0.83

[a]SPV was performed under conditions of 250 μL/min, 1 mM phosphate buffer, pH 7, containing 0.15 M of NaCl, and 1.0 mg cell wet weight. The substrate was 100 mg/L, pH 7 in the buffer.
[b]Calculated data from Li and Tan.[18]

metabolism stage. For example, a microbe metabolizes glucose to produce acidic compounds such as lactic acid through glycolysis and pyruvic acid. Glycolysis is generally faster than the TCA cycle, and the major factor in the short response time might be glycolysis. There is no consumption of oxygen in glycolysis, and production of only two molecules of NADH requires a small amount of oxygen in the ATP production reaction that follows. Compared to the TCA cycle, glycolysis makes only a small contribution to the oxygen consumption in the oxygen electrode method, although it produces a certain quantity of acidic compounds. Thus, a relatively weak response might be obtained for compounds that are only metabolized in the TCA cycle. The mechanism suggests that the method can be used under anaerobic conditions. Microorganisms that metabolize certain materials specifically under anaerobic conditions would be suitable for use in a microbial sensor combined with SPV.

Basically, SPV measures not the "oxygen demand," or the consumption of oxygen, but the production of acidic compounds, while the oxygen electrode method measures the consumption of oxygen. Acidic compounds produced by the microbe are mainly carbon dioxide and carbonic acid. Oxidation of the substrate produces acidity, so the production of acidic compounds is strongly related to oxygen consumption. This difference in principle can be responsible for the different substrate specificities even though the two methods use the same microbe. Strictly speaking, BOD_{SPV} has different characteristics than BOD_S and BOD_5. There is also a difference between BOD_S and BOD_5. The 5-day method utilizes any kind of microorganism allowing it to metabolize for long time, while the sensor method utilizes the designated microorganism allowing it to metabolize for only short time and we have to be aware of this when comparing different BOD values for samples with compounds of different specificities. Fortunately, environmental monitoring often utilizes a sample from permanent sites or the same kind of sample, and one can take account of the difference prior to analysis.

SPV-based BOD values of some wastewater samples were determined. The 5-day BOD value of BOD solution and various wastewater samples was determined by the Japanese Industrial Standard (JIS) method[10] as follows. An incubator flask including the BOD regent, the sample, and the microbe was cultured at 20°C for 5 days. After cultivation, concentration of the dissolved oxygen was measured by an oxygen electrode.

The wastewater samples tested were treated and untreated wastewater collected from a septic tank in our institute and a brewery. Each wastewater sample was diluted with the carrier solution prior to use. As shown in Fig. 10.18, good agreement between the SPV-based sensor and BOD_5 methods was obtained for the test sample. The correlation coefficient between BOD_{SPV} and BOD_5 was 0.979 and 0.976 for the institute sample and the brewery, respectively. Though the sample from the brewery might have contained redox compounds such as ascorbic acid, the sensitivity without an extra reference electrode is almost the same as that of the sample from the institute.

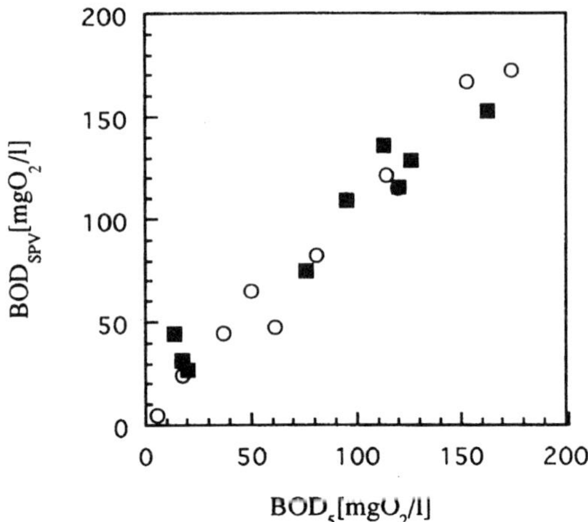

Figure 10.18. Relationship between the SPV-based sensor and 5-day method for measuring wastewater. The sensor method was used under conditions of 250 μL/min, 1 mM phosphate buffer, pH 7, and 1.0 mg cell in wet weight. The 5-day method was followed as described in JIS K0102, and an oxygen electrode was used for the measurement of dissolved oxygen. The wastewater was sampled at a septic tank of an institute (■), and a brewery (○).

10.6. CONCLUSION

The SPV technique was used to develop different kinds of biosensors. The detection mechanism and technique of immobilization of biological sensing materials were also discussed. It was shown that SPV detects not only surface-potential change coming from, e.g., adsorbed charge density or a change in pH on the electrode surface but also from a change in surface impedance owing to the adsorbed substance, which made it possible to develop different kinds of highly sensitive biosensors, such as an enzyme sensor, an immunosensor, and a microbial sensor. The simple structure of an SPV electrode also provides a flexibility for immobilization of sensing materials on a semiconductor electrode leading to stable immobilization of different kinds of biosensitive materials such as enzymes, antibodies, and microorganisms. The results expand the applicability of the SPV method in the biosensor field, although this report only treated a limited number of sensors. The method should also be applicable to other microbial sensors by replacing the immobilized microorganism *T. cutaneum* with various others that are specific and optimal to certain analytes.

REFERENCES

1. Hafeman DG, Parce JW, McConnell HM. Light addressable potentiometric sensor for biochemical systems. *Science* 1988;240:1182–1185.
2. Katsube T, Uchida H. High sensitive immuno-sensor with a surface photo-voltage technique. *Proc. 1st World Wide Congr. Bio-Sensors*, Singapore, 1990; pp 170–171.

3. Sasaki Y, Kanai Y, Uchida H, et al. Highly sensitive taste sensor with a new differential LAPS method. *Sens Act B* 1995;24–25:819–822.

4. Sato T, Shimizu M, Uchida H, et al. Light-addressable suspended-gate gas sensor. *Sens Act B* 1994;20:213–216.

5. Shimizu M, Kanai Y, Uchida H, et al. Integrated bio-sensor employing a surface photo-voltage technique. *Sens Act B* 1994:24–25:187–192.

6. Uchida H, Zhang WY, Katsube T. High speed chemical image sensor with digital LAPS system. *Sens Act B* 1994;34:446–449.

7. Matsuda I, Inoue K. Novel photoreactive surface modification technology for fabricated devices. *Trans Am Soc Artif* Intern Organs, July-September 1990;36(3):M161–M164.

8. Katsube T, Araki T, Hara M, et al. A multi-species biosensor with extended-gate field effect transistors, *Proc. 6th Sensor Symp., Tokyo, Japan,* 1986; pp 211–214.

9. Japanese Industrial Standard Committee, JIS K 3602. Apparatus for the estimation of biochemical oxygen demand (BOD_S) with microbial sensor. Japanese Standards Association, Tokyo, 1990.

10. Japanese Industrial Standard Committee, JIS K 0102. Testing methods for industrial waste water. Japanese Standards Association, Tokyo, 1974.

11. Karube I, Matsunaga T, Mitsuda S, et al. Microbial electrode BOD sensors. *Biotechnol Bioeng* 1977;19:1535–1547.

12. Karube I, Mitsuda S, Matsunaga T, et al. A rapid method for estimation of BOD by using immobilized microbial cells. *J Ferment Technol* 1979;55:243–248.

13. Hikuma M, Suzuki H, Yasuda T, et al. Amperometric estimation of BOD by using living immobilized yeasts. *Euro J Appl Microbiol Biotechnol* 1979;8:289–297.

14. Tan TC, Li F, Neoh KG. Measurement of BOD by initial rate of response of a microbial sensor. *Sens Act B* 1993;10:137–142.

15. Tanaka H, Nakamura E, Minamiyama Y, et al. BOD biosensor for secondary effluent from wastewater treatment plants. *Water Sci Tech* 1994;30:215–227.

16. Ohki A, Shinohara K, Ito O, et al. A BOD sensor using *Klebsiella oxytoca* AS1. *Intern J Environ Anal Chem* 1994;56:261–269.

17. Li F, Tan TC, Lee YK. Effects of pre-conditioning and microbial composition on the sensing efficacy of a BOD biosensor. *Biosens Bioelectron* 1994;9:197–205.

18. Li F, Tan TC. Monitoring BOD in the presence of heavy metal ions using a poly(4-vinylpyridine)-coated microbial sensor. *Biosens Bioelectron* 1994;9:445–455.

19. Iranpour R, Straub B, Jugo TJ. Real time BOD monitoring for wastewater process control. *Environ Eng* 1997;123:154–159.

20. Hyun CK, Tamiya E, Takeuchi T, et al. A novel BOD sensor based on bacterial luminescence. *Biotech Bioeng* 1993;41:1107–1111.

21. Li XM, Ruan FC, Ng WY, et al. Scanning optical sensor for the measurement of dissolved oxygen and BOD. *Sens Act B* 1994;21:143–149.

22. Comber SDW, Gardner MJ, Gunn AM. Measurement of absorbance and fluorescence as potential alternatives to BOD. *Environ Technol* 1996;17:771–776.

11

Surface Plasmon Resonance Biosensors

Ralf W. Glaser

11.1. INTRODUCTION

Surface plasmon resonance (SPR) is a real-time, label-free, optical detection method for studying the interaction of soluble analyte with immobilized ligand. SPR occurs when light is internally reflected at the interface between a medium of high refractive index, a thin layer with good electric conductivity, and a medium of low refractive index.[1] The evanescent wave that develops at the interface interacts with free electrons in the conductive layer and gives rise to so-called plasmons. The energy for this interaction is lost from the reflected light, resulting in a minimum of the reflected intensity at the resonance angle. The evanescent wave extends only a few hundred nanometers into the medium of lower refractive index and decays exponentially over a fraction of the wavelength. The propagation of the evanescent wave along the interface and thus the resonance angle depend on the refractive index in this thin layer adjacent to the interface.

This principle can be used to measure refractive index changes due to adsorption of material to the surface of a sensor (high-refractive-index material coated with a metal layer) from a fluid or gas phase.[2] The change of refractive index near the interface is approximately proportional to the mass of the molecules that enter the interfacial layer. This allows a label-free measurement of the interaction of large biomolecules with immobilized ligands.[3] The changes at the interface are monitored in real time, whereby the rate at which molecules enter and leave the interface is itself the limiting factor for the rate of data collection.

Highly sensitive SPR measurements are a great technological challenge.[4] Hence the availability of a commercial instrument, the BIAcore from Pharmacia Biosensor AB in 1990, was a major breakthrough for the application of real-time, label-free SPR measurements in biomedical research. The BIAcore combined an autosampler

---

Ralf W. Glaser • Institut für Molekularbiologie, Friedrich Schiller Universität D-07708 Jena, Germany.

Biosensors and Their Applications, edited by Yang and Ngo, Kluwer Academic/Plenum Publishers, New York, 1999.

and an integrated fluid-handling system[5] with the high-resolution sensor on top of a microflow chamber in a temperature-stabilized environment. The instrument is controlled by a PC, which is also used for on-line display and later evaluation of the results (Fig. 11.1). First applications of the instrument focused on the determination of kinetic interaction parameters in immunological research and protein engineering.[6,7] Within a few years, however, SPR has become a standard technique in many areas of biological and medical research; for a review see Szabo et al.[8] It has been used for applications as diverse as real-time characterization of genetic engineering manipulations of DNA,[9] investigation of the function of heat-shock proteins GroEL/GroES,[10] determination of oligosaccharide composition of a glycoprotein,[11] and detection of chemotherapeutics in food.[12]

An enormous amount of technological expertise has been accumulated through these experiments (see http://www.biacore.com/scientific/reflist.html). Hardware and software have been improved continuously in close cooperation with the users. Four different instruments of the BIA series with a range of different sensor surfaces were made available from Biacore AB (former Pharmacia Biosensor AB) in 1997 (http://www.biacore.com). They satisfy the needs of different types of applications and a different sample throughput, including one instrument that is equipped for continuous measurements of analyte concentration in chromatography or process control. In biomedical research, surface plasmon resonance has practically become a synonym for measurements with an instrument of the

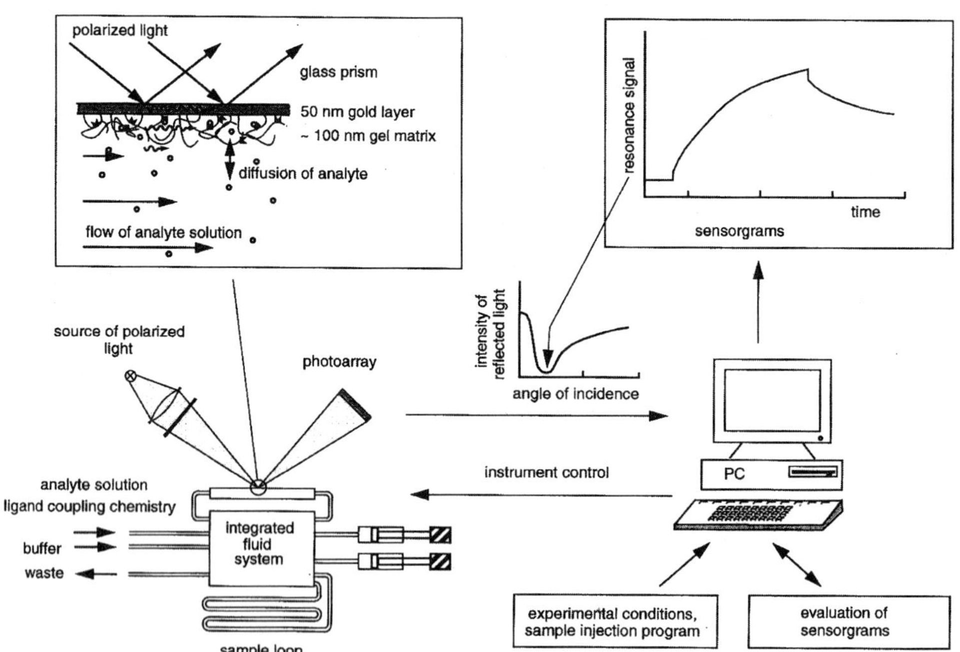

Figure 11.1.   Basic scheme of surface plasmon resonance measurements in a BIAcore

BIA series, and most applications discussed in this chapter were measured with these instruments.

Another SPR instrument, IBIS (Intersens, Amersfoort, Netherlands), has recently become commercially available, but as yet has only a few users in the Netherlands.[13] SPR fiber-optic biosensors are based on a completely different design.[14] Light intensity is measured as a function of wavelength instead of angle of incidence. The miniaturized sensor can be placed at some distance from the detection unit, but the sensitivity is much lower than in BIAcore.

Unlike many other types of biosensors, SPR sensors are used not only to measure the activity or concentration of an analyte but also for qualitative and quantitative characterization of biomolecular interactions. This chapter presents some basic considerations for the preparation and evaluation of SPR experiments.

11.2. RELATED TECHNIQUES

Propagation of an evanescent wave along an interface to a lower-refractive-index medium can be detected on the basis of different physical principles.[15] Two instruments (IAsys and BIOS) that use waveguides to measure charges in the refractive index at a sensor surface are commercially available.[16] The BIOS instrument (Artificial Sensing Instruments ASI AG, Zurich, Switzerland) uses a diffraction grating to couple laser light to a waveguide of high refractive index, which is deposited on a glass or plastic slide.[17] In the IAsys instruments (Affinity Sensors, Cambridge, U.K. — formerly Fisons Applied Sensor Technology), light is fed into the waveguide through a low-refractive-index coupling layer.[18] Most of the issues discussed in this chapter are relevant to these detection principles as well. In theory, waveguide sensors could even achieve a slightly higher sensitivity than SPR sensors. In practice, however, sensitivity is limited by temperature stability, mass transport rate, ligand immobilization, and instrument design rather than by the physical principle employed to detect the refractive-index changes at the sensor surface.

For the user, the major difference between the two instruments is sample handling and design of the measurement chamber. Fison's IAsys uses a stirred microcuvette, which is both a sensor and a solution reservoir. This design has an advantage for certain types of equilibrium experiments[19] but is at a disadvantage for many other experiments. In contrast to the BIAcore microflow chamber, analyte concentration in the stirred cuvette is gradually reduced by binding the sensor surface; and dissociation is incomplete owing to analyte accumulation in the buffer. The rate of mass transport is the same order of magnitude for BIAcore and IAsys. IAsys is available as a research instrument with manual sample loading and as a routine instrument with autosampler (http://www.affinity-sensors.com). Options for surface immobilization are also similar to those of the BIAcore, but the instruments have not found such a broad range of applications. More emphasis is placed on flat, two-dimensional ligand immobilization to study interactions with cells or other large particles. ASI's BIOS-1 is a modular instrument that offers several options for sample handling and can also be used for measurements in the gas phase. It has few users, but the company continues to develop new application areas. The fact that BIAcore is clearly the market leader seems to be a matter of instrument design, user contacts, application expertise, and marketing rather than the difference in sensing principle.

11.3. IMMOBILIZATION OF LIGANDS

The most common method for covalent immobilization of proteins to the sensor surface uses chips with a carboxymethyldextran gel matrix. In an activation step carboxyl groups of the gel form N-hydroxysuccinimide esters via reaction with N-hydroxysuccinimide (NHS) and N-ethyl-N'-(dimethylaminopropyl) carbodiimide (EDC) in water. In the coupling step these NHS esters react with nucleophilic groups on the surface of the protein and form covalent bonds. The remaining esters are inactivated with a high concentration of ethanolamine. The whole procedure takes 20–30 min and is performed in the instrument, so that the density of immobilized protein can be easily adjusted.

The coupling step is usually performed at low ionic strengths and at a pH below the isoelectric point of the protein. Under these conditions, the negative charges of the carboxymethyl groups create a large Donnan potential in the matrix, leading to a considerable enrichment of macromolecules that are positively charged at low pH. The procedure can be applied to a wide range of proteins with small modifications of the parameters. IgG antibodies can be immobilized at surface concentrations up to 10^{-5} kg/m^2; some 30–80% of the molecules remain accessible for biospecific binding. To obtain a low surface concentration of ligands, activation of carboxy groups should be reduced rather than the concentration of ligands or the duration of the coupling step. This reduces the average number of covalent bonds and, consequently, the percentage of inactivated ligands. It is also possible to perform several activation/inactivation cycles before eventually coupling the protein, which also reduces the concentration of negative charges in the matrix.

Immobilization of ligands in a three-dimensional gel layer which has approximately the thickness of the layer that is probed by the surface plasmon resonance sensor has a number of advantages over immobilization on a two-dimensional surface. More ligands can be immobilized, and more analyte can bind without steric hindrance between neighboring molecules. Ligands are not in contact with a surface, so interaction can take place under conditions similar to those in solution. Indeed most analytes show very low nonspecific adsorption to the sensor chip as long as the interactions are studied at physiological or higher ionic strength. The percentage of denatured ligand molecules showing nonnative epitopes is very small compared to solid-phase immunoassays that use noncovalent adsorption of proteins to a plastic surface.

Large particles such as viruses or cells cannot enter the matrix, and thus they bind at a distance to the sensor surface, where their contribution to the signal is reduced. In addition, it is difficult to estimate the density of ligands that are accessible to a binding site on the surface of the particle. For such experiments immobilization of ligands on a two-dimensional surface may be an advantage.

However, even binding of phage-displayed antibody libraries was studied with antigen immobilized in a carboxymethyl dextran matrix.[20]

Alternatives to the direct covalent immobilization of ligands are noncovalent binding to antibodies, binding of biotin-labeled ligands to streptavidin, or binding of ligands with a polyhistidine tail via nickel ions. Site-specific immobilization prevents the possible interference of ligand–analyte binding with covalent matrix attachment of the ligand in the vicinity of the analyte binding site. An attractive way to present lipids or membrane-anchored proteins on the sensor surface is the formation of a hybrid bilayer directly on the gold surface.[21]

11.4. QUALITATIVE CHARACTERIZATION OF MOLECULAR INTERACTIONS

Real-time measurements with surface plasmon resonance biosensors are well suited for studying the sequential binding of several molecules to the immobilized ligand and to each other. The signal resulting from the individual components is roughly proportional to molecular weight — so the stoichiometry of binding can be determined directly.

Sequential binding is routinely used to obtain "epitope maps" of an antigen for a panel of monoclonal antibodies; the antibodies are tested for their ability to bind to a complex of one of the antibodies with antigen. Thus groups of antibodies are found that recognize identical epitopes or epitopes that are near to each other on the surface of the antigen.[6,22] This can be done in two ways: (1) One antibody is immobilized, antigen is bound to it, and the second antibody is tested for binding to the complex; this sequence does not work properly if the antigen contains several identical epitopes. (2) The antigen is immobilized, and binding of the antibodies is compared before and after saturation of an epitope with one of the antibodies; this technique can also be used if the antigen is not a monomer, or if it is heterogeneous and displays different epitopes on different molecules.

We used sequence (1) to identify enhancement of binding between different neutralizing monoclonal antibodies directed against human interleukin-10.[23] A higher avidity of antibodies in the presence of other antibodies is quite common. However, this is usually due to the formation of complexes that lead to bivalent binding of the antibodies; in this case the dissociation rate constants of the antibodies is reduced. For the neutralizing anti-IL-10 antibodies, the SPR measurements showed that the association rate of several antibodies was increased when other antibodies were bound to the protein (Fig. 11.2). The most probable explanation for this effect is a conformational change induced by antibody binding.

Complex molecular systems can be studied with this approach. One example was the assembly and dissociation of a signal transduction complex which controls chemotaxis in *E. coli*.[24] The sequence in which this quaternary protein complex is built and the induced dissociation in the presence of ATP were studied with SPR, and this led to an understanding of the function of the signal transduction complex *in vivo*.

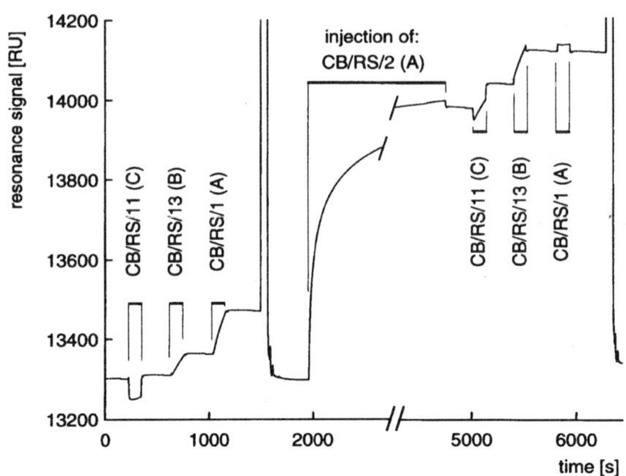

Figure 11.2.   Experiment to measure the association rate of three antibodies, CB/RS/11, CB/RS/13, and CB/RS/1 to IL-10 and to the complex of IL-10 and antibody CB/RS/2. 30 μg/mL CB/RS/11, 13.7 μg/mL CB/RS/13, and 6.2 μg/mL CB/RS/1 are injected for 120 s each. The matrix is regenerated by a 60-s injection of 0.5 M citrate buffer, pH 2.0. 150 μg/mL CB/RS/2 is injected for 48 min and injection of the three test antibodies is repeated at the same concentrations. Finally the matrix is regenerated again. The measurement was made on a BIAcore 2000™ at a flow rate of 3 μL/min. CB/RS/1 and CB/RS/2 bind the same epitope, so CB/RS/1 does not bind to the complex. CB/RS/11 and CB/RS/13 bind different epitopes on IL-10; CB/RS/11 binding is nearly unaffected and CB/RS/13 binding to the complex is significantly enhanced compared with binding to free interleukin.

11.5.   MEASUREMENT OF ANALYTE CONCENTRATION

The classical application for biosensors is the quantitative detection of a particular analyte in a solution of unknown composition. SPR sensors are well suited for this purpose, as long as only a moderate sample throughput is required. The noise level in the BIAcore is below 10^{-9} kg/m^2, which corresponds to 1 RU (resonance unit) or a shift in the intensity minimum of 0.0001°.[25] A signal of about 50 RU is necessary for reliable measurements.

Analytes with a molecular weight of more than about 5 kD can be measured directly from the signal they give when they are bound to the immobilized ligand. Smaller analytes do not give rise to a very strong signal. A more accurate measurement can be achieved by allowing the analyte to interact with a soluble ligand of greater molecular weight, and measuring the reduction of free soluble ligand using the SPR sensor.

Several parameters of the sensorgrams can be used to calculate the analyte concentration. For analyte–ligand interactions with low affinity and high dissociation rate, the level of binding at equilibrium is measured. For high-affinity interactions it is more appropriate to measure the initial binding rate. The initial increase in the SPR signal usually shows a linear dependence on analyte concentration over a wide concentration range, even under conditions of bivalent binding, mass transport limitation, and other factors that lead to nonexponential sensorgrams. If the signal during analyte injection is disturbed by differences in the composition between the buffer and the analyte solution, analyte can be injected for a fixed time, and the amount of bound analyte can be determined from the signal detected after injection.

The maximum amount of information is obtained from the experiment if the complete sensorgrams are analyzed as described below for the determination of kinetic constants. With known $R_{\max}$ and k_a, c is then obtained from the experimental fit. Under suitable conditions it is even possible to determine both c and $R_{\max}$ from the fit: this is necessary to correct for loss of binding activity of the sensor chip after repeated measurements, or to measure on single-use sensor chips, where the amount of immobilized ligand is slightly different from chip to chip.

In the equation that describes the ideal sensor response the association rate constant k_a and the analyte concentration c are always a product $c \times k_a$, so it is not possible to calculate both c and k_a from sensorgrams of a pseudo-first-order interaction; k_a must be known explicitly, or it must be implicitly contained in a calibration curve. However, this is not necessarily so for all types of kinetics. A method for determining analyte concentration under conditions of mass transport control that does not rely on knowledge of the association rate constant is discussed in Section 11.10 below.

Concentration measurements with unpurified solutions require careful control experiments to preclude errors due to interference from specific or nonspecific interaction with the sensor chip by substances other than the analyte of interest. The dissociation part of the sensorgram can give us additional information about the substance that has bound to the sensor chip during injection of the test solution; it should be compared with the expected dissociation rate of the analyte. If necessary, pH or other parameters may be changed to test for a typical dissociation behavior of the analyte examined.

Figure 11.3 illustrates one way in which quantitative information can be obtained from the dissociation signal. We studied rearrangement of protein domains in a solution of HPLC-purified dimeric single-chain variable domain antibody fragments (scFv) at elevated temperature (W. Höhne et al., submitted). The concentrations of two forms of the analyte, monovalently and bivalently binding scFv, were determined. They were distinguished by their characteristic time constant of dissociation, measured after very short periods of analyte injection. With the knowledge of k_a and k_d for both forms of the analyte, a quantitative calculation of the concentrations was possible; $R_{\max}$ could be determined from the signal after longer injections of analyte solution (not shown). Unexpectedly, a rapid increase in bivalently binding scFv (emerging from noninteracting dimeric molecules) was followed by a slower conversion into monovalently binding molecules.

11.6. DETERMINATION OF KINETIC AND THERMODYNAMIC INTERACTION CONSTANTS

For a reaction $A + B \leftrightarrow AB$ between analyte A and the immobilized ligand B the change of the SPR signal, R, is proportional to the formation or dissociation of the complex AB. The ideal response of the sensor would be

$$R(t) = \frac{c \times k_a}{c \times k_a + k_d} R_{\max}(1 - e^{-(c \times k_a + k_d)t}) + R_b \tag{1}$$

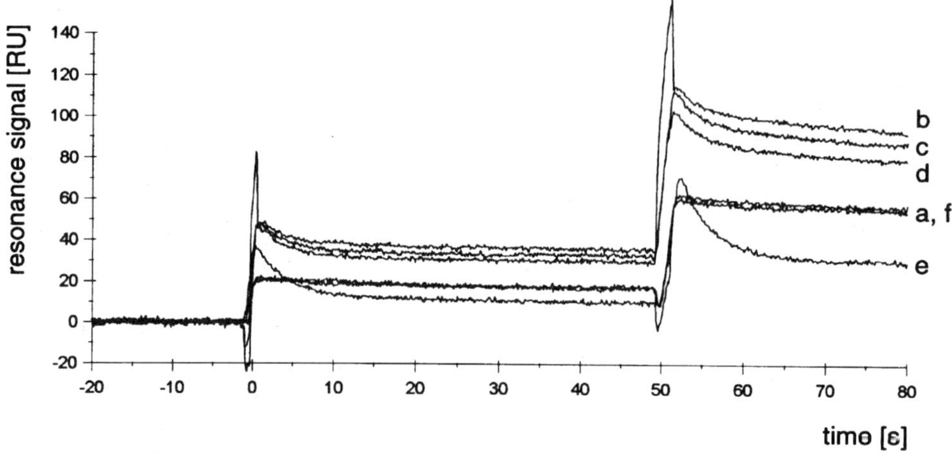

Figure 11.3. Sensorgrams of successive analyte injections of 1 and 2 s duration to immobilized turkey egg lysozyme at a flow rate of 50 μL/min in a BIAlite instrument. Analyte is a 6 μg/mL solution of dimeric single-chain 1F9 antibody variable fragments before (a); after 30 min (b); 1 h (c); 2 h (d); and 15.5 h at 40°C (e); or 16 h at 0°C (f).

and

$$R(t) = R_0 \times e^{-k_d \times t} \tag{2}$$

for association and dissociation, respectively.[26] R_{max} is the signal that would be measured if all possible ligand binding sites are occupied by analyte, R_b is the shift of the baseline during injection of analyte solution, which results from free analyte in solution and from possible differences in buffer composition, and R_0 is the signal at the start of dissociation.

The concentration of analyte in solution, c, should be known. The amount of ligand B immobilized in the matrix is known from the increase in the signal during immobilization, but usually it is not known what percentage of the ligand molecules is active, i.e., able to bind analyte. During immobilization by NHS/EDC as described above, covalent bonds are formed in the neighborhood of the binding site in a considerable fraction of the molecules, preventing interaction with the analyte.

The value of k_d is determined by an exponential fit of the dissociation phase. R_b, the difference in the signal when the injected buffer is changed for analyte solution and vice versa, is determined either at the beginning or at the end of analyte injection. The association part of the curve is then shifted relative to the rest of the sensorgram, so that R_b can be ignored. If the time constant of the exponential association is significantly faster than the time constant of dissociation, both R_{max} and k_a are derived from an exponential fit of the association phase, and all parameters can be obtained from a single injection of analyte. If $c \times k_a \ll k_d$, association and dissociation have approximately the same time constant and k_a can only be determined if R_{max} is known. This can be obtained from an experiment with higher analyte concentration, from binding of another analyte, or from the signal during ligand immobilization if 100% of the binding sites are accessible. In practice, several injections of analyte solution with different concentrations must be measured to get reliable results and to test the validity of the kinetic model used. It must be possible to fit all of the curves with the same R_{max}, k_a, and k_d. The data manipulations — nonlinear fits, offset corrections, etc. — can be done in a convenient way with the BIAevaluation software that is provided with the instruments.

The sensorgrams can also be evaluated by linearization. For the pseudo-first-order reaction the plot of dR/dt vs. R gives a linear curve, the slope of which is the time constant of association, $c \times k_a + k_d$. This slope is determined from experiments with different analyte concentrations, and the plot of $c \times k_a + k_d$ vs. c provides both association and dissociation rate constants.[3] The value of k_d is also obtained from linearization of the dissociation curve as $\ln(R_0/R)$ vs. t or $\ln(dR/dt)$ vs. t. With the availability of suitable nonlinear fitting software these linearized plots are no longer used. Nonlinear fits provide a better basis for recognizing the extent of systematic deviations from the expected kinetic behavior, and they are also less sensitive to random noise.

11.7. NONEXPONENTIAL BINDING BEHAVIOR

In a bimolecular reaction $A + B \leftrightarrow AB$, where the analyte concentration [A] is kept constant by a continuous flow, pseudo-first-order kinetics are expected for binding and dissociation. The SPR signal, which is proportional to [AB], should show exponential kinetic behavior. In practice, however, a number of systematic deviations from an exponential function are observed. There is much discussion about nonideal behavior of interacting molecules in SPR biosensors, which might lead to the impression that there are more artifacts than in other methods. I think this is not the case. The high signal/noise ratio makes deviations visible that would not be detected otherwise, e.g., nonlinearities in the instrument of a few resonance units. In many cases deviations are small, but certain experiments are very difficult to analyze because the parameters calculated from different parts of the curves or from data obtained under different conditions vary by orders of magnitude.

Deviations from idealized kinetic behavior can result from various effects. Some are undesired artifacts while others can provide valuable information about the interacting molecules. It depends on the situation whether a particular effect prevents us from obtaining specific information from an experiment or whether it gives us a better understanding of the interaction that we are studying. Bivalent binding, e.g., often prevents the quantitative measurement of interaction parameters for binding of bivalent IgG antibodies to immobilized antigen. The same effect could be used to quantify the concentration of molecules with one or two functional binding sites, as described above.

Bivalent or multivalent binding of the analyte always complicates or prevents the quantitative determination of dissociation rate constants and thermodynamic equilibrium constants. Bivalent binding cannot be efficiently suppressed by low

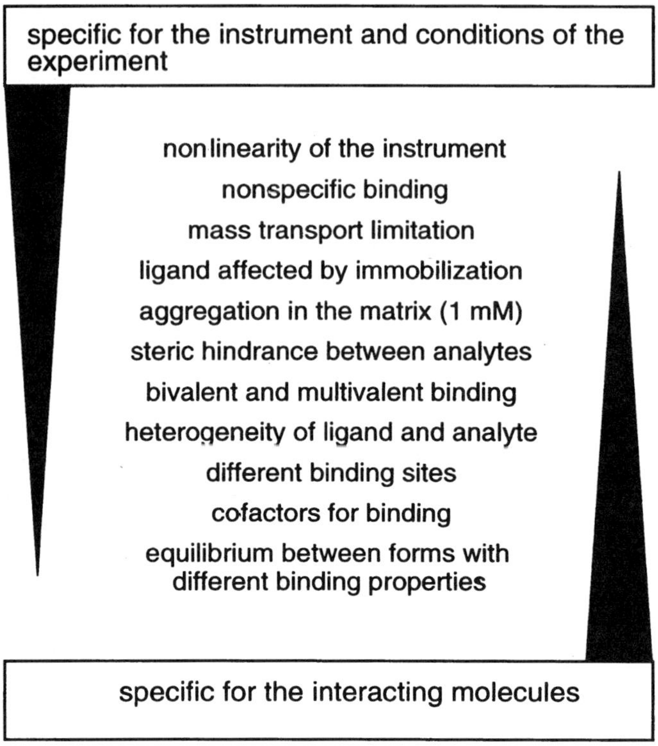

specific for the instrument and conditions of the experiment

nonlinearity of the instrument

nonspecific binding

mass transport limitation

ligand affected by immobilization

aggregation in the matrix (1 mM)

steric hindrance between analytes

bivalent and multivalent binding

heterogeneity of ligand and analyte

different binding sites

cofactors for binding

equilibrium between forms with different binding properties

specific for the interacting molecules

Figure 11.4

ligand density because the flexibility of the carboxymethyldextran matrix allows bivalent binding even at ligand concentrations where the average distance between ligands is large. Apparent dissociation rate constants depend on both the ligand density and on the amount of analyte bound. Sometimes the dissociation rate constant for monovalent binding can be determined by extrapolation to high analyte saturation and low ligand density, but this is not always the case. The best way to get interaction parameters for monovalent binding is to immobilize the multivalent molecule and measure interaction with the monovalent binding partner as analyte in solution. In the case of antibodies the interaction can be studied with Fab fragments.

Bivalent or multivalent binding can also be the result of aggregation. The concentration of immobilized ligand in the matrix, and thus the concentration of analyte bound to the ligand, can be higher than 1 mM. Rather weak interactions are sufficient to cause reversible formation of dimers or oligomers under these conditions. A different situation is the presence of long-lived aggregates in the analyte solution. They may result, e.g., from misfolding or proteolysis in a small percentage of the molecules. If dissociation of monovalent analyte is fast, dimers or higher aggregates can accumulate on the sensor and eventually dominate the signal, even if they represent only a very small percentage of all molecules. Careful experimentation is necessary to recognize such artifacts and to characterize the interaction of molecules with low affinity.[27]

Another intriguing possibility is a ligand–ligand or an analyte–analyte interaction that hides the binding site and prevents the analyte–ligand interaction of interest. In the immobilized ligand, dimerization can result in complex kinetics owing to competition between analyte and other ligand molecules for the ligand binding sites. In the analyte, noninteracting dimers do not prevent pseudo-first-order kinetic behavior, but they do produce a nonlinear relation between the free analyte concentration that is effective for binding and the total analyte concentration that is determined by other methods.

The presence of two or more different binding sites on one molecule or the existence of several types of molecules with different interaction properties is certainly an important feature of the ligand or analyte studied. However, the heterogeneity may have been introduced artificially, e.g., by insufficient purification or partial denaturation. Irrespective of whether we are interested in a complete description or only in the parameters for one particular aspect of the interaction, a more complex model has to be used and validated. The next section describes some considerations concerning consistency of data that are essential for choosing the correct model. They apply not only for complex kinetic schemes but also for the simple pseudo-first-order model.

11.8. CONSISTENCY AND CHOICE OF THE RIGHT MODEL

Although many biomolecular interactions take place on a surface with at least one immobilized partner, it is the aim of most quantitative kinetic and thermodynamic measurements to find parameters for idealized analyte–ligand interactions at low concentration in solution. If the interaction depends on ligand density, surface charge, and other parameters, then it is easier to predict the behavior under certain biological conditions from interaction parameters in solution than from the behavior observed under the different conditions on the sensor surface. Thus it is important to determine how parameters derived from the sensorgrams relate to the parameters of the interaction in solution. The most reliable way to characterize the interaction in solution is a competition measurement in solution as described below. Where this is not possible, one minimally requires the following checks for the data obtained:

1. Control experiments for "non-specific" binding: Analyte solution is injected onto a sensor chip without a ligand, or — better — with an immobilized molecule of similar physicochemical properties for which no specific interaction with the analyte is expected. On modern instruments this can be done on a separate channel, in parallel with the specific binding experiment.
2. Check of consistency between kinetic and thermodynamic behavior: The dissociation constant calculated from the kinetic rate constants, $K_D = k_d/k_a$, has to be equal to the dissociation constant calculated from the

amount of analyte bound at equilibrium,

$$R_{eq}(c) = R_{max} \frac{c}{c + K_D} \tag{3}$$

3. Consistency of data obtained under different conditions: It must be possible to describe experiments at different analyte concentrations with the same set of parameters, i.e., not only with the same k_a and k_d, but also with the same R_{max}, if there is no obvious reason to assume that the binding capacity of the sensor changes between experiments.[28]
4. Exchange of ligand and analyte: The same parameters should be obtained whether one or the other molecule is immobilized.

The situation is particularly complex if we have to assume that the deviations from pseudo-first-order kinetics reflect intrinsic properties of ligand or analyte. We often do not have *a priori* knowledge about the type of interaction that has to be assumed or about heterogeneity. So the first step is to find the origin of the deviations from pseudo-first-order kinetic behavior and thus the correct mathematical model to describe the sensorgrams. Although single sensorgrams can mostly be described quite well as the sum of two exponential components, most models fail to describe the dependence on analyte concentration or other aspects for all data.

It is only when the correct model has been defined unambiguously by a qualitative and semiquantitative evaluation of results obtained under different conditions that it makes sense to calculate the parameters of the model. The correct model and a consistent interpretation of possible artifactual deviations from it allow us to identify such conditions and those parts of the curves that best reflect the properties of the analyte–ligand interaction of interest.

For the interaction of Fab fragments of the monoclonal antibody CB-4/1 with recombinant HIV core protein p24 (rp24) we found a fast association of the antibody to a small percentage of the rp24, followed by a slow phase in which additional antigen was bound. Dissociation, however, was uniform. Moreover, a "memory effect" of rp24 was observed, i.e., immediately after antibody was dissociated from rp24, the fast association was enhanced.[29] The following measurements were made to find a model for the interaction: (a) association and dissociation kinetics of CB-4/1 Fab at different concentrations (Fig. 11.5) and for different times of injection with immobilized rp24; (b) interaction of rp24 at different concentrations with immobilized CB-4/1; (c) injection of CB-4/1 Fab at high concentration on immobilized rp24 until the binding sites were nearly saturated, injection of buffer for various short intervals, followed again by Fab injection; (d) interaction of CB-4/1 Fab with the immobilized peptide epitope; (e) interaction of Fab and rp24 in solution, measurement of free Fab at different times after mixture; (f) association and dissociation of Fab on immobilized rp24 in the presence of different concentrations of competing peptide epitope; (g) interaction of another anti-p24 antibody with immobilized rp24. It was concluded from these measurements that most rp24 exists in a form that cannot bind to CB-4/1 antibody. The protein has to undergo a conformational

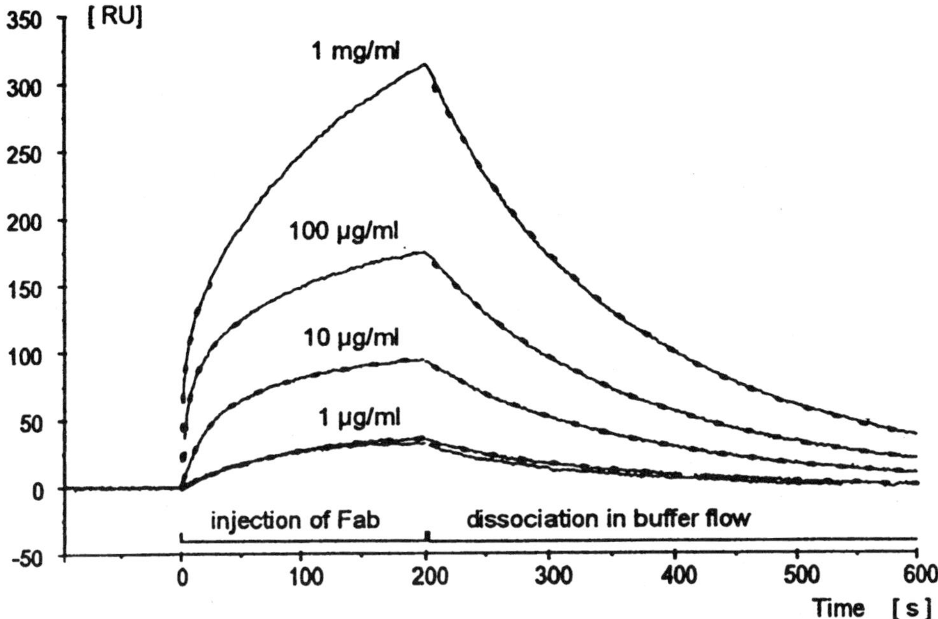

Figure 11.5. Sensorgram of the binding of 2×10^{-8}, 2×10^{-7}, 2×10^{-6}, and 2×10^{-5} M Fab fragment of antibody CB-4/1 to immobilized recombinant HIV core protein rp24. The 2×10^{-8} M injection was measured twice, at the beginning and at the end of the experiment. The bulk effect, i.e., the signal from different refractive indexes of buffer and solution, was subtracted. The sensorgrams reveal large deviations from an $A + B \leftrightarrow AB$ interaction model. ---- simulated curves: association $R = R_{eq1} \times (1 - e^{-t/\tau_1}) + R_{eq2} \times (1 - e^{-t/\tau_2})$; dissociation $R = R_0 \times e^{-kd \times (t - 200\,s)} +$ offset.

change for the epitope to become exposed. This process itself does not follow first-order kinetics, which suggests that more than two conformations exist. The possibility that the behavior was due to dimerization of rp24 hiding the epitope could be excluded. Dimerization would have been a much more obvious reason for the kinetics observed in experiments (a) and (c), but it was qualitatively incompatible with the result of (b), where only a fraction of soluble rp24 interacted with the antibody, but this fraction was independent of rp24 concentration.

The interpretation of the kinetic interaction measurements as reflecting a conformational change in rp24 was confirmed by the three-dimensional NMR structure of the N-terminal domain of HIV p24.[30] In this structure, several of the residues in helix III and at their N-terminal end, which were identified as important for recognition by CB-4/1,[31] were not accessible to the antibody.

Although in this case a model of interaction and conformational changes that provides a quantitative global fit to all data from the different modes of measurement could not be defined, fascinating qualitative results were obtained, and at least a semiquantitative description of the processes could be given.

11.9. STUDYING INTERACTIONS IN SOLUTION

Immobilization of ligand in a carboxymethyldextran matrix may influence the interaction with analyte, as discussed above. These problems can be avoided if the two molecules interact in solution, and the surface plasmon resonance sensor is used to determine the concentration of free analyte — analogous to the method for affinity determination with enzyme immunoassays described by Friguet et al.[32] The use of SPR sensor technology for this type of measurement has a number of advantages compared with an enzyme immunoassay: (1) The time for a single concentration measurement is short (10–100 s), so the kinetics of the interaction in solution can be measured. (2) The time constant for exchange between the bulk analyte solution and free analyte molecules in the matrix of the sensor chip is less than 1 s; the bulk solution is constantly replaced, so the immobilized ligand does not disturb the thermodynamic equilibrium in solution. (3) In an enzyme immunoassay monovalently and bivalently bound analytes may give different responses according to the extent of dissociation during the conjugation step. The label-free SPR measurement gives exactly the same signal independent of the binding mode.

Determination of affinity by these competition assays is more reliable than calculation of affinity from the kinetic rate constants k_a/k_d.[33] Bivalency of one of the molecules can be taken into account, and problems such as mass transport limitation or artifactual changes of interaction properties by immobilization do not influence the result, because they do not disturb a reliable measurement of free-analyte concentration.

For the competition assay the ligand immobilized on the SPR sensor can be different from the ligand that is studied in solution — it only has to distinguish between free and bound analyte. To study the interaction between HIV core protein p24 and the Fab fragment of monoclonal antibody CB-4/1 in solution [experiment (e) in the example described above] we immobilized a mutated peptide from the p24 epitope, which was much more suitable for measuring free Fab concentration than the antigen p24 itself.[29]

Measurement of dissociation kinetics in solution is also a way to get reliable dissociation rate constants below $3 \times 10^{-5}\,\mathrm{s}^{-1}$ with the SPR sensor. The reproducibility of the concentration measurement has to be checked very carefully during such experiments, since regeneration and storage can affect the binding properties of the ligand on the sensor chip.

11.10. MASS TRANSPORT LIMITATION

Mass transport limits the dynamic range of SPR instruments toward high association rate constants. Analyte is transported to the sensor surface by convection parallel to the sensor surface, and by diffusion. Two different processes are related to mass transport: (1) replacement of buffer with analyte solution and vice versa after valves have been switched in the instrument; and (2) replacement of free analyte that has bound to the sensor by analyte from bulk solution, or transport of dissociated analyte back to the bulk solution.

The first process gives rise to short transition periods at the beginning and at the end of analyte injection. This process can be studied with a solution of high refractive index that does not interact with the sensor surface, e.g., sucrose or bovine serum albumin solution. Its time constant is approximately proportional to the flow rate, and at a flow of $10\,\mu\text{L/min}$ the exchange between two solutions is more or less complete after about $2\,\text{s}$. The signal resulting from the transition is quite obvious, and process (1) will not be discussed further.

The second process is less obvious, because it is tightly connected to the interaction between analyte and ligand and often shows the same type of kinetics. When free analyte is bound to the sensor, it must be replaced by analyte from the bulk solution. The diffusional transport needs a concentration gradient, so there is a concentration difference between the bulk solution and the surface. As analyte concentration at the sensor surface is not known, analyte concentration in the bulk solution is used to calculate the association rate constant, and k_a is underestimated. When the sensor is rinsed with buffer, the concentration of free analyte at the sensor surface is not zero, because dissociated analyte must be transported away from the sensor surface. Some of the analyte can rebind, so the dissociation rate constant k_d is also underestimated. Some theoretical knowledge is essential in order to recognize these effects and avoid misinterpretation of experimental data.

A simple approximation considers mass transport and binding or dissociation as two identical net fluxes j that are in line and share their driving force — the difference between analyte concentration in bulk solution c_0 and the equilibrium concentration c_{eq}, which is the theoretical analyte concentration that would correspond to equilibrium at a certain level of analyte binding

$$c_{\text{eq}} = K_D \frac{g_b}{g_{\text{fr}}} \tag{4}$$

where g_b and g_{fr} are the surface concentrations of bound and free analyte, respectively.[34] The concentration of free analyte at the surface, c_{sf}, divides the driving force into the two components, and the flux is

$$j = L_m(c_0 - c_{\text{sf}}) = L_r(c_{\text{sf}} - c_{\text{eq}}) \tag{5}$$

where the Onsager coefficients for mass transport and for binding/dissociation are $L_r = k_a \times g_{\text{fr}}$ and $L_m = \sqrt[3]{(D^2\,\text{flw}/h^2 b1)}$, respectively. Apart from the theoretical calculation, L_m can easily be determined from association rate measurements under conditions where the signal is determined only by mass transport, e.g., during the accumulation of negatively charged protein at low ionic strength.

For $L_m \gg L_r$ we get $C_{\text{sf}} \approx C_0$, and calculation of the kinetic rate constants from analyte concentration in bulk solution is correct. If L_m and L_r are the same order of magnitude, both binding and mass transport determine the observed kinetics. The relative error in the calculation of apparent kinetic constants (using the bulk analyte concentration c_0) depends on k_a, g_{fr}, $D^{-2/3}$, $\text{flw}^{-1/3}$, and the geometry of the flow chamber. It is the same for association and dissociation. To improve the situation, the ligand density g_{fr} can be reduced (at the expense of signal/noise) and the flow rate, flw, can be increased (the effect is small because $L_m \sim \text{flw}^{1/3}$). To some extent

k_a can be extrapolated from plots of k_a vs. g_{fr} or k_a^{-1} vs. $flw^{-1/3}$. In any case, mass transport has to be considered. The highest association rate constants that can be determined reliably depend on the signal/noise at low ligand densities, i.e., on the molecular weight of the analyte. About $10^6\,M^{-1}\,s^{-1}$ can be determined for 50-kD molecules.

As long as only a small percentage of ligand binding sites is occupied, i.e., g_{fr} is nearly constant, we still see pseudo-first-order kinetics. It is only for $c \gg K_D$ that some characteristic features of mass transport limitation become obvious in the association curves. In the dissociation phase, rebinding can be suppressed by the addition of low-molecular-weight molecules that compete with the analyte studied for ligand binding sites.

At $L_m \ll L_r$, mass transport alone determines the observed kinetics, and no information about interaction parameters can be obtained from the experiment except for thermodynamic parameters at equilibrium and a lower limit for k_a. At very high association rate constants, the buildup of the concentration gradient along the flow chamber[34] and the distribution of the analyte between the different layers of the gel matrix[35] modify the kinetics compared with the approximation above. Both aspects give rise to characteristic effects in association kinetics that can be confused with cooperative binding.

Mass transport limitation is not only a potential source of error — in a positive sense the effect can be used to measure the analyte concentration without knowing the association rate constant. Concentration of antibodies could be measured by adsorption to antigen immobilized at high density, although the binding properties of different antibodies to the antigen were not identical.[35] If the concentration of free ligand and the association rate constant are sufficiently high, the initial rate of association is limited by mass transport and depends only on analyte concentration in bulk solution, diffusion constant of the analyte, and flow rate.

11.11. CONCLUSIONS

Commercial instruments for SPR studies were available for less than 7 years when this review was written. Within this time SPR has become a standard technique in many areas of biology, medicine, and biotechnology. Applications go far beyond the determination of analyte concentration and kinetic rate constants of analyte–ligand interactions. Real-time label-free measurements at high signal/noise ratios led to a better understanding of the mechanisms of biomolecular interaction in the time frame of seconds to hours and enabled functional studies of complex molecular assemblies. The range of applications for SPR measurements is still growing rapidly.

REFERENCES

1. Kretschmann E, Raether H. Radiative decay of nonradiative surface plasmons excited by light. *Z Naturforsch* 1968;23A:2135–2136.
2. Liedberg B, Nylander C, Lundström I. Surface plasmon resonance for gas detection and biosensing. *Sens Act* 1983;4:299–304.

3. Karlsson R, Michaelsson A, Mattsson L. Kinetic analysis of monoclonal antibody–antigen interaction with a new biosensor based analytical system. *J Immunol Meth* 1991;145:229–240.

4. Liedberg B, Nylander C, Lundstrom I. Biosensing with surface plasmon resonance—how it all started. *Biosens Bioelectron* 1995;10:i–ix.

5. Sjölander S, Urbaniczky C. Integrated fluid handling system for bimolecular interaction analysis. *Anal Chem* 1991;63:2338–2345.

6. Fägerstam LG, Frostell A, Karlsson R, et al. Detection of antigen-antibody interactions by surface plasmon resonance: application to epitope mapping. *J Mol Recogn* 1990;3:208–214.

7. Borrebaeck CAK, Malmborg A-C, Furebring C, et al. Kinetic analysis of recombinant antibody-antigen interactions: relation between structural domain and antigen binding. *Bio/Tech* 1992;10:697-698.

8. Szabo A, Stolz L, Granzow R. Surface plasmon resonance and its use in biomolecular interaction analysis (BIA). *Curr Opin Struct Biol* 1995;5:699–705.

9. Nilsson P, Persson B, Uhlen M, et al. Real-time monitoring of DNA manipulations using biosensor technology. *Anal Biochem* 1995;224:400–408.

10. Hayer-Hartl MK, Martin J, Hartl FU. Asymmetrical interaction of GroEL and GroES in the ATPase cycle of assisted protein folding. *Science* 1995;269:836–841.

11. Hutchinson AM. Characterization of glycoprotein oligosaccharides using surface plasmon resonance. *Anal Biochem* 1994;220:303–307.

12. Sternesjo A, Mellgren C, Bjorck L. Determination of sulfamethazine residues in milk by a surface plasmon resonance-based biosensor assay. *Anal Biochem* 1995;226:175–181.

13. van Leeuwen HC, Strating MJ, Rensen M, et al. Linker length and composition influence the flexibility of oct-1 DNA binding. *EMBO J* 1997;16:2043–2053.

14. Jorgenson RC, Yee SS. A fiber optic chemical sensor based on surface plasmon resonance. *Sens Acts* 1993;B7:213–220.

15. Garland PB. Optical evanescent wave methods for the study of biomolecular interactions. *Quart Rev Biophys* 1996;29:91–117.

16. Hodgson J. Light, angles, action. *Biotechnology* 1994;12:31–35.

17. Tiefenthaler K. Grating couplers as label-free biochemical waveguide sensors. *Biosens Bioelectron* 1993;8:xxxv–xxxvii.

18. Cush R, Cronin JM, Steward WJ, et al. The resonant mirror: a novel optical biosensor for direct sensing of biomolecular interactions. *Biosens Bioelectron* 1993;8:347–353.

19. Hall DR, Winzor DJ. Use of a resonant mirror biosensor to characterize the interaction of carboxypeptidase A with an elicited monoclonal antibody. *Anal Biochem* 1997;244:152–160.

20. Malmborg AC, Borrebaeck CA. BIAcore as a tool in antibody engineering. *J Immunol Meth* 1995;183:7–13.

21. Plant AL, Brigham-Burke M, Petrella EC, et al. Phospholipid/alkanethiol bilayers for cell-surface receptor studies by surface plasmon resonance. *Anal Biochem* 1995;226:342–348.

22. Daiss JL, Scalice ER. Epitope mapping on BIAcore: theoretical and practical considerations. *Methods* 1994;6:143–156.

23. Sabat R, Seifert M, Glaser RW, et al. Neutralizing murine monoclonal antiinterleukin-10 antibodies enhance binding of antibodies against a different epitope. *Mol Immunol* 1997;33:1103–1111.

24. Schuster SC, Swanson RS, Alex LA, et al. Assembly and function of a quaternary signal transduction complex monitored by surface plasmon resonance. *Nature* 1993;365:343–347.

25. Stenberg E, Persson B, Roos H, et al. Quantitative determination of surface concentration of protein with surface plasmon resonance using radiolabeled proteins. *J Colloid Int Sci* 1991;143:513–526.

26. O'Shannessy DJ, Brigham-Burke M, Soneson KK, et al. Determination of rate and equilibrium binding constants for macromolecular interactions using surface plasmon resonance: use of nonlinear least squares analysis methods. *Anal Biochem* 1993;212:457–468.

27. Morton TA, Myszka DG, Chaiken IM. Interpreting complex binding kinetics from optical biosensors: a comparison of analysis by linearization, the integrated rate equation, and numerical integration. *Anal Biochem* 1995;227:176–185.

28. Glaser RW, Hausdorf G. Binding kinetics of an antibody against HIV p24 core protein measured with real-time biomolecular interaction analysis suggest a slow conformational change in antigen p24. *J Immunol Meth* 1996;189:1–14.

29. Gitti RK, Lee BM, Walker J, et al. Structure of the amino-terminal core domain of the HIV-1 capsid protein. *Science* 1996;273:231–235.
30. Höhne WE, Küttner G, Kiessig S, et al. Structural base of the interaction of a monoclonal antibody against p24 of HIV with its peptide epitope. *Mol Immunol* 1993;30:1213–1221.
31. van der Merwe PA, Barclay AN, Mason DW, et al. Human cell-adhesion molecule CD2 binds CD58 (LFA-3) with a very low affinity and an extremely fast dissociation rate but does not bind CD48 or CD59. *Biochemistry* 1994;33:10149–10160.
32. Friguet B, Chaffotte AF, Djavadi-Ohaniance L, et al. Measurement of the true affinity constant in solution of antigen–antibody complexes by enzyme-linked immunosorbent assay. *J Immunol Meth* 1985;77:305–319.
33. Nieba L, Krebber A, Pluckthun A. Competition BIAcore for measuring true affinities: large differences from values determined from binding kinetics. *Anal Biochem* 1996;234:155–165.
34. Glaser RW. Antigen–antibody binding and mass transport by convection and diffusion to a surface. *Anal Biochem* 1993;213:152–161.
35. Schuck P. Kinetics of ligand binding to receptor immobilized in a polymer matrix, as detected with an evanescent wave biosensor: I. A computer simulation of the influence of mass transport. *Biophys J* 1996;70:1230–1249.
36. Karlsson R, Roos H, Fägerstam L, et al. Kinetic and concentration analysis using BIA technology. *Methods* 1994;6:99–110.

12

Luminescent Biosensors

L. J. Blum and P. R. Coulet

12.1. INTRODUCTION

Luminescence is a widespread phenomenon that occurs when molecules emit light while returning to the ground state after having been excited. Several types of luminescence can be distinguished depending on the source of the energy that leads to this excited state. Of special interest is chemiluminescence, wherein light emission depends on chemical reactions. Bioluminescence can be considered as a special form of chemiluminescence associated with living organisms and reactions catalyzed by enzymes. The presence of enzymes confers molecular recognition properties on the system that are unique in terms of specificity, especially when chiral molecules have to be considered. In the design of biosensors, where a sensing layer involving immobilized enzymes able to selectively recognize an analyte in a complex medium is associated with a transducer converting this recognition event into an electrical signal, luminescent reactions appear to be extremely attractive.

Similar mechanisms of light emission occur in numerous living organisms including bacteria, planktonic species, fishes, insects, and fungi, the best known among them probably being those of glow worms or fireflies, which depend on ATP.

Our group has used luminescence reactions catalyzed by peroxidase, firefly luciferase, or the oxidoreductase–luciferase system from marine bacteria for the preparation of bioactive layers in the design of luminescent biosensors. The principle of such sensors is illustrated in Fig. 12.1. The specific recognition of an analyte by the sensing layer triggers a light emitting reaction (LER). The light is then transmitted through a waveguide (generally fiber-optic) to the light detector, which is the photomultiplier tube (PMT) of a luminometer.

L. J. Blum and P. R. Coulet • Laboratoire de Génie Enzymatique, UPRESA CNRS 5013, Université Claude Bernard Lyon 1, F-69622 Villeurbanne Cedex, France.

Biosensors and Their Applications, edited by Yang and Ngo, Kluwer Academic/Plenum Publishers, New York, 1999.

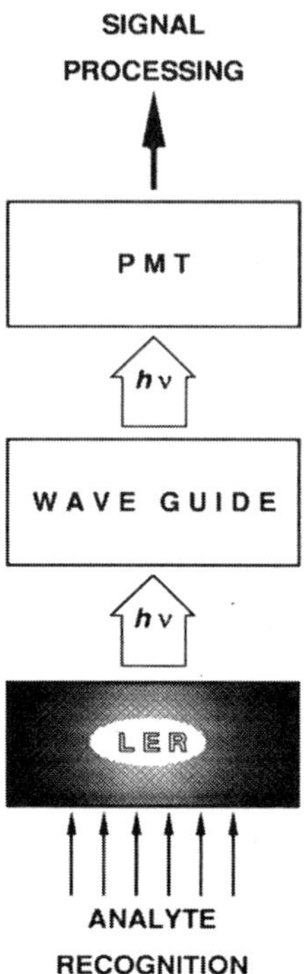

Figure 12.1. Principle of a fiber-optic biosensor based on a light-emitting reaction.

12.2. ENZYME REACTIONS

A first level of detection is achieved by the selected luminescent enzymic system targeted toward H_2O_2, ATP, or NAD(P)H. Depending on the type of LER involved, maximum emission occurs at a characteristic wavelength. The detection of other analytes is obviously desirable and can be achieved through more complex systems involving auxiliary enzymes working sequentially with the luminescent system. This idea is what led to the design of the multienzyme-based biosensors presented in this chapter.

12.2.1. Basic Reactions

12.2.1.1. Detection of H_2O_2

The most commonly used chemiluminescent compounds are luminol (5-amino-2,3-dihydro-1,4 phthalazinedione) and related hydrazides. Luminol-mediated chemiluminescence in the presence of horseradish peroxidase (HRP, EC 1.11.1.7), enables the detection of hydrogen peroxide at a very low level according to the following reaction.

$$2H_2O_2 + \text{Luminol} + OH^- \xrightarrow{\text{HRP}} \text{3-aminophthalate} + N_2 + 3H_2O + h\nu$$

$$(\lambda_{\max} = 430\,\text{nm}) \qquad (1)$$

12.2.1.2. Detection of ATP and NAD(P)H

Light emission may depend on intracellular ATP or NAD(P)H concentration and thus can serve for the detection of these analytes, which can be performed *in vitro* with a very high selectivity and sensitivity. For ATP determination, luciferase (EC 1.13.12.7) from the firefly *Photinus pyralis* provides a light emission in the presence of luciferin, Mg^{2+}, and molecular oxygen:

$$\text{ATP} + \text{Luciferin} + O_2 \xrightarrow[\text{Mg}^{2+}]{\text{Luciferase}} \text{AMP} + \text{Oxyluciferin} + \text{PPi} + CO_2 + hv$$

$$(\lambda_{max} = 560 \, \text{nm}) \qquad (2)$$

For NADH or NADPH determination, one can use the luminescent marine bacteria systems from *Vibrio fischeri* or *Vibrio harveyi*, which consist of two consecutive enzymatic reactions. $FMNH_2$, a reaction product of the first enzyme, NAD(P)H:FMN oxidoreductase (EC 1.6.8.1) is a substrate for the second enzyme, luciferase (EC 1.14.14.3), generating the light emission when molecular oxygen and a long-chain aldehyde (R–CHO), generally decanal, are present as cosubstrates. Thus, bacterial luciferase associated with NAD(P)H:FMN oxidoreductase enables NAD(P)H measurements in the presence of FMN and a long-chain aldehyde (R–CHO) according to the following reactions:

$$\text{NAD(P)H} + H^+ + \text{FMN} \xrightarrow{\text{Oxidoreductase}} \text{NAD(P)}^+ + FMNH_2 \qquad (3)$$

$$FMNH_2 + \text{R–CHO} + O_2 \xrightarrow{\text{Luciferase}} \text{FMN} + \text{R–COOH} + H_2O + hv$$

$$(\lambda_{max} = 490 \, \text{nm}) \qquad (4)$$

12.2.2. Extension Through Oxidoreductases as Auxiliary Enzymes

12.2.2.1. Oxidases

Several oxidases lead to the production of H_2O_2. They can be associated with the luminol–peroxidase reaction for H_2O_2 detection, thus enabling measurement of the oxidase substrate:

$$A + O_2 + H_2O \xrightarrow{\text{Oxidase}} B + H_2O_2 \qquad (5)$$

12.2.2.2. Dehydrogenases

$$AH_2 + \text{NAD(P)}^+ \xrightarrow{\text{Dehydrogenase}} A + \text{NAD(P)H} + H^+ \qquad (6)$$

A dehydrogenase specific to a target substrate AH_2 or A, depending on the sense the reaction achievable, provides or consumes NAD(P)H in a stoichiometric amount in which variations can be monitored through the bacterial luminescent system thus correlated to the dehydrogenase substrate concentration.

12.3.   DESIGN OF THE SENSING LAYER

In a biosensor, the design of the sensing layer is of prime importance for the performance of the system. Numerous methods of enzyme immobilization have been described in the literature and many types of supports are available. Polymeric membranes are suitable for the design of sensing layers but the primary problems to be faced are still the enzyme activity retained on the membrane and the ratio of activities achievable when a multienzyme system has to be considered taking account of the overall operational and storage stability. An additional difficulty is the use of entrapped cosubstrates in the design of reagentless biosensors.

12.3.1.   Basic Procedures for Enzyme Immobilization

Two methods were used to prepare bioactive membranes with immobilized enzymes suitable for biosensors, one based on chemically activated collagen films and the other on polyamide membranes commercially available in a preactivated form. Both have specific advantages and drawbacks to be evaluated prior to selection for a specific purpose.

12.3.1.1.   Immobilization on Collagen Membranes

The original method for the development of electrochemical biosensors in our laboratory, which is now used for optical biosensors, consists of an acyl-azide activation of lateral carboxyl groups of aspartate and glutamate residues in reconstituted collagen films.[1] Briefly, it involves three consecutive steps: after a 3-day mild esterification in a 0.2 M HCl–MeOH mixture and a hydrazine treatment at room temperature, the action of nitrous acid at 4°C leads to formation of the acyl-azide ($-CON_3$). After a rapid washing the enzyme coupling is carried out immediately by immersing the activated film in the enzyme solution for 2 h. The coupling occurs mainly through the available amino groups on lateral chains of the lysine residues of the enzyme. The bioactive membranes thus obtained can be used immediately or stored either at 4°C in buffer or dried at $-20°C$. This method was successfully applied to the immobilization of both the luminescent bienzyme system of bacteria[2] and to firefly luciferase.[3] This coupling procedure, originally developed with a local production of collagen films, is now used with films that are supplied by ICN (normally devoted to serving as a support for cell cultures) and are available worldwide.

12.3.1.2.   Immobilization on Preactivated Polyamide Membranes

The second method is based on the use of polyamide membranes supplied in a preactivated form by Pall Co, which enable an enzymic membrane to be prepared in a few minutes.[4] For coupling, disks 10 mm in diameter matching the biosensor tip are cut out of a preactivated sheet (15 × 15 cm) stored in a sealed bag under vacuum.

Routinely, 10 μL of the selected enzymatic solution is applied on each side. Covalent binding occurs spontaneously at room temperature within 1 min. All the coupling solutions were prepared in 0.1 M phosphate buffer, pH 7.

12.3.2. Coimmobilization of Multienzyme Systems on the Same Membrane

When several enzymes are involved in the detection sequence, they can be mixed in the coupling solution and then randomly bound on the same membrane. Attention must be given to the fact rarely mentioned that the ratio of activities in solution is not maintained after immobilization. Nevertheless, active membranes have been easily prepared with the luminescent enzymes associated with auxiliary ones. It must be noted that a separate preparation of oxidoreductase from either *V. harveyi* or *V. fischeri* can be coimmobilized with the luciferase of one or the other microorganism.

12.3.3. Compartmentalization

In order to enhance the response of multienzyme-based biosensors, we have developed a new approach to make use of coupled reactions with immobilized enzymes, which consists of a compartmentalized system including different membranes on which only one enzyme was immobilized.[5] In a first step, a bienzymatic system was studied and the coupled reaction lactate oxidase–peroxidase was chosen as the model system. The oxidation of lactate catalyzed by lactate oxidase (LOD) leads to the production of H_2O_2, which in turn serves as a cosubstrate for the chemiluminescence reaction of luminol catalyzed by peroxidase. The compartmentalized system was obtained by stacking membranes bearing peroxidase and LOD, respectively, in such a way that the former was in close contact with the fiber-optic bundle and the latter interfaced with the reaction medium.

12.3.4. Cosubstrate Confinement

Most of enzyme sensors require the presence of cosubstrates that have to be added to the reaction medium. This can be solved using a flow injection analysis design, but there is now increasing demand for self-contained reagentless biosensors. The approach we have developed involves confining the coreactants in a porous matrix that is inserted between the surface of the fiber-optic bundle and the bioactive membrane, as illustrated in Fig. 12.2.[6–8] Light is emitted when the confined coreactants are released at a sufficient rate to react with the immobilized enzymes. However, the matrix must offer optical properties compatible with the light transmission, which is the case with poly(vinyl)alcohol (PVA), giving a translucent matrix after polymerization. When the coreactants differ in their affinity for water (e.g., when considering the bacterial luminescent system, FMN is hydrophilic whereas decanal is hydrophobic), their simultaneous release rate must be sufficient to ensure that the respective concentrations in the microenvironment are optimum for enzymatic

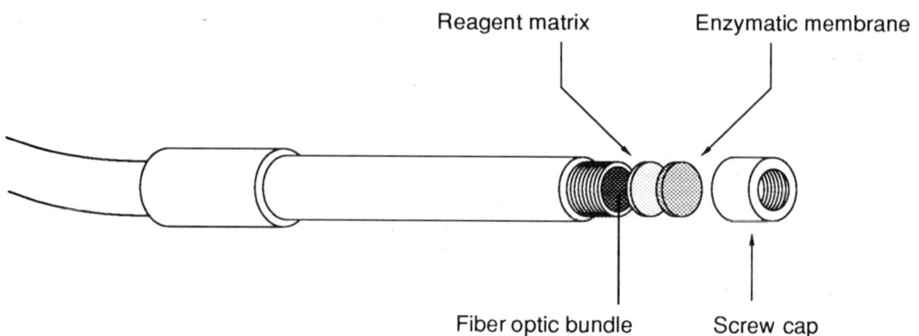

Figure 12.2. Sensing tip of the reagentless fiber-optic sensor, including a polymeric matrix with entrapped reagents for their continuous release in the vicinity of the enzyme membrane.

catalysis. Also, the release of the coreactants must enable the use of the sensor over a long period as well as prevent enzymatic inhibition by an excess of coreactants and/or an accumulation of reaction products.

12.5. SENSOR DESIGN

The sensor, originally designed for batch analysis, consisted of an immobilized enzyme membrane secured by a screw cap in close contact with one end of a 1-m-long optical glass fiber bundle 8 mm in diameter connected to a PMT associated with a signal-processing system.[9] Continuous-flow bioluminescent methods were also developed using a flow cell adapted to the sensing tip of the fiber optic sensor.[10,11] Both batch and flow injection analysis (FIA) systems are shown in Fig. 12.3.

For batch analysis, the biosensor was immersed in 4.5 ml of a stirred and thermostated (25°C) reaction medium. To avoid interference from ambient light, the reaction vessel was surrounded by a black lighttight jacket, and a lighttight septum allowed the sample to be injected into the reaction medium with a syringe.

For FIA, a specially designed flow cell made of black polyvinyl chloride and including a minute reaction chamber of 125 μL stirred with a small magnetic bar was used. The reagent solution was delivered by a peristaltic pump and samples were injected into the flowing stream with a low-pressure injection valve. When the detection limit and the dynamic range were considered, similar performances were obtained in both cases; however, the combination with FIA techniques facilitates automation and leads to better reproducibility.

12.6. APPLICATIONS

1. H_2O_2: H_2O_2 measurements could be achieved with peroxidase in the sensing membrane in the range 2×10^{-8} to 2×10^{-5} M and the maximum light intensity was obtained within 1 min.

2. ATP: With firefly luciferase immobilized in the sensing tip, ATP measurements could be performed over a wide linear dynamic range from 2×10^{-11} to 1×10^{-6} M.

3. NADH: With the bacterial oxidoreductase–luciferase system, the detection limit of the biosensor was as low as 3×10^{-10} M, and the linear dynamic range fell between 1×10^{-9} and 3×10^{-6} M.

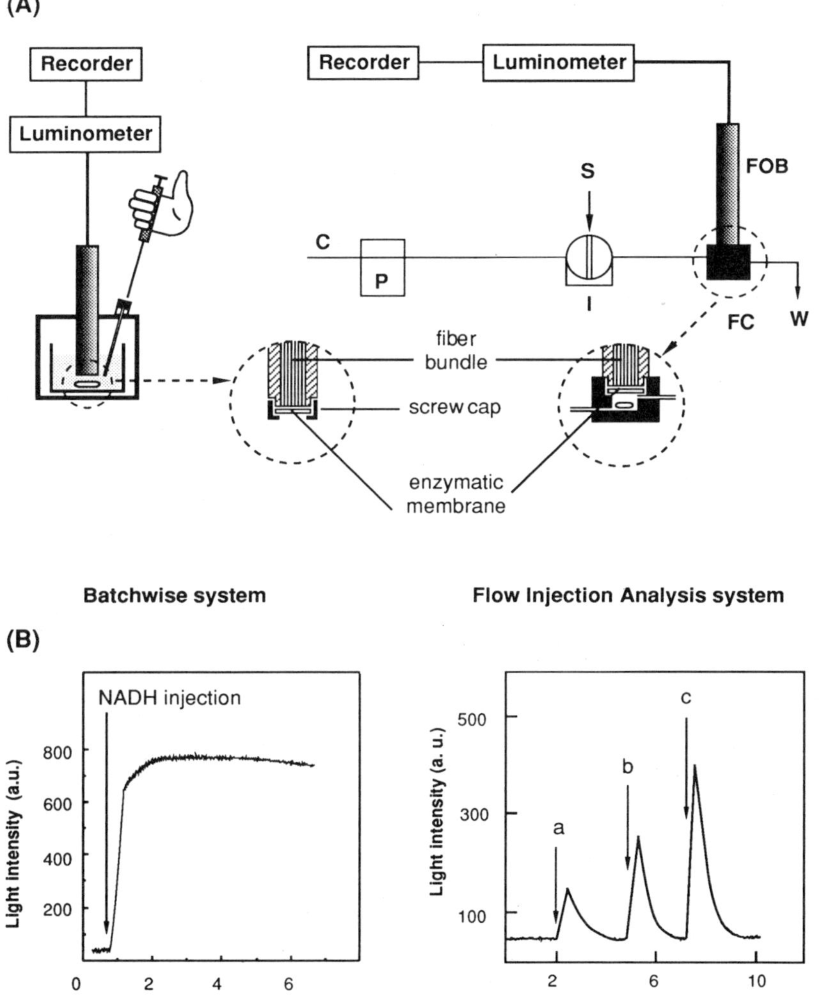

Figure 12.3. (A) Schematic diagram of the batchwise and flow injection analysis systems: C, continuous flowing stream or carrier stream; P, peristaltic pump; I, injection valve; S, sample; FC, flow-cell; FOB, fiber-optic biosensor; W, waste. (B) Traces obtained for NADH measurements; for batchwise analysis, the final NADH concentration was 1×10^{-7} M; with the flow system, samples containing different amounts of NADH were injected: (a) 1×10^{-10} mole; (b) 2×10^{-10} mole; (c) 5×10^{-10} mole (a.u.: arbitrary unit).

12.6.1. Determination of Other Analytes with Auxiliary Enzymes

12.6.1.1. Sorbitol, Ethanol, and Oxaloacetate

The determination of various target analytes was achieved with suitable auxiliary enzymes such as dehydrogenases associated with the luminescent system in the sensing layer. With sorbitol dehydrogenase coimmobilized with the bacterial bienzymatic system, D-sorbitol could be measured from 2×10^{-8} to 2×10^{-5} M, with coimmobilized alcohol dehydrogenase, ethanol from 4×10^{-7} to 7×10^{-5} M, and with coimmobilized malate dehydrogenase, oxaloacetate from 1×10^{-9} to 3×10^{-6} M.[12]

A sorbitol fiber-optic biosensor was also designed by making use of a compartmentalized sensing layer.[13] Sorbitol dehydrogenase was immobilized on one membrane stacked with another bearing the bacterial oxidoreductase–luciferase system. The response time of the sensor involving the compartmentalized layer was 4–6 min and the relative standard deviation, calculated from ten cumulative injections of 2×10^{-7} M D-sorbitol was 5.2%. D-Sorbitol in foodstuffs and pharmaceutical samples (chocolate, sweets, and toothpaste) have been assayed with the sensor and the results obtained compared well with a colorimetric method of enzymatic analysis. Moreover, the sorbitol sensor offered an advantage in that it provided the results faster than the colorimetric method since the result was obtained about 6 min after the sample injection instead of after more than 35 min with the colorimetric procedure, which required an incubation step prior to the measurement. With the chemiluminescence system, specific oxidases can also be combined with peroxidase to extend the potential of the sensor.

12.6.1.2. Glucose in Drinks

The feasibility of chemiluminescent flow analysis of glucose with the fiber-optic sensor involving coimmobilized glucose oxidase and peroxidase was demonstrated by our group.[14] H_2O_2 generated by the glucose oxidase reaction in the presence of glucose was detected using the chemiluminescence reaction of luminol catalyzed by peroxidase. Using the FIA system, the detection limit was 0.25 nmole with a coefficient of variation less than 4% for repeated measurements of 2.5 nmole glucose. The method was applied to glucose analysis in soft drinks on diluted and degassed samples. The results were in good agreement with those obtained with a standard spectrophotometric method, i.e., the end-point method using a reagent kit based on the hexokinase/glucose-6-phosphate dehydrogenase reaction with detection at 340 nm of NADH.

12.6.1.3. Lactate in Foodstuffs

As noted above, a chemiluminescence-based lactate sensor including a compartmentalized sensing layer was developed.[5] The lactate oxidase membrane interfaced with the reaction medium and the peroxidase membrane was in close contact with the fiber-optic bundle. FIA of lactate standards and samples was performed with the fiber-optic sensor operated with the compartmentalized enzymatic system. The coefficient of variation for 10 replicates of 6.25×10^{-9} mole lactate standards was 1.7%. Samples from the dairy industry (acid-type and sweet-type whey solutions) were assayed by the present method and the results compared well with measurements done with a lactate oxidase amperometric electrode developed in our laboratory.

A bioluminescence-based fiber-optic sensor for L-lactate was also designed making use of the same approach as that employed for the sorbitol sensor, i.e., with a compartmentalized sensing layer.[15] For this purpose, L-lactate dehydrogenase from rabbit muscle was immobilized alone on a preactivated polyamide membrane while the oxidoreductase from *V. fischeri* and the luciferase from *V. harveyi* were coimmobilized on another preactivated polyamide membrane. The sensing layer was obtained by stacking these two membranes, the one with dehydrogenase interfacing with the reaction medium and the oxidoreductase–luciferase membrane against one end of the glass fiber bundle. Here again, the performances of the biosensor in terms of sensitivity, detection limit, and dynamic linear range were markedly improved by compartmentalization of the sensing layer. Compared with the classical approach in which the three enzymes were coimmobilized on the same membrane, the compartmentalization led to a fivefold increase in the lactate biosensor sensitivity. Under optimum reaction conditions, i.e., 6.3 mM NAD^+ and pH 7.2, the detection limit was 2×10^{-7} M L-lactate and the calibration graph was linear from this value up to 2×10^{-6} M, whereas with the coimmobilized system, the biosensor response was linearly related to lactate concentration only from 1×10^{-6} to 5×10^{-6} M, i.e., half a decade. No loss of activity was detected after 8 h of intensive use, which is a clear indication of the excellent operational stability of the lactate sensor.

12.6.1.4. Enzyme Activity

Most biosensor applications concern the analysis of metabolites and only a few are devoted to the monitoring of enzyme activity, so we have investigated the possibility of measuring enzyme activity with the NADH bioluminescence-based sensor.[16] The study was focused on L-lactate dehydrogenase (EC 1.1.1.27) because of its importance in clinical analysis. The enzymatic oxidation of L-lactate into pyruvate proceeds through the reduction of NAD^+ into NADH. The assay of lactate dehydrogenase was thus achieved by monitoring the appearance of NADH with the bacterial bioluminescence-based fiber-optic sensor.

The flow system was slightly different from that used for NADH measurements and consisted, in this case, of one carrier stream and two reagent streams. The sample was injected into the carrier stream; merged with the lactate dehydrogenase reagent stream, which contained 1 mM NAD^+ and 210 mM lactate; passed through a mixing coil 90 cm long; and finally merged with the bioluminescence reagent stream, which contained 10 mM DTT, 0.15 mM FMN, and 45 μM decanal. Under these conditions, a linear relationship was obtained between the peak light intensity and the lactate dehydrogenase activity of serum samples in the range from 5 to at least 250 U/L.

12.6.2. Stability

As already emphasized for amperometric enzyme electrodes,[17] the stability of the sensing layer is one of the most important factors determining the reliability of biosensors and we focused on this parameter. Practical applications of biosensors are highly dependent on the actual stability of the sensing element, i.e., of the enzymatic system used. Intrinsic properties of the enzymes, varying with their origin, and storage conditions are important factors for preserving high analytical capacities of the sensing tip, and we focused on these two key points. For instance, when storing the bacterial bienzymatic sensing membrane at $-20°C$ in the presence of 20% glycerol, the exhibited activity of the bioactive membrane prepared with enzymes from *V. fischeri* was higher than that measured prior to freezing and then remained stable for more than 4 months.[18]

12.7. CONCLUSIONS AND TRENDS

Combining optical transduction with the sensitivity of bioluminescence and the convenience of easy-to-handle immobilized enzymes allowed us to design sensitive and selective enzyme sensors. As for every type of biosensor, stability of the biological sensing layer remains a key parameter to be taken into account. The use of more stable recombinant enzymes, now under investigation in our group, appears very promising and is expected to provide reliable luminescent biosensors with highly improved performances in the near future.

REFERENCES

1. Coulet PR, Julliard JH, Gautheron DC. A mild method of general use for covalent coupling of enzymes to chemicaly activated collagen films. *Biotechnol Bioeng* 1974;16:1055–1068.
2. Blum LJ, Coulet PR. Bioluminescent determination of nicotinamide adenine dinucleotide with immobilized bacterial luciferase and flavin mononucleotide oxidoreductase on collagen films. *Anal Chim Acta* 1984;161:355–358.
3. Blum LJ, Coulet PR, Gautheron DC. Collagen strip with immobilized luciferase for ATP bioluminescent determination. *Biotechnol Bioeng* 1985;27:232–237.
4. Assolant-Vinet CH, Coulet PR. New immobilized enzyme membranes for tailor-made biosensors. *Anal Lett* 1986;19:875–885.
5. Berger A., Blum LJ. Enhancement of the response of a lactate oxidase/peroxidase-based fiberoptic sensor by compartmentalization of the enzyme layer. *Enzyme Microb Technol* 1994;16:979–984.
6. Gautier SM, Blum LJ, Coulet PR. Bioluminescence-based fiber optic sensor with entrapped co-reactant: an approach for designing a self-contained biosensor. *Anal Chim Acta* 1991;243:149–156.
7. Gautier SM, Blum LJ, Coulet PR. Cofactor-containing bioluminescent fiber optic sensor: new developments with poly(vinyl)alcohol matrices. *Anal Chim Acta* 1991;255–253–258.
8. Gautier SM, Michel PE, Blum LJ. Reagentless bioluminescent sensor for NADH. *Anal Lett* 1994;27:2055–2069.
9. Blum LJ, Gautier SM, Coulet PR. Luminescence fiber optic biosensor. *Anal Lett* 1988;21:717–726.
10. Blum LJ, Gautier SM, Coulet PR. Continuous-flow bioluminescent assay of NADH using a fibre-optic sensor. *Anal Chim Acta* 1989;226:331–336.

11. Blum LJ, Gautier SM, Coulet PR. Design of bioluminescence-based fiber optic sensors for flow-injection analysis. *J Biotechnol* 1993;31:357–368.
12. Gautier SM, Blum LJ, Coulet PR. Fibre-optic biosensor based on luminescence and immobilized enzymes: microdetermination of sorbitol, ethanol and oxaloacetate. *J Biolumin Chemilumin* 1990;5:57–63.
13. Michel PE, Gautier SM, Blum LJ. A high-performance bioluminescent trienzymatic sensor for D-sorbitol based on a novel approach of the sensing layer design. *Enzyme Microb Technol* 1997;21:108–116
14. Blum LJ. Chemiluminescent flow injection analysis of glucose in drinks with a bienzyme fiberoptic biosensor. *Enzyme Microb Technol* 1993;15:407–411.
15. Michel PE, Gautier SM, Blum LJ. Effect of compartmentalization of the sensing layer on the sensitivity of a multienzyme-based bioluminescent sensor for L-lactate. *Anal Lett* 1996;29:1139–115.
16. Gautier SM, Blum LJ. Coulet PR. Dehydrogenase activity monitoring by flow injection analysis combined with luminescence based fibre-optic sensors. *Anal Chim Acta* 1992;266:331–338.
17. Coulet PR. Enzyme electrodes: from the self-contained probe to the design of an automatic analyzer. In Schmid RD, Guilbault GG, Karube I., et al. , eds. Biosensors International Workshop 1987, *GBF Monographs*. Weinheim: VCH, 1987, vol. 10, pp 75–80.
18. Blum LJ, Gautier SM, Coulet PR. Highly stable bioluminescence-based fiber optic sensor using immobilized enzymes from *Vibrio harveyi. Anal Lett* 1989;22:2211–2222.

13

Micromachining for Biosensors and Biosensing Systems

S. Shoji

13.1. INTRODUCTION

Miniaturization of chemical sensors and biosensors has been developed by three-dimensional microfabrication technologies based on photolithography commonly used in integrated circuits — called micromachining. Biosensors based on ion sensitive field effect transistor (ISFET) are examples of such devices.[1] Microflow control devices of microvalves, micropumps, and microflow sensors have also been fabricated by micromachining.[2] Such microdevices open new possibilities for the miniaturization of conventional chemical and biochemical analysis systems. Micro total analysis systems (μTAS) including microfabricated detectors (silicon-based electrochemical or optical sensors), microflow control devices, and even control/detection circuits comprise one of the recent main areas of interest in the microelectromechanical systems (MEMS) field. Miniaturized flow injection analysis (FIA) and liquid chromatography systems have been demonstrated using microfabricated elements.[3,4] An analytical system based on enzyme-linked immunosorbent assay (ELISA) using ISFET was also demonstrated.[5] Micromachining is also used to fabricate DNA analysis systems using a polymerase chain reaction (PCR) process.[6,7]

The mechanical elements of μTAS should have three-dimensional structures, and the sensors and detecting circuits must be integrated. The materials used for the elements should be chosen with a view toward biocompatibility, and micromachining techniques must be diversified to develop sophisticated μTAS for biomedical applications. Basic micromachining techniques of etching and bonding and recent progressive techniques in MEMS that will be useful for microbiosensors and μTAS fabrication are reviewed in this chapter. Silicon etching techniques to fabricate

S. Shoji • Department of Electronics, Information and Communication Engineering, Waseda University, Shinjuku-ku, Tokyo 169-8555, Japan.

Biosensors and Their Applications, edited by Yang and Ngo, Kluwer Academic/Plenum Publishers, New York, 1999.

three-dimensional bulk microstructures are described, and surface micromachining and the lost wafer process, which are applicable for fabricating free-standing microstructures, are introduced. These techniques are suitable for the fabrication of circuit-integrated microsensors and microactuators. The LIGA and HEXSIL processes, which enable very high aspect ratio microstructures are introduced. Mass production of microstructures will be obtained by a molding or a moldinglike process. Various microchannel microfabrication methods that are often required in the μTAS are reviewed. Combination processes of etching and thin-film deposition, etching and bonding are used for this purpose. A selective etching of a sacrificial layer, i.e., the application of surface micromachining is also applied. Finally the wafer-bonding method, useful for packaging of microbiosensors and assembling of the μTAS, are also described.

13.2. ETCHING

Four silicon etching methods have been used to fabricate three-dimensional silicon microstructures — wet and dry isotropic etching and wet and dry anisotropic etching. The typical etching profiles of these methods are shown in Fig. 13.1. Nitric acid, fluoric acid, and acetic acid systems are used for the wet isotropic etching.[8] In

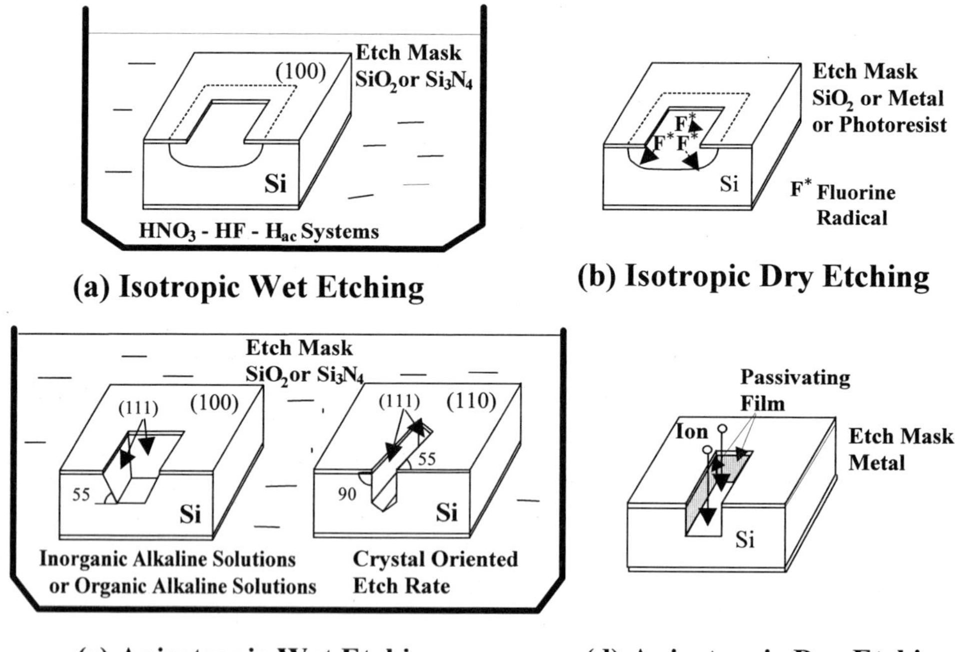

Figure 13.1.   Silicon etching methods.

isotropic dry etching, an F* radical generated in RF-excited CF_4 plasma is used as the etching reactor. Vaporized XeF_2 in a vacuum chamber is another useful etchant for the new dry isotropic etching without the RF-excited plasma.[9] Since isotropic etching shows large undercut and poor controllability of etching properties, wet and dry anisotropic etching techniques have been widely used when precise control of a three-dimensional structure is necessary.

13.2.1. Wet Etching

Crystal-plane-dependent anisotropic wet etching is widely used in micromachining to fabricate three-dimensional bulk Si structures.[10] Some kinds of alkaline hydroxides of KOH, NaOH, CsOH–water, NH_4OH–water, organic aqueous solutions of ethylenediamine–pyrocathechol–water (EPW), hydrazine–water, and quaternary ammonium hydroxide–water (QAHW) have been used as etchants. The important properties of this etching are etch rate, etch ratio between the crystal planes, smoothness of the etched plane, etch rate of SiO_2 as the etch mask, toxicity, and process compatibility to integrated circuits. Characteristics of Si anisotropic etching are listed in Table 13.1.

The most popular alkaline hydroxide for this purpose is KOH, and the etching procedures have been studied in detail. Uniform and constant etching and the smooth etching surface can be realized with highly concentrated KOH of above 20 wt.% at above 60°C. Alkaline hydroxide–water has large etch rate ratios of the (100) plane/(111) plane and the (110) plane/(111) plane.

Organic aqueous solutions have small etch rate ratios of the (100) plane/(111) plane and the (110) plane/(111) plane and a low etch rate of SiO_2 compared to the alkaline hydroxide–water. EPW should be used with an inert gas to purge the etching chamber in order to avoid oxidation of the etchant. Adding pyrazine to the solution makes the etch rate constant. Since this etchant has a very low etch rate for boron-doped silicon, it is widely used in the selective etching process as described below. Hydrazine–water, normally a 1:1 solution, has similar etching properties to an EPW solution. However, the etch rate of the (100) plane is two to three times higher than that of EPW. The pyramidal-shaped etch pits are sometimes observed in this etching. Ag, Au, Ti, and Ta withstand this etching solution whereas Al, Cu, and Zn are attacked. EPW and hydrazine are carcinogens. TMAH–water shows very good process compatibility with integrated circuits owing to the fact that there is no contamination. The Al etching rate is reduced when the solution is doped with Si or silicic acid [$Si(OH)_4$], which is is very important for circuit-integrated microsensor fabrication. Addition of ammonium persulfate improves the etch rate and the smoothness of the etched surface.[11] Etch stop methods by doping-dependent and bias-controlled selective etching are carried out in wet anisotropic etching as shown in Fig. 13.2. All etching solutions exhibit a strong reduction of their etch rate at high boron concentrations as is shown in Table 13.1. There are some methods of p–n junction etch stop using 2, 3, 4 electrodes. The reverse-biased p–n junction can provide a large selective etching of p-type over n-type Si.

Table 13.1. Features of Anisotropic Wet Si Etching[10]

Etching solution	Temperature (°C)	Etch rate of (100) plane	Anisotropy (100)/(111)	Anisotropy (110)/(111)	Selectivity Si(100)/SiO$_2$	Selectivity Si(100)/p^+Si	Remarks
KOH (41 wt.%)	80	1.0	400	600	700	-20 ($>10^{20}$ cm^{-3})	p–n Etch stop (low concentration)
KOH + Iso propanole (26% + 4%)	80	1.1					
CsOH (50 wt%)	70	0.3	50		2000		
NH$_4$OH (4 wt%)	50	0.7		200	875		Pyramidal pits
EPW + pyrazine	115	1.25	35		6250	-50 ($>7 \times 10^{19}$ cm^{-3})	p^+ Etch stop carcinogens
Hydrazine (50%)	100	3.0	16	9	17650	-100 ($>2.2 \times 10^{21}$ cm^{-3})	p^+ Etch stop carcinogens
TMHA (22 wt%)	90	1.0	33		4200	-10 ($>10^{20}$ cm^{-3})	IC compatibility
TMAH + ammonium persulfate	85	1.3					p–n Etch stop

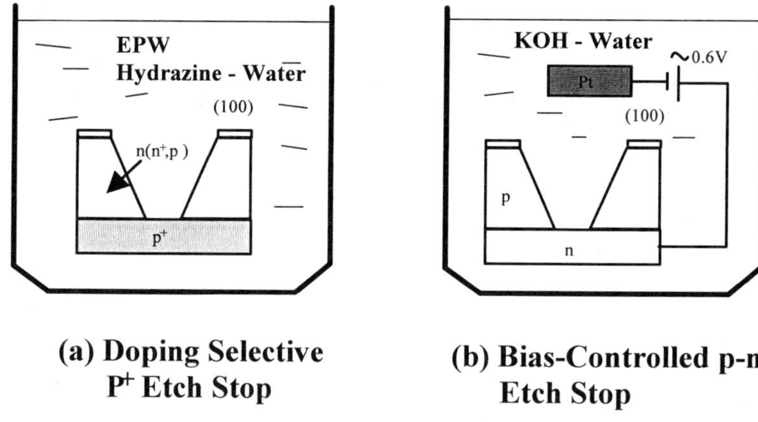

**(a) Doping Selective
P⁺ Etch Stop**

**(b) Bias-Controlled p-n
Etch Stop**

Figure 13.2.   Si etch stop method (selective etching).

13.2.2. Dry Etching

Anisotropic dry etching is obtained by reactive ion etching (RIE). Many types of plasma sources have recently become available, including magnetron and electron cyclotron resonance (ECR). A low dry etch rate precluded use of deep Si etching. However, high etch rate, high aspect ratio and high silicon/mask selectivity are obtained using recently improved RIE equipment. The RIE system employs inductively coupled plasma (ICP), which is commercially available and can achieve high etch rates of several micrometers per minute, geometrical aspect ratio of at least 15:1, and etch selectivity higher than 200:1 to thermal oxide and 50:1 to photoresist.[12] This system enables fabrication three-dimensional bulk microstructures having high aspect ratios which is difficult with anisotropic wet etching.

13.3.   FREE-STANDING MICROSTRUCTURE FABRICATION

13.3.1. Surface Micromachining

Surface micromachining is widely used for fabricating microsensors and micro-actuators. In many cases poly-Si formed by chemical vapor deposition (CVD) is employed as the structure, and SiO_2 or PSG (phosphor-silicate glass) as the sacrificial layer.[13] The typical process is shown in Fig. 13.3a. The sacrificial layer is deposited and patterned first. The poly-Si layer is formed by CVD and patterned as the structure. A multilevel poly-Si structure can be fabricated by repeating deposition and patterning of the sacrificial and the poly-Si layers. In the last step, the sacrificial oxide is removed by HF etching. Residual stress control of poly-Si and careful drying of the rinse liquid after sacrificial etching are required to fabricate free-standing microstructures reproducibly.[14] This process is often used for capacitor-type microsensors, electrostatic comb drive microactuators, and microresonators. The compatibility of this process with an integrated circuit such as a complementary metal oxide semiconductor integrated circuit (CMOS IC) or a bipolar CMOS

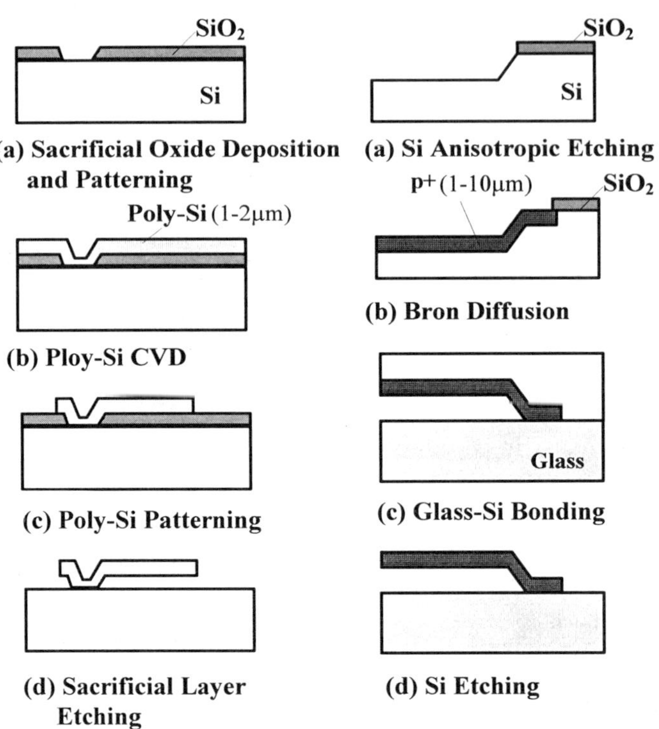

Figure 13.3.   Fabrication methods for free-standing microstructures.[13,16]

integrated circuit (BiCMOS IC) enables fabrication of integrated sensor–circuit systems. The thickness of the poly-Si is typically limited to a few micrometers because of the stress problem. Dielectrics of CVD SiO_2, Si_3N_4, and Al can be used both as structural materials and as a sacrificial layer.[15]

13.3.2.   Lost Wafer Process

Combining a p^+ etch stop and wafer bonding is another process for fabricating free-standing microstructure.[16] Figure 13.3b shows the sequence of the so-called "lost wafer process." Anisotropic etching is used to define a gap between the free-standing structure and the substrate. Boron is then diffused through a thick SiO_2 mask. A two-step diffusion process is often used to create thick and thin p^+ microstructures. Then the Si wafer is anodically bonded to the glass substrate. Finally, silicon is anisotropically etched everywhere except on the p^+ layer. Free-standing microstructures made of bulk Si can be fabricated on the glass substrate. This structure is useful for capacitor-type microsensors because of the reduction of the parasitic capacitance and electrostatic microactuators.

13.4. HIGH ASPECT RATIO MICROSTRUCTURE FABRICATION

13.4.1. LIGA

LIGA (German: Lithographie, Galvanoformung, Abformung) was developed to fabricate very high aspect ratio microstructures.[17] It involves three basic processing steps: lithography, electroplating, and molding as shown in Fig. 13.4. A thick photoresist [typically polymethyl methacrylate PMMA)] ranging from a few hundred micrometers to a few millimeters in thickness was exposed by highly collimated x-rays radiated from a synchrotron through an x-ray mask. The eventual thickness of the structure depends on the wavelength, i.e., the energy of the x-rays. The metal is electroplated on the substrate, filling the spaces and the top surface of the

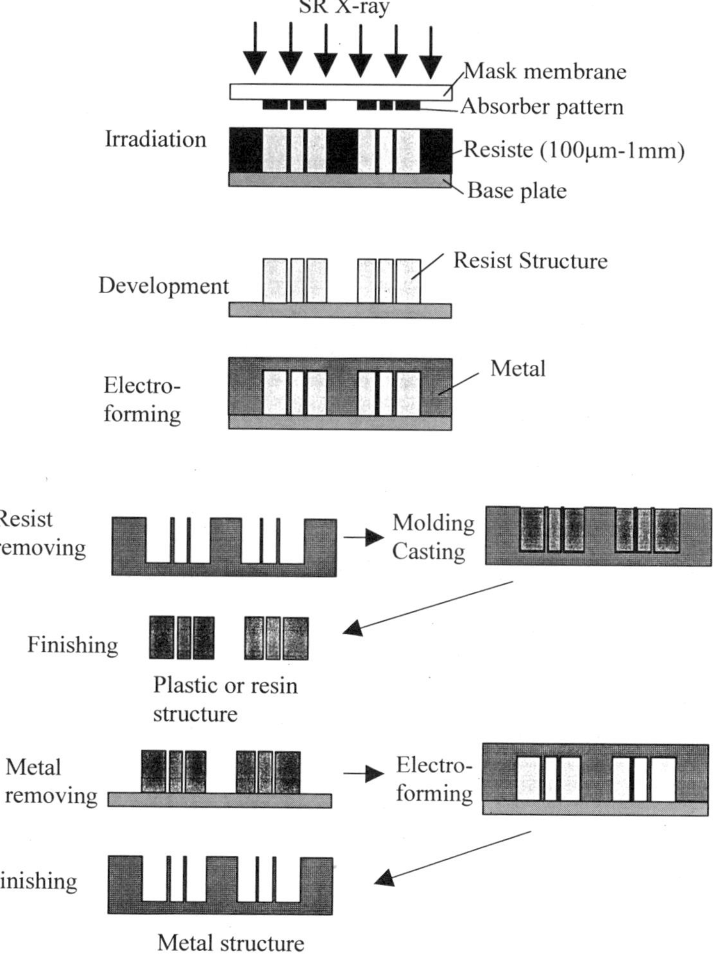

Figure 13.4.   LIGA process.[17]

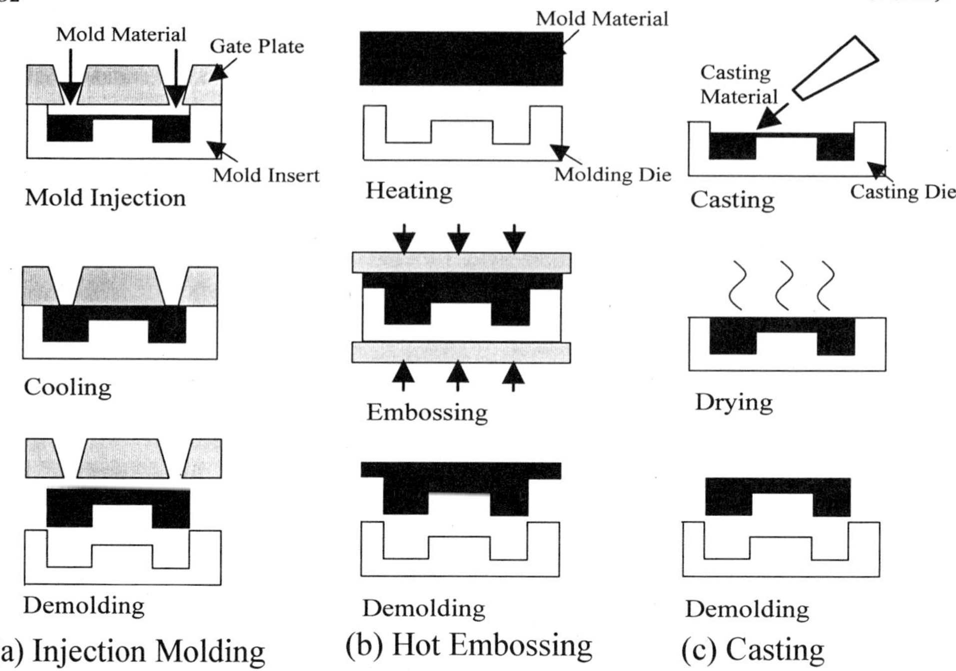

Figure 13.5.   Plastic microstructure fabrication process.[17]

photoresist mask, and the mask is then removed. The metal microstructure can be used repeatedly as a mold insert for injection molding, a molding die for hot embossing, or a die for casting, as shown in Fig. 13.5. Flexibility of the plastic material is one of the advantage of LIGA process.[18] Plastic microfluidic elements that can be used in μTAS are fabricated by plastic-molding.[19] The metal structure can also be reformed by electroplating using a plastic die. An aspect ratio higher than 100 is obtained by this process. By using a metal sacrificial layer, a movable microstructure can also be fabricated. This process is called SLIGA (sacrificial LIGA). However, it needs synchrotron equipment, which makes fabrication expensive.

LIGA-like processing methods using RIE or ECR etching of polymer instead of x-ray lithography have been also developed,[20,21] but the available aspect ratio ranges from 10 to 15. A new type of photoresist mask (IBM SU-8) has been developed to achieve high aspect ratio photolithography with near-UV light (400 nm).[22] An aspect ratio higher than 18 is obtained in the thickness range between 80 and 1200 μm, which will be very useful in actual MEMS fabrications.

13.4.2.   HEXSIL

High aspect ratio microstructures can be fabricated by a combination RIE and poly-Si CVD, so-called HEXSIL process (hexagonal silcon).[23] The basic HEXSIL process is shown in Fig. 13.6a. The first step is to etch the deep narrow trenches into the Si wafer by deep RIE. After deposition of a sacrificial oxide layer, the trenches are filled with CVD poly-Si. The top surface of the substrate is lapped and polished.

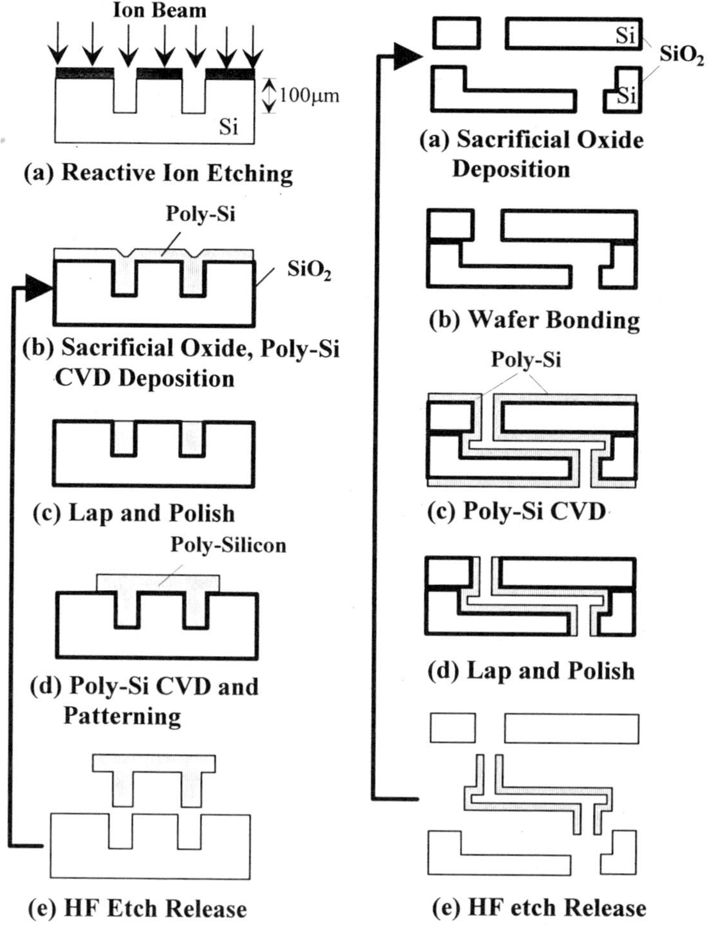

Figure 13.6. HEXSIL process.[23]

Then the second poly-Si is deposited and patterned. The sacrificial layer is removed using HF solution with a surfactant to remove the structure from the substrate. The wafer can be reused and similar microstructures can be fabricated repeatedly, like molding. An aspect ratio of up to 15 can be obtained by the HEXSIL process using ICP DRIE. Figure 13.6b shows another HEXSIL process that uses wafer bonding. Making throughput holes and a cavity on the Si substrate causes poly-Si micro-vessels having inlets and outlets to be fabricated by molding.

13.5. MICROCHANNEL FABRICATION

13.5.1. Application of Bulk Micromachining

Fabrication methods of microchannels using combined micromachining methods have been developed. The combination of a p^+ etch stop and CVD

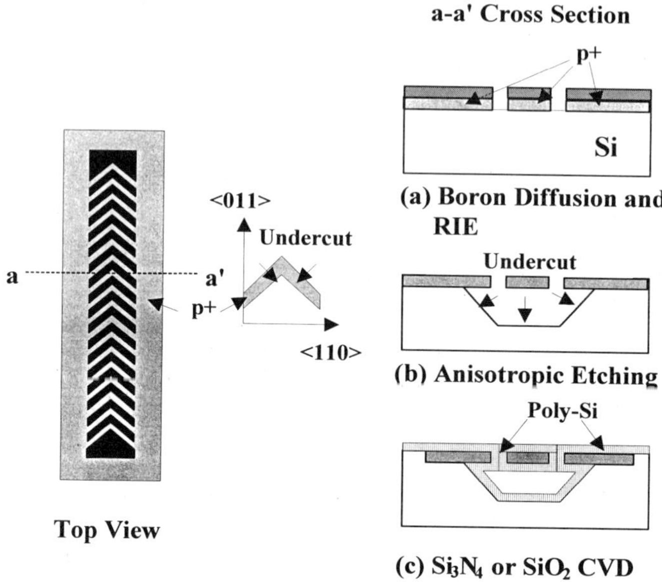

Figure 13.7.   Microchannel fabrication process combined p^+ selective etching and CVD.[24]

deposition is applied to make buried microchannels as shown in Fig. 13.7.[24] A shallow boron diffused layer is formed on the front side of the wafer. Chevron-shaped openings that are perpendicular to the $\langle 100 \rangle$ direction are etched into the p^+ layer by RIE. This layer is undercut by EPW anisotropic wet etching to create a channel aligned with the $\langle 110 \rangle$ plane. The gaps in the chevron-shaped opening are filled with CVD dielectric. The channel surface is covered with dielectric at the same time. This process has been adapted for use in making a nozzle array for an inkjet printer. Microchannels are also formed on a quartz substrate by a similar method using isotropic wet etching with poly-Si overlayer and subsequent CVD of TEOS oxide.[25] Figure 13.8a illustrates the combined process of Si etching, CVD, and bonding. First channels are formed in the Si substrate by any type of Si etching. Si_3N_4 and SiO_2 are then deposited on the Si by CVD.[26] After the Si and a glass are anodically bonded, the Si is etched away to leave a free-standing Si_3N_4 microchannel. Buried microchannels are formed by the combination of Si anisotropic etching and CVD, as shown in Fig. 13.8b.[26] At first, a narrow deep trench is etched by RIE. After the whole surface of the wafer is covered with CVD Si_3N_4, the Si_3N_4 at the bottom of the trench is removed by RIE. Then the Si around the bottom of the trench is etched either anisotropically or isotropically to create the channel structure. The Si_3N_4 mask is etched away and Si_3N_4 (SiO_2 or poly-Si) formed by CVD covers the channel and closes the trench.

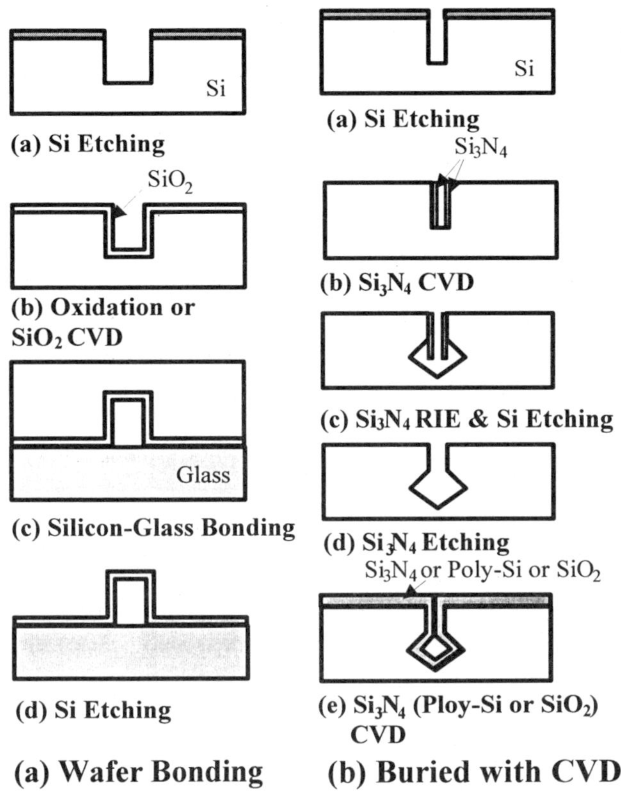

Figure 13.8.   Microchannel fabrication process by bulk micromachining.[26]

13.5.2 *Application of Surface Micromachining*

Some methods for fabricating the microchannel using a positive photoresist mask as the sacrificial layer have been developed.[27] Polymer microchannels can be fabricated on the Si wafer as shown in Fig. 13.9a.[28] After the access holes are formed from the back side of the wafer, poly-*p*-xylyrene (parylene C) is deposited on the Si₃N₄ barrier layer at the top of the wafer by CVD. Then the thick sacrificial photoresist mask is coated and patterned and the top parylene layer is deposited. A thick photosensitive polyimide overlay is formed by spin-coating and photolithography. After etching the barrier and parylene layers from the back side, the sacrificial mask is removed with acetone at 40–50°C. Microchannels can be formed on any substrate using a simple three-mask and low-temperature process. This IC-compatible process achieves an integration of microfluidic and circuit elements. Figure 9b shows a similar process for making metal microchannels in which electroplating is used for fabricating the metallic channel cover.[29]

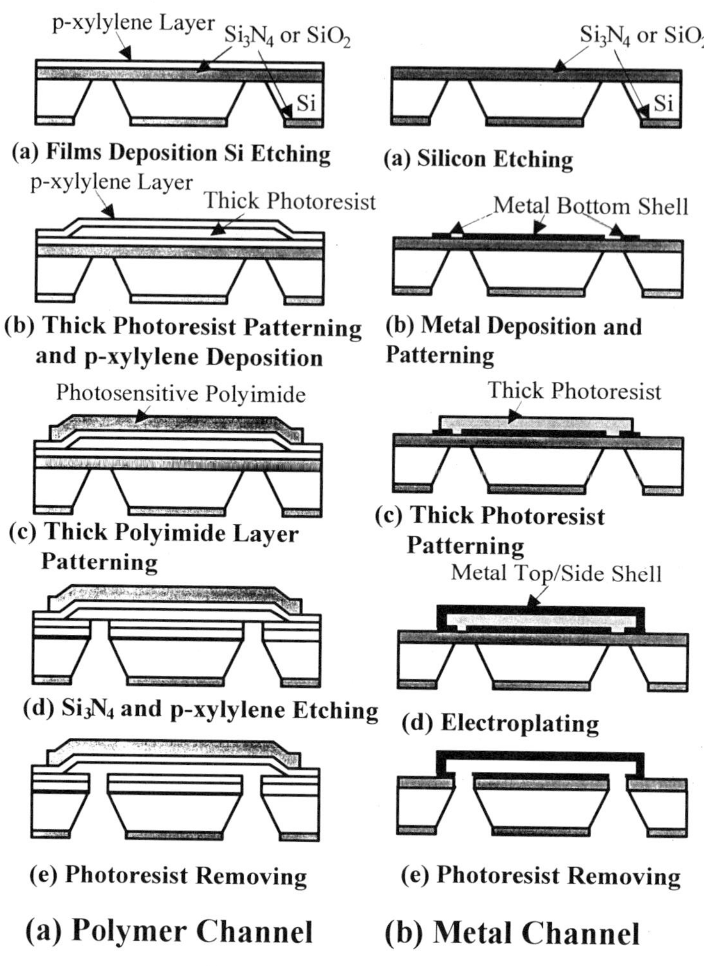

Figure 13.9. Microchannel fabrication process by surface micromachining.[28,29]

13.6. BONDING

Wafer-bonding techniques play a very important role in packaging of biosensors and assembly of the μTAS. Many kinds of bonding methods have been reported.[30] Typical bonding methods for glass–Si, Si–Si, glass–glass are shown in Fig. 13.10. The features of these bonding methods are listed in Table 13.2.

13.6.1. Gluing

The simplest bonding methods involve gluing the wafers with an adhesive. A thin adhesive layer can be formed by spin-coating with a controlled viscosity adhesive. Many available combinations of wafers can be bonded with a low-

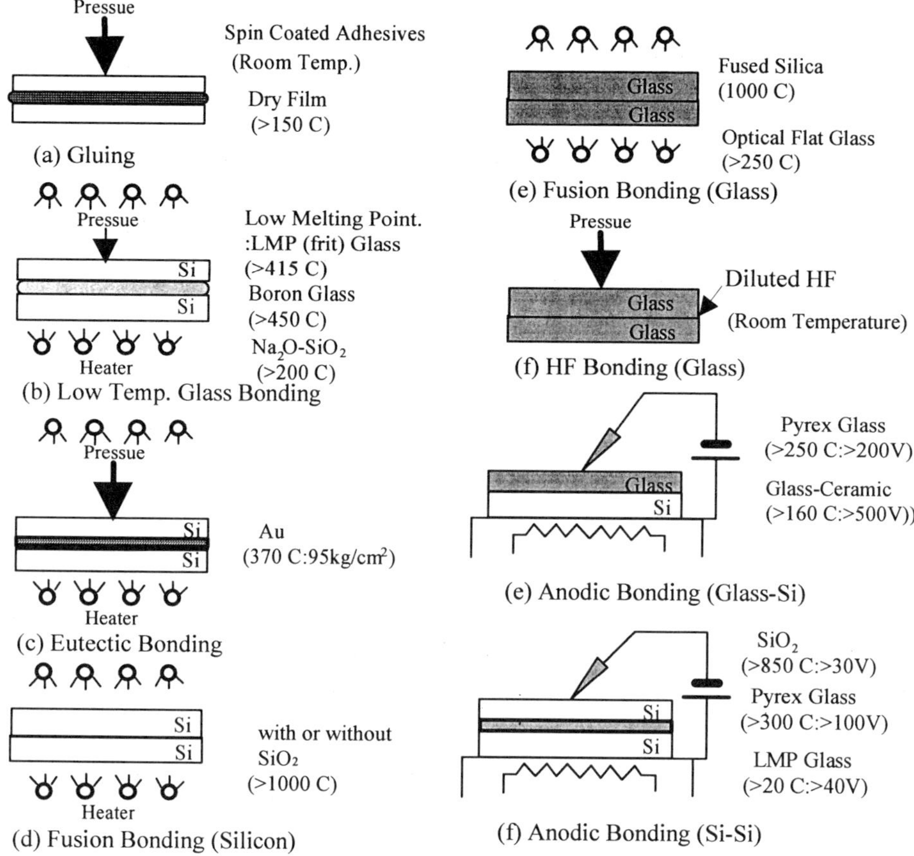

Figure 13.10. Setups of the wafer bonding methods.[30]

temperature process by the proper choice of adhesive. A photosensitive dry film is also used for selective bonding.

13.6.2. Low-Temperature Glass Bonding

Some kinds of low-melting-temperature glass are used for the thin intermediate layer between the bonded wafers. A typical method using screen-printed frit glass is the so-called "frit seal." Boron glass deposited by a solid source can also be used as the intermediate layer in Si–Si bonding. Bonding under 200°C can be achieved using spin-coated sodium silicate gel on the wafer.

13.6.3. Eutectic Bonding

Si–Si bonding is performed using a thin gold intermediate layer. Bonding is observed at 370°C, which is the eutectic temperature of Si and Au. This method needs a clean Si–Au interface.

Table 13.2.   Features of Wafer Bonding Methods[30–33]

Bonding method	Substrates	Intermediate layer	Temperature (°C)	Applied pressure (kg/cm^2)	Applied voltage (V)	Selective bonding (patterning)	Remarks (requirements)
Gluing	Any substrates	Spin coated adhesives	Room temperature	>0.2		x	
		Photosensitive dry film	>150	>0.5		Photolithography	
Low-temperature-melting glass bonding	Si–Si	Frit glass	>415			Screen printing	
		Liquid glass (Na$_2$O–SiO$_2$)	>200			x	
		Boron glass	>450			x	
Eutectic bonding	Si–Si	Au	370	95		Photolithography	
Fusion bonding	Si–Si		>1000			x	
	Glass–glass (fused silica)		1000			x	
	Glass–glass		>200			x	Optical flat
HF bonding	Glass–glass	(Diluted HF)	Room temperature	>0.04		x	
Anodic bonding	Glass–Si (Pyrex glass)		>250		>200	SiO$_2$(>1 μm) TiW/Au	
	Glass–Si (glass–ceramic)		>160		>300		
	Si–Si	SiO$_2$	>850		>30	Photolithography	
	Si–Si	Sputtered Pyrex glass	>300		>100	Lift off	
	Si–Si Si–Al/Glass Si–ITO/Glass	Sputtered low-melting-point glass	Room temperature	>0.8	>40	Lift off	Particle free

13.6.4. Fusion Bonding

Both Si–Si and glass–glass (fused silica) bonding are carried out at temperatures above 1000°C without any intermediate layers. At such high temperatures, wafers are thermally "fused." A hydrophilic surface pretreatment of wafers is necessary and careful elevation of the temperature is required to avoid voids. Glass–glass bonding below 250°C is also obtained if the glasses have an optically flat surface.[31]

13.6.5. HF Bonding

Room temperature glass–glass bonding has been reported.[32] The method is a simple one in which 1% diluted HF solution is dropped into the gap between the glass wafers and a pressure of 40 gf/cm^2 is applied. This method is also suitable for bonding between Si wafers having SiO_2 on their surfaces.

13.6.6. Anodic Bonding

Glass–Si anodic bonding is a most popular method in micromachining. Glass and Si wafers are placed in intimate contact and heated to the temperature at which alkaline ions in the glass are mobile. A high positive dc voltage is applied to the glass and the alkaline ions are depleted from the glass interface. The high field induces a very high electrostatic pressure at the glass–Si interface so as to bring the glass and Si into strong intimate contact. Finally a chemical bond is formed between the glass and the Si. There is a dependence between the bond strength and the total charge density integration of the current density during bonding that is useful for monitoring bonding behavior. Pyrex glass (Corning #7740) is widely used because of its good thermal expansion coefficient, which matches that of Si, as a thermal expansion mismatch causes bowing owing to residual stress after cooling. In the case of Pyrex glass, the bonding temperature is typically above 350°C and the applied voltage above 500 V. Anodic bonding using a glass (HOYA #SD-2) with a thermal expansion coefficient perfectly matched to Si is useful to reduce the residual stress. A glass ceramic (HOYA prototype #PS-100) that has high alkaline ion mobility at low temperatures and a similar thermal expansion coefficient to Si is also useful for anodic bonding.[33] With this glass ceramic bonding temperature can be reduced to 160°C.

Si–Si anodic bonding methods that make use of a thin glass intermediate layer have been also developed. Two Si wafers covered with thin oxide layers are anodically bonded at 850°C with an applied voltage of 30 V. A sputtered Pyrex glass is often used for the intermediate layer. The residual stress in the glass layer can be controlled by the gas pressure during sputtering. The subsequent steam annealing under 565°C renders the anodic bonding reproducible. A sputtered low-melting-point glass (Corning #7570) can be used instead of Pyrex glass. The low viscosity and the high relative permittivity of the glass is useful to reduce the bonding temperature.

13.7. CONCLUSION

Various micromachining techniques that can be used for the fabrication of microsensors, microactuators, and micrototal analysis systems have been reviewed. Fabricating biosensors and μTAS for biosensing applications requires a low-temperature process. Biocompatibility of the materials of microelements should be taken into account. Many micromachining processes in combination are necessary to achieve μTAS.

REFERENCES

1. Dewa AS, Ko WH. Biosensors In: Sze SM. *Semiconductor Sensors*. New York: John Wiley & Sons, Inc, 1994, pp. 415–472.
2. Shoji S, Esashi M. Microflow devices and systems. *J Micromech Microeng* 1994;4:157–171.
3. Manz A, Verpoote E, Raymond DE, et al. μ-TAS: Miniaturized total chemical analysis systems. In: van den Berg A, Bergveld P, eds. *Micrro Total Analysis Systems*. Dordrecht: Kluwer Academic Publishers. 1995, pp 5–27.
4. Miyake R, Tsuzuki K, Takagi T, et al. 1997, A highly sensitive and small flow-type chemical analysis system with integrated absorptiometric micro-flowcell. *Proc. 10th IEEE Micro Electro Mechanical Systems Workshop,* 1997:102–107.
5. Yacoub G, Wolf H. A miniaturized ISFET-ELISA system with a pretreated fused silica capillary as reaction cartridge. *Tech. Digest 8th Int. Conf. on Solid-State Sensors and Actuators: Transducers'95* 1995;2:898–901.
6. Woundenberg TM, Winn-Deen ES, Albin M. High-density PCR and beyond, *Proc. 2nd Int Symp. Miniaturized Total Analysis Systems* 1996:55–59.
7. Northrup MA, Beeman B, Hadley D, et al. Integrated miniature DNA-based analytical instrumentation. *Proc. 2nd Int Symp. Miniaturized Total Analysis Systems* 1996:153–157.
8. Gardner JW. Conventional silicon processing, microsensors: principles and applications. In: Sze SM. *Semiconductor Sensors*, New York, John Wiley & Sons Ltd, 1994, pp 36–58.
9. Chang FI, Yeh G, Lin P, et. al. Gas-phase silicon micromachining with xenon difluoride. *SPIE Proc.* 1995;2641:117–128.
10. Shoji S, Esashi M. Microfabrication and microsensors. *Appl Biochem Biotech* 1993;41:21–34.
11. Klaassen EH, Reay RJ, Storment, AJ, et al. Micromachined thermally isolated circuit. *Proc. Solid-State Sensors and Actuators Workshop* 1996:127–131.
12. Klaassen EH, Petersen K, Noworolski JM, et al. Silicon fusion bonding and deep reactive ion etching: a new technology for microstructures. *Sens Act.* 1996:A52:132–139.
13. Howe RT. Surface micromachining for microsensors and microactuators. *J. Vac. Sci Technol B* 1988;6:1809–1813.
14. Kim JY, Kim C-J. Comparative study of various release methods for polysilicon surface micromachining. *Proc. 10th IEEE Micro Electro Mechanical Systems Workshop* 1997:442–447.
15. Westberg D, Paul O, Andersson GI, et al. Surface micromachining by sacrificial aluminum etching. *J Micromech Microeng* 1996;6:376–384.
16. Gianchandai YB, Najafi K. A bulk silicon dissolved wafer process for microelectromechanical devices. *J Microelectromech Syst* 1992;1:77.
17. Rogner A, Eicher J, Munchmeyer D, et al. The LIGA technique: what are the new opportunities. *J Micromech Microeng* 1992;2:133–140.
18. Ehrfeld W, Abraham M, Ehrfeld U, et al. Materials for LIGA products. *Proc. 7th IEEE Micro Electro Mechanical Systems Workshop* 1994:86–90.
19. Kamper K-P, Ehrfeld W, Dopper J, et al. Microfluidic components for biological and chemical microreactors. *Proc. 10th IEEE Micro Electro Mechanical Systems Workshop* 1997:338–343.
20. Murakami K, Minami K, Esashi M. High aspect ratio fabrication method using O_2 RIE and electroplating. *Proc. Micro System Technologies'94* 1994:143–152.

21. Juan WH, Pang SW, Selvakumar A, et al. Using electron cyclotron resonance (ECR) source to etch polyimide molds for fabrication of electroplated structures. *Tech Digest Solid-State Sensors and Actuator Workshop*, 1994:82–85.
22. Despont M, Lorenz H, Fahrni N, et al. High-aspect-ratio, ultrathick, negative-tone near-UV photoresist for MEMSD applications. *Proc. 10th IEEE Micro Electro Mechanical Systems Workshop* 1997:518–522.
23. Keller C, Ferrari M. Milli-scale polysilicon structures. *Tech Digest IEEE Solid-State Sensors and Actuators Workshop* 1994:132–137.
24. Chen J, Wise KD, 1995, A high-resolution silicon monolithic nozzle array for inkjet printing. *Tech. Digest 8th Int. Conf. on Solid-State Sensors and Actuators: Transducers'95* 1995;2:321–324.
25. Kaplan W, Elderstig H, Vieider C. A novel fabrication method of capillary tubes on quartz for chemical analysis applications. *Proc. 7th IEEE Micro Electro Mechanical Systems Workshop* 1994:63–68.
26. Tjerkstra RW, de Boer M, Berenschot E, et al. Etching technology for microchannels. *Proc. 10th IEEE Micro Electro Mechanical Systems Workshop* 1997:147–152.
27. Shoji S, van der Schoot BH, de Rooij NF, et al. Smallest dead volume microvalves for integrated chemical analyzing systems. *Tech. Digest 6th Int. Conf. Solid-State Sensors and Actuators* 1991:1052–1055.
28. Man PF, Jones DK, Mastrangelo CH. Microfluidic plastic capillaries on silicon substrates: a new inexpensive technology for bioanalysis chips. *Proc. 10th IEEE Micro Electro Mechanical Systems Workshop* 1997:311–316.
29. Papautsky I, Frazier AB, Swerdlow H. A low temperature IC compatible process for fabricating surface micromachined metallic microchannels. *Proc. 10th IEEE Micro Electro Mechanical Systems Workshop* 1997:317–326.
30. Shoji S, Esashi M. Bonding and assembling methods for realizing a μTAS. In: van den Berg A and Bergveld P. *Micro Total Analysis Systems*. Dordrecht: Kluwer Academic Publishers, 1995, pp 165–179.
31. Ando D, Oishi K, Nakamura T, Umeda S. Glass direct bonding technology for hermetic seal package. *Proc. 10th IEEE Micro Electro Mechanical Systems Workshop* 1997:186–190.
32. Nakanishi H, Nishimoto T, Nakamura N., et al. Fabrication of electrophoresis devices on quartz and glass substrates using a bonding with HF solution. *Proc. 10th IEEE Micro Electro Mechanical Systems Workshop* 1997:299–304.
33. Shoji S, Kikuchi H, Torigoe H. Anodic bonding below 180°C for packaging and assembling of MEMS using lithium alminosilicate-β'-quartz glass-ceramic. *Proc. 10th IEEE Micro Electro Mechanical Systems Workshop* 1997:482–487.

14

Simultaneous Determination of Glucose and Analogous Disaccharides by Dual-Electrode Enzyme Sensor System

Xian-En Zhang

14.1. INTRODUCTION

Glucose and disaccharides, such as maltose, sucrose, and lactose, are found in a wide range of foods and beverages. Determination of these substances is important for at least three reasons: nutritional evaluation of natural foods, quality control of food and beverage production, and bioprocess control when the substances serve as microbial substrates. There are several conventional techniques for determining these substances. One of the most widely used is the Fehling titration method, which measures the total reducing sugar. Others include: the densometer method, which measures the density of a specific sugar in a pure sample; the colorimetric method, which measures individual sugars; and optical rotational analysis, which measures a pure sample.

Although these methods are fast and easy, their precision, sensitivity, and specificity are not always adequate and cannot meet the requirements of on-line process control. Moreover, they cannot be used when one sugar has to be distinguished from another, which is becoming a matter of interest to the biological industry and researchers studying microbial metabolism. For example, in some dairies, milk products are pretreated with enzymes to convert lactose to glucose and β-galactose for the benefit of lactose-intolerant adults, and the present method for determining lactose in dairy products is not convenient.

Another example is perhaps more complex. Molasses and starch are the most common raw materials used in the fermentation industry. Molasses contains a few kinds of fermentable sugars such as glucose and fructose but has mainly sucrose,

Xian-En Zhang ● Wuhan Institute of Virology, Chinese Academy of Sciences, Wuchang, Wuhan 430071, P.R. China.

Biosensors and Their Applications, edited by Yang and Ngo, Kluwer Academic/Plenum Publishers, New York, 1999.

243

which is not a reducing sugar. Neither the Fehling method nor other conventional methods can be used to determine the total fermentable sugar in molasses. Starch is usually not a ready carbon source for microorganisms. Under normal conditions, it has to be hydrolyzed before fermentation. The composition of a starch hydrolysate depends on the enzymes employed during hydrolysis. The enzymatic hydrolyzed products can be limited to glucose, or be a mixture of dextrin and glucose or maltose, or all three. If glucose and maltose coexist, there is likely to be confusion when the Fehling method is used. The results would show a lower value than the true one if the reducing sugar is expressed as glucose, as the molecular weight of glucose is one-half that of maltose, or higher than the true value if the reducing sugar is expressed as maltose, since each substrate contains one semiacetal hydroxyl group. This ambiguity may exist worldwide. Using molar concentration (mmole/L) to replace percentage concentration (g/L) is inconvenient for determining fermentation indexes and kinetics because the biomass is measured by dry weight. Having arrived at this point, we conclude that the only solution is simultaneous determination of coexisting fermentable sugars.

Of the variety of quantifying methods currently available, enzyme sensors are the most attractive because of their speed, convenience, and low cost. Glucose-sensing electrodes are well developed and widely used, while sensors for disaccharides have been reported frequently.[1-8] A new approach using what is known as a sequence electrode has been developed for determining an individual disaccharide that contains at least one glucose residual as part of its molecular makeup.[9] The principle of the sequence electrode is as follows:

$$\text{Disaccharide} + H_2O \xrightarrow{\text{Appropriate enzyme}} \text{Glucose} \tag{1}$$

$$\text{Glucose} + O_2 \xrightarrow{\text{Glucose oxidase}} \text{Gluconate} + H_2O_2 \tag{2}$$

Disaccharides can thus be determined by amperometrically measuring either the oxygen consumed or the hydrogen peroxide produced in reaction (2).

However, sequence electrodes suffer from interference of coexisting glucose when they are exposed to mixtures in samples such as some soft drinks, starch hydrolysate, and molasses. Many attempts have been made to solve this problem, e.g., eliminating the coexisting glucose by pretreatment with glucose oxidase (GOD)[10] or using a manifold comprised of columns of appropriate enzyme(s) and GOD in sequence coupled to amperometric detection of the hydrogen peroxide produced.[11] For reliability, we recommend the dual-electrode method, which involves the simultaneous use of a GOD electrode and a sequence electrode. With the GOD electrode measuring only glucose and the sequence electrode measuring the sum of glucose and disaccharide, the concentration of both substances can be calculated. The principle of this method was proposed by Pfeiffer et al.[1] in 1980 but has only recently become reliable. This chapter describes the results we have obtained in the past few years.

Table 14.1. Dual Functional Enzyme Sensors with Different Working Electrodes

Analyte	Electrode type	Working system	Ref.
Glucose and maltose	Clark oxygen electrode	FIA	12
	Hydrogen peroxide electrode	FIA	13
	Screen-printed electrode	Disposable	15,16
Glucose and sucrose	Carbon-paste electrode	FIA	14
	Hydrogen peroxide electrode	FIA	13
Glucose and lactose	Carbon-paste electrode	Batch	20

14.2. DUAL-ELECTRODE ENZYME SENSOR SYSTEM

Several different dual-electrode enzyme sensors have been developed, but, basically, they all consist of a GOD electrode and a sequence electrode either in a flow-injection analysis (FIA) manifold or in a type of disposable sequence electrode strip. The chemical electrode used to construct the enzyme sensor can be a Clark oxygen electrode, a hydrogen peroxide electrode (Pt electrode), carbon paste electrode, or screen-printed electrode (Table 14.1). Ag–AgCl electrodes were used as reference electrodes except in the case of disposable types. The components of the enzyme membrane of a sequence electrode depend on the disaccharide to be detected. Table 14.2 lists the components of selected enzyme membranes.

Temperature affects the sensor response in two ways. First, most of the sensors described herein have temperature coefficients from 1 to 5% per °C, and the background current changes with a change in temperature. Second, air bubbles are produced and released from the solution when the temperature increases because the solubility of air varies with the temperature, which leads to an increase in electrode noise. The greater the difference between the working and environmental temperatures, the more seriously the noise will interfere with the sensors. Thus biosensor setups are usually equipped with a temperature control unit, but for dependable measurement results the thermostat must be precise to ca. 1% per °C at room temperature, of course, the need for such high precision raises the cost of manufacture.

Our idea was to develop an automatic temperature-compensation module to be used in place of an expensive thermostat. A thermistor whose resistance varies

Table 14.2. Enzymes in the Sequence
Enzyme Electrodes

Analyte	Enzyme
Glucose	GOD
Maltose	GOD + amyloglucosidase, GOD + maltose
Sucrose	GOD + INV + MUT
Lactose	GOD + β-galactosidase

sharply with the temperature was installed in the measuring cell to detect temperature changes in the system. The information was sent to a built-in computer that had an equation derived from the temperature coefficient of the sensor stored within it. By this device, the temperature effect on the sensor response was negligible in the range of $\pm 3°C$ of the starting temperature and the bubble problem was eliminated as the temperature of the buffer stream was in equilibrium with the room temperature so that the dissolved air was released very slowly.

14.2.1. *Preparation of the Oxygen-Electrode-Based Sequence Electrode*

A cross-linking reaction was initiated by spreading a mixture of enzymes, bovine serum albumin (BSA), and glutaraldehyde onto a porous Teflon or polypropylene membrane with a fine glass rod. The mixture was allowed to stand for 30 min at room temperature to form an enzyme membrane, which was then covered with a dialysis membrane. This "sandwich" was fixed over the tip of an oxygen electrode to make the enzyme electrode, with the dialysis membrane toward the outside of the electrode. The dialysis membrane helped to hold the enzyme molecules in place but allowed the analytes to penetrate freely. For more details see Qu et al.[12]

14.2.2. *Preparation of the Hydrogen-Peroxide-Electrode-Based Sequence Electrode*

A cellulose acetate (CA) membrane was used to prevent the Pt anode from interfering with other electroactive substances. To prepare the CA membrane, CA solution (15% in acetone) was cast on a glass slide and polymerization was performed at room temperature for 4 h. A mixture of enzyme and BSA was spread onto the CA membrane, after which glutaradehyde was added to initiate the cross-linking reaction. The membrane was then fixed on the Pt anode and covered with the dialysis membrane. A working potential of 0.7 V was applied to the Pt electrode vs. the Ag–AgCl reference electrode.[13]

14.2.3. *Preparation of the Carbon-Paste (CP)-Electrode-Based Sequence Electrode*

Two cavities in a thin flow cell cartridge were connected with copper conducting wires. Each was packed tightly with the CP mixture (60% graphite powder and 40% light mineral oil, w/w) and smoothed by polishing the tips of the electrodes on weighing paper (Fisher). A GOD membrane was prepared on each electrode surface by cross-linking GOD, BSA, and glutaraldehyde. The mixture was spread on the oil surface of the CP by carefully placing a piece of commercial Kimwipe that had been cut to the size of the electrode surface. Direct incorporation of enzymes into the CP may be a kind of waste as most of the enzyme molecules are entrapped and the substrate cannot pass through because of the mineral oil surface. The use of Kimwipes as the surface layer accommodates both the enzyme suspension and the oil phase in the same region. After the GOD membranes formed, the mixture of appropriate enzyme(s), BSA, and glutaraldehyde was deposited on the K2 electrode to form an outer layer. The working potential for the CP electrode was 0.9 V vs. the Ag–AgCl reference electrode.[14] A sequence electrode can also be made by a one-step method, i.e., by immobilizing the GOD and other enzyme(s) simultaneously.

14.2.4. Preparation of Disposable Sequence Electrode Strips

The sequence-electrode sensor consisted of a PVC substrate upon which three layers were printed — all through polyester screens. Conducting tracks of silver ink were applied to the substrate and allowed to dry for 24 h. Carbon ink was applied to each conducting track and allowed to dry for at least 48 h. An insulation shroud was then applied, leaving terminals and active surfaces of the electrode exposed. After washing and drying a mixture of GOD and $K_3Fe(CN_6)$ solution or ferrocene derivative was dropped onto the working area of the pair of carbon electrode strips that constitute a single electrode. Forty-five pairs of such strips are made in each printing batch. The electrodes were then dried over silica gel under reduced pressure at room temperature overnight. The working voltage applied to dual screen-printed electrodes was 0.25–0.3 V.[15,16]

14.2.5. FIA Configuration and Measuring Procedure

Several biosensor FIA systems have been proposed each one slightly different than the others. Such a system consists principally of a system of tubing, a manual or an automatic sample injection valve, a peristaltic pump, a thermostat or temperature compensation module, dual-electrode enzyme sensor, multichannel amplifiers, and a chart recorder. One of the features of such a system is the use of an automatic temperature compensation method instead of a conventional thermostat (Fig. 14.1), which will be discussed below.

Before measurements are taken, the carrier stream (phosphate buffer or distilled water) is pumped through the flow line at a rate of 2–3 mL/min. The sample solution is injected into the carrier stream when the background current stabilizes. When the sample flows through the surfaces of the electrodes, consumption of oxygen or production of hydrogen peroxide in the enzymatic reaction generates current changes, the peaks of which are measured for responses. The output of each electrode is processed by the electronic system and displayed on the screen or on a meter. The amplified analogue signals can be read through the chart recorder.

The calibration of both electrodes is carried out with standard glucose and disaccharide solutions at the beginning of each experiment and should be rechecked during the experiment. The measurement results can be calculated using the following equations:

$$C_g = R_{g'}/A_{g'} \qquad (3)$$

$$C_d = (R_t - A_g \times R_{g'}/A_{g'})/A_d \qquad (4)$$

where C_g and C_d are, respectively, the concentrations of glucose and disaccharide in

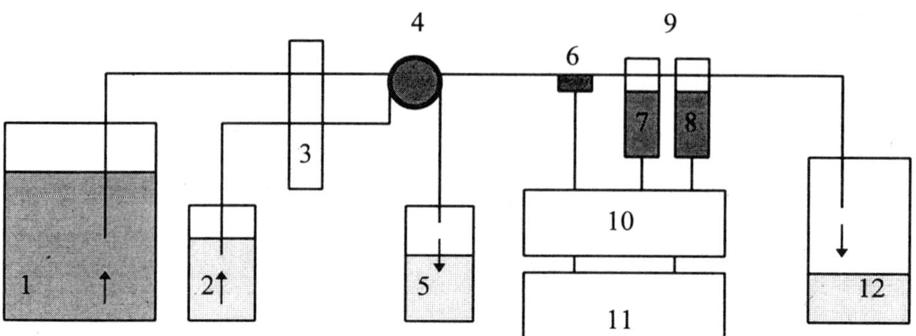

Figure 14.1. Diagram of the typical flow injection dual-electrode enzyme sensor system: (1) buffer supply, (2) sample or calibration solution, (3) peristaltic pump, (4) sample injection valve, (5) sample effluent, (6) temperature probe, (7) GOD electrode (8) sequence electrode, (9) measuring flow cell, (10) multichannel amplifier with digital display, (11) chart recorder, (12) waste.

the mixture; $R_{g'}$ and R_t are the respective responses of the GOD and the sequence electrodes; $A_{g'}$ and A_g are, respectively, the calibration coefficients of the GOD and the sequence electrodes to glucose; and A_d is the calibration coefficient of the sequence electrode to disaccharide.

14.3. SIMULTANEOUS DETERMINATION OF GLUCOSE AND SUCROSE

Glucose mutarotation in water is a slow process. It is known that the acceleration of mutarotation from α-isomer to β-isomer can be achieved using either mutarotase (MUT) or phosphate ions.[17] To compare the efficiencies of these means of acceleration, two sucrose electrodes were prepared with and without MUT. When sucrose was injected into the buffer stream, the response activity of the electrode with MUT was about three times that of the electrode without MUT (Fig. 14.2). MUT was thus coimmobilized with invertase (INV) and GOD in the enzyme membrane. As the CP electrode is hydrogen-peroxide-dependent, hydrogen peroxide produced in the enzymatic reaction of the upstream electrode may interfere with the downstream electrode. One way to prevent the downstream electrode from being affected by this interference is to deposit a catalase (CAT) membrane onto the thin channel between the two electrodes to prevent hydrogen peroxide from passing through the channel.[14]

The effect of pH on the working electrode was evaluated using the ratios of the response to glucose and hydrogen peroxide (R_G/R_H), as the CP electrode is sensitive to a change in pH. The electrode response increased with increasing pH to some extent. The maximum response ratios for sucrose and glucose were observed at pH 6.8 and above. A phosphate buffer (0.02 mmole/L, pH 6.8) was used as the carrier stream.

The linear range of the sensor system was dependent on the sample injection volume. With 50-μL injections, the linear ranges were 0.1–5 mmole/L for glucose

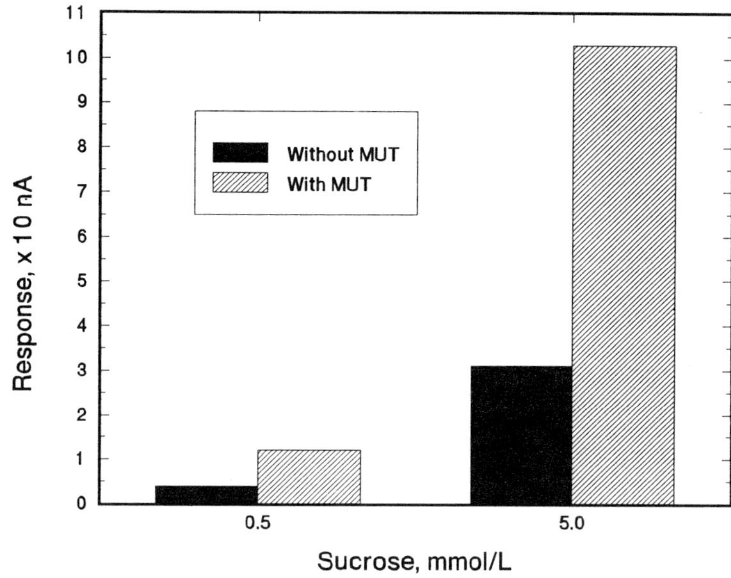

Figure 14.2. Comparison of glucose mutarotation acceleration between the phosphate buffer method and mutarotase method. (Zhang et al.,[14] with permission of VCH publishers, Inc.)

and 0.5–7.5 mmole/L for sucrose. The precision of the system was checked using glucose and sucrose solutions and a mixture of both. For the GOD and sequence electrodes the coefficients of variation (CVs) were, respectively, 2.7% and 1.3% in the determination of glucose and 2.9% and 1.5% for glucose and sucrose in the mixture. The precision of the sensor is comparable to results reported elsewhere.

One point of interest is the investigation of individual enzyme activity, which is an important factor in working with sequence electrodes. The decay in activity of any one of the coimmobilized enzymes can lead to loss of the electrode response. The mutarotation rate of glucose solution was monitored with a GOD electrode that responds only to β-glucose. As the mutarotation processes take more than 60 min to complete, and the initial mutarotation speed is high, INV (MUT) activity for α-glucose must be measured immediately after the α-glucose is dissolved in the phosphate buffer. The activities of the immobilized GOD, INV, and MUT were evaluated based on the ratio of the responses to appropriate substrates and the response to hydrogen peroxide (R_X/R_H). Here we define R_S as the response to sucrose, $R_{\alpha-G}$ as the response to α-glucose, and $R_{\beta-G}$ as the response to β-glucose. R_X/R_H fluctuated in the initial days but did not decrease for at least 1 month. On the 36th day, R_S/R_H dropped to a minimum while $R_{\alpha-G}/R_H$ and $R_{\beta-G}/R_H$ remained constant. This suggests that the stability of immobilized INV is the limiting factor for the stability of the sucrose sensor under the experimental conditions.

In contrast to the biosensor based on a hydrogen peroxide electrode, the CP-electrode-based biosensor has some drawbacks, e.g., a long precondition time (as much as 1 h), a high working potential that may cause a nonspecific electrode

Table 14.3. Comparison of the Dual-Electrode Enzyme Sensor Method with the
Acidic-Hydrolysis–GOD Electrode Method for Determinations of Glucose and Sucrose

| | Sucrose (mmole/L) | | Glucose (mmole/L) | |
Sample	Dual-electrode method	Acidic-hydrolysis– GOD electrode method	Dual-electrode method	Acidic-hydrolysis– GOD electrode method
Carotene juice #1	47.4	58.5	59.6	56.8
Carotene juice #2	57.4	68.9	67.8	61.9
Carotene juice #3	157.4	136.4	57.1	62.2
JianLiBao	211.0	227.8	70.5	80.0
Pepsi	0	0	325.3	330.7
Coca Cola #2	0	0	304.5	292.5
Sprite	212.4	198.4	164.3	176.6

current generated by some electroactive species, the short lifetime of the PC
electrode, and a pH dependence. We constructed a dual-electrode enzyme sensor
based on a hydrogen peroxide electrode, which was not sensitive to pH and ascorbic
acid (an electroactive species).[13] The linear range was 1–30 mmole/L with a CV of
less than 5%. With the sensor system, glucose and sucrose in selected beverages were
determined simultaneously. A comparable method involved measuring the glucose
with the GOD electrode first, treating the sample with sulfate to decompose the
sucrose, measuring the total glucose in the treated sample, and calculating the
amount of sucrose. The two methods agreed well with one another (Table 14.3) but
the dual-electrode technique was much faster.

14.4. SIMULTANEOUS DETERMINATION OF GLUCOSE AND MALTOSE

It is important to know that not only maltose but also starch and those of its
degradation products that contain more than two glucose moieties can serve as
substrates of amyloglucosidase. A low-molecular-weight-cut dialysis membrane was
placed in front of the A–G (amyloglucosidase–GOD) electrode to separate these
substrates from immobilized amyloglucosidase. Theoretically, as was described
above, the enzymatic reaction at the upstream electrode consumes oxygen and
produces hydrogen peroxide, which may interfere with the response of the downstream electrode. However, in an oxygen-electrode-based system or a parallel
installed-electrode system with a larger measuring cell, this interference is too slight
to be detected, so incorporating catalase is unnecessary.

For the sensor system constructed with oxygen electrodes,[12] distilled water
(pH 6.0) was used as carrier stream for the following reasons: (1) the oxygen
electrode is not sensitive to pH; (2) the optimum pH for measurement of either
substrate was around 5.8; and (3) under slightly acid conditions the responses were

Table 14.4.   Linear Ranges and CVs of the Maltose Sequence Electrodes Made with a Hydrogen Peroxide Electrode

Sample	Concentration (mmol/L)	Batch response values (digits)	Average (digits)	CV (%)	Linear regression coefficients
Glucose	5	58, 60, 64, 59, 60	60.2	3.8	
	15	192, 187, 192, 189	190.0	1.29	0.99998
	30	374, 384, 384, 381	380.8	1.24	
Maltose	5	68, 66, 66, 66	66.5	1.50	
	15	227, 228, 218, 224	224.2	2.01	0.99986
	30	459, 449, 442, 447	446.8	0.76	

quite constant with changes in pH. The linear range of the system was 0.3–30 mmole/L of sugars, and the CVs were 3.2% for the GOD electrode and 1.6% for the A–G electrode in measuring the mixture of glucose and maltose. The results attained with the dual-electrode system agreed well with those of Fehling titration at the molar concentration scale.

The sensor system was used for an on-line study of starch hydrolyses catalyzed by various amylases.[18] Determinations could be made every 2 min. A 10 μL sample with recycled backflow minimized any loss of the reaction medium. The production, growth, and decay of glucose/maltose concentrations during starch hydrolysis under various enzymatic conditions were closely monitored, making this system a powerful tool for the study of the catalytic kinetics of amylases and screening and analyzing enzyme systems in nature.

An alternative sensor system was constructed based on hydrogen peroxide electrodes. The linear range and error analysis of the sequence electrodes are shown in Table 14.4. The linear range of the maltose sequence electrode like that of the sucrose electrode strongly depends on the concentration of the coexisting glucose. A

Table 14.5.   Lifetime of Glucose and Maltose Sequence Electrodes

Measurement No.	Response (digits)		Relative residual activity (%)	
	Maltose	Glucose	Maltose	Glucose
1	9	25	100	100
500	9	22	100	88
1000	12	21	133	84
1500	11	19	122	76
2000	9	11	100	44
2500	9	11	100	44
3000	7	10	78	40

linear range up to 30 mmole/L of the total glucose and maltose could be expected. The lifetime of the enzyme membrane was studied and is shown in Table 14.5. After 3000 measurements over a 20-day period, the response activity of the sequence electrode remained 78% for maltose and 40% for glucose. These results clearly suggest that the loss of sequence electrode activity for maltose is due to the partial inactivation of GOD.

The sensor system was used to monitor the saccharification process in a local brewery (Dongxihu Beer Company). The saccharification was catalyzed by α-amylase followed by the addition of malt extract containing β-amylase. The main products of saccharification were maltose and glucose. With successful simultaneous determination of glucose and maltose, the ratio of the two sugars could be adjusted toward the desired fermentation, and the process could be controlled. Utilization of the substrates was increased by 5% and the quality of the dry beer was improved, which will be of great benefit.

A disposable type of sensor for the determination of glucose and maltose was developed using screen-printing technology, which was introduced for biosensor manufacture in the middle of 1980s (see, e.g., Mattews et al.[19]), based on the principle of mediated enzyme electrodes.[20] Figure 14.3 shows the surface structure of a single electrode.[15] Each electrode contains two carbon electrodes, working as a dual "half-cell" without the Ag–AgCl reference electrode. Potassium ferricyanide, a mediator that transfers electrons between the enzyme molecules and the electrode at a low working potential (0.25 V) was incorporated into the sensor configuration. The Faradaic current of the electrode is dependent on the concentration of the potassium ferricyanide(III).

A preliminary study indicated that the enzyme electrodes were quite stable at lower pH. The linear range of the system was up to 35 mmole/L of glucose or 25 mmole/L of maltose, with CVs ranging from 2.35 to 5.29%. Because of low working potential, the interference of other electroactive species is minimized. Another advantage is that neither calibration nor washing steps are needed since the electrodes can be made cheaply and homogeneously. This being the case calibration is done by the manufacturer from batch to batch, each electrode being used only once. It would then be possible to be more sure of the value of the half-cell potential of the working electrode. A working model based on a new design is being developed for commercial use.

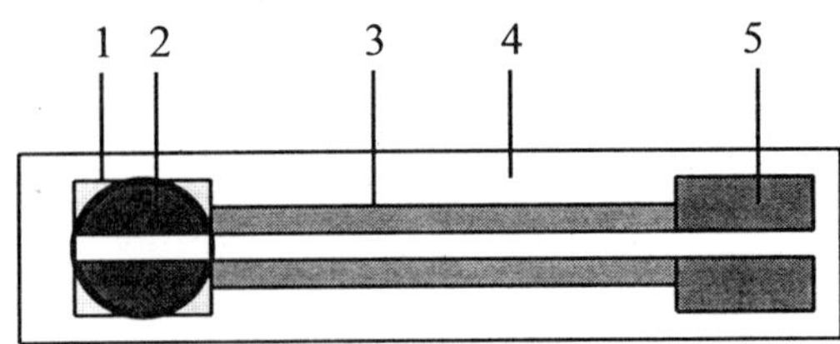

Figure 14.3. Configuration of screen-printed enzyme electrode: (1) printed carbon electrode, (2) adsorbed enzyme(s) and mediator, (3) conducting track, (4) PVC substrate, (5) electric connector.

14.5. SIMULTANEOUS DETERMINATION OF GLUCOSE AND LACTOSE

The sequence electrode was made of β-galactosidase and GOD. The dual-electrode system now being discussed is based on a CP electrode.[21] This sensor was not designed to be incorporated into a FIA system but rather to be used with a stirring measuring cell. The linear range was 0.1–2.5 mmole/L. The system enabled the determination of lactose concentration in milk in the presence of glucose to be carried out more precisely and with a higher degree of sensitivity than the conventional colorimetric method. Several different commercial milk products were selected for the experiment. In regular milk, low-fat milk, and lactose-reduced milk, concentrations of lactose were determined to be 149, 167, and 3 mmole/L, respectively, using the biosensor method, and 138, 154, and <30 mmole/L, respectively, using the colorimetric method. Concentrations of glucose in these samples were simultaneously determined to be 1.5, 1.5, and 157 mmole/L, respectively, using the biosensor method, while the colorimetric method could not be used for this measurement.

Determinations of recovery percentages using the dual-electrode method were carried out with mixtures of the same samples. The average recoveries for lactose were 100.7% (the mixture of milk and lactose-reduced milk) and 99.6% (the mixture of low-fat milk and lactose-reduced milk). For glucose, determination recovery for both mixtures was 105.6%.

14.6. DISCUSSION

The dual-electrode enzyme sensor method enables a speedy and accurate determination of concentrations of glucose and disaccharides simultaneously from a single measurement. The technique was derived from the single-sequence-electrode method in combination with sample pretreatment by either acidic or enzymatic hydrolysis. As predicted, no effect of disaccharides on glucose determination was observed, while the maximum concentration of disaccharides that could be determined was dependent on the concentration of coexisting glucose. The higher the glucose concentration, the lower the maximum disaccharide concentration that could be measured accurately.

Coordination of enzymes in a sequence electrode is one of the key points. However, the kinetics of the sequence enzymatic reaction in a heterogeneous phase has been less studied. Theoretically, equal quantities (in units) of related enzymes should be equally functional in a sequence enzymatic reaction in solution (homogeneous phase), but this does not usually prove to be the case with immobilized enzyme membranes or biosensors. Apart from pH and temperature, many other factors can affect enzyme activity. For example, an enzyme can be partially inactivated by chemical or physical modification during immobilization, and immobilization orientation, distance between immobilized enzyme molecules, and mass transfer rate in the sensor membrane can also make significant contributions to the final recovery activity of an enzyme. This illustrates why the optimal ratios of enzymes in the similar sequence electrodes have varied from report to report by different authors. In our experience, the enzyme catalyzing advanced step should predominate over the enzyme catalyzing the following step, so that all the enzymes are saturated by their substrates. This would enable performance of all the steps in the sequence reaction in first or pseudo-first order.

ACKNOWLEDGMENTS: The author would like to give sincere thanks to colleagues for their contribution to the research work described in this chapter. Support from National Natural Science Foundation of China and Chinese Academy of Sciences is greatly appreciated.

REFERENCES

1. Pfeiffer D, Scheller F, Janchen M, et al. Bienzyme electrodes for ATP, NAD$^+$, starch and disaccharides based on a glucose sensor. *Anal Lett* 1980;9:65–69.
2. Garton L., Appelquist R, Johansson G, et al. Determination of starch and maltose using immobilized amyloglucosidase and glucose electrode in a flow injection system. *J Chem Technol Biotech* 1989;46:327–333.
3. Varadit M, Adanyi N, Nagy G, et al. Studying the bienzyme reaction with amperometric detection for measuring maltose. *Biosens Bioelectron* 1993;8:339–345.
4. Scheller F, Renneberg R. Glucose eliminating enzyme electrode for direct sucrose determination in glucose containing samples. *Anal Chim Acta* 1983;152:265–269.
5. Xu YH, Guilbault GG, Kuan SS. Sucrose enzyme electrode. *Anal Chem* 1989;61:782–785.
6. Hu WP, Zhang XE, Zhang X, et al. Study on multiple-enzyme electrode for sucrose determination. *Chinese J Biotech* 1991;7:293–300.
7. Hu WP, Zhang XE, Zhang ZP, et al. A multi-functional enzyme sensor for sucrose and glucose determination. *Chinese Biochem J* 1993;9:741–747.
8. Pilloton R, Mascini M, Casella IG, et al. Lactose determination in raw milk with a two enzyme based electrochemical sensor. *Anal Lett* 1987;20:1803–1814.
9. Satoh I, Karube I, Suzuki S. Enzyme electrode for sucrose. *Biotech Bioeng* 1976;18:269–272.
10. Dullar R, Reinhadt B, Schugerl K. High reliability and stability of enzyme cartridges in flow injection analysis. *Anal Chim Acta* 1989;225:253–262.
11. Massom M, Townshend A. Simultaneous determination of sucrose and glucose in mixtures by flow injection analysis with immobilized enzymes. *Anal Chim Acta* 1985;171:185.
12. Qu HB, Zhang XE, Zhang SZ. Simultaneous determination of maltose and glucose using a dual-electrode flow injection system. *Food Chem* 1995;52:187–192.
13. Zhang XE, Wei HP, Tang JK, et al. Study on an multifunctional enzyme sensor system. Summary Report of the National Key Project No. 85-722-13-02, 1996.
14. Zhang XE, Rechnitz GA. Simultaneous determination of glucose and sucrose by a dual working electrode multienzyme sensor flow injection system. *Electroanalysis* 1994;6:361–367.
15. Ge F, Zhang XE, Zhang ZP, Zhang XM. Potassium ferricyanide mediated disposable biosensor for determination of maltose and glucose. *Anal Lett* 1998;31:383–394.
16. Ge F, Zhang XE, Zhang ZP, et al. Simultaneous determination of maltose and glucose using a screen-printed electrode system, *Biosens Bioelectron* 1998;13:333–339.
17. Barlikova A, Svorc J, Miertus S. Hybrid biosensor for the determination of sucrose. *Anal Chim Acta* 1991;247:83–87.
18. Qu HB, Zhang XE, Chui WW, et al. Kinetic study on starch hydrolysis using a glucose/maltose sensing system. *Biotech Tech* 1995;9:445–450.

19. Matthews, DR, Bown E, Watson A, et al. Pen-sized digital 30-second blood-glucose meter. *Lancet* 1987;1:778–779.
20. Cass AEG, Davis G, Francis GD, et al. Ferrocene-mediated enzyme electrode for amperometric determination of glucose. *Anal Chem* 1984;56:667–671.
21. Takashi K, Zhang XE, Rechnitz GA. Simultaneous determination of lactose and glucose in milk using two working enzyme electrodes. *Talanta* 1994; 41:843–848.

15

Application of Biosensors to the Measurement of Neurotransmitter Function

C. A. Marsden, S. R. Beckett, and A. Waterfall

15.1. INTRODUCTION

The 1990s is the "decade of the brain" and it has seen considerable advances in the methodology available for investigating brain function in relation to behavioral and physiological responses. The most obvious of such approaches is the use of imaging techniques (e.g., position emission tomography and functional nuclear magnetic resonance) to probe brain region structure and neurochemistry. An alternative approach is to develop biosensors that can be implanted into brain tissue to measure changes in extracellular levels of various neurochemicals including neurotransmitters, neuromodulators, products of receptor occupation (e.g., cyclic AMP), products of intracellular metabolism, products of oxidative stress, and drugs. The development of these *in vivo* probe techniques has resulted in new understanding of the action of drugs; one of the best examples is the establishment of a link between drug dependence and activation of dopamine mesolimbic function. Equally important are the new data becoming available on neurotransmitter control of behavior and the processes involved in neurodegeneration.

Several approaches have been used to develop appropriate "biosensors'" for use in the brain in the living animal and human.

1.   The Microdialysis Probe: A semipermeable hollow membrane inserted into the brain and continuously perfused with artificial cerebrospinal fluid. Substances in the extracellular space cross the membrane in a concentration-dependent

C. A. Marsden, S. R. Beckett, and A. Waterfall ● School of Biomedical Sciences, Medical School, Queen's Medical Centre, Nottingham NG7 2UH, United Kingdom.

Biosensors and Their Applications, edited by Yang and Ngo, Kluwer Academic/Plenum Publishers, New York, 1999.

257

manner and are collected in the perfusate; the levels are then measured using an appropriate analytical method of high sensitivity (usually HPLC). Microdialysis probes are not true biosensors but offer a convenient highly adaptable method for collecting a wide range of extracellular substances in the brain in both unanesthetized animals[1,2] and man[3,4] (Table 15.1).

2. The Carbon-Fiber Electrochemical (Voltammetric) Probe: These are used to measure the current produced by oxidation of electroactive substances in the extracellular space at the surface of a carbon-based electrode.[5,6] Again these probes are not selective in their own right — except that they will only detect electroactive substances. Some selectivity can be provided by the potential that is applied and/or the voltammetric measurement method used.[6,7]

3. The "Semiselective" Voltammetric Probe: In this situation the electrode is both electrically and chemically modified to improve selectivity towards individual compounds or groups of compounds, e.g., the use of Nafion to coat the carbon fiber to exclude acid metabolites of amine neurotransmitters, so making the electrodes "amine selective,"[6,8] and the Nafion-coated porphyrin/carbon fiber electrode to measure nitric oxide.[9]

4. The Enzyme Selective Electrochemical Probe: These probes incorporate an immobilized enzyme that converts nonelectroactive compounds into an electroactive product that can then be measured. An example is the glutamate biosensor using L-glutamate oxidase.[10,11] This approach also has been successfully applied to the development of sensors for measuring glucose, using glucose oxidase, in brain.[12] Similar approaches are being adopted in the development of other enzyme-based biosensors (e.g., acetylcholine involving the use of three enzymes — acetylcholine esterase, choline oxidase, and horseradish peroxidase). In all cases specific methods for immobilizing the different enzymes have to be carefully considered and all sources of interference with the signal minimized to ensure that the biosensors achieve the required high sensitivity. A variant of this approach is to immobilise an enzyme on the electrode surface that converts an interfering electroactive compound into a nonelectroactive product so removing the interference; an example is the ascorbate oxidase "killer" probe to remove ascorbic acid, which is abundant in the brain and oxidises at a potential similar to that of the important catecholamine neurotransmitters (dopamine and noradrenaline).[13]

5. The Antibody-Selective Probe: In this situation a suitable surface is coated with a selective antibody (to date these have concentrated on neuropeptide neurotransmitters or modulators), and the microprobe is inserted into a brain region allowing the endogenous peptide to bind to the antibodies. The probe is then removed, exposed to the specific *labeled* peptide *in vitro* and this binding is compared to the unlabeled endogenous binding using x-ray film.[14]

In this chapter these various options will be reviewed briefly to highlight the major advantages and disadvantages of each with particular emphasis on the ways

Biological group	Chemical	Assay[b]	References
1. Classical Neurotransmitters and precursors	DOPA	HPLC + ECD	1, 14, 25, 26
	Dopamine	HPLC + ECD	
	Noradrenaline	HPLC + ECD	
(a) Amines	Adrenaline	HPLC + ECD	
	5-Hydroxytryptophan	HPLC + ECD	
	Serotonin	HPLC + ECD	
	Acetylcholine	HPLC + ECD	
(b) Amino acids	GABA	HPLC + ECD or FD	2, 15, 24, 27
	Glutamate	HPLC + ECD or FD	
	Aspartate	HPLC + ECD or FD	
	Taurine	HPLC + ECD or FD	
	(other amino acids)		
2. Neuropeptides	Neurotensin	RIA	28
	Cholecystokinin	RIA	28, 29
	Enkephalins	RIA	30
	CGRP	RIA	31
	Angiotensin II	RIA	32
	Endothelin-I	RIA	32
	Insulin	RIA	33
3. Other neuromodulators			
(a) Nitric oxide formation	Citrulline	HPLC + FD	34
	NO_2	HPLC + UV	34
	NO_3	HPLC + UV	34
	cGMP	ELISA	35
(b) Purines	Adenosine	HPLC + FD	36
	(other purines)		
4. Products of receptor activation	cAMP	RIA or binding assay[c]	1
	cGMP	RIA or binding assay[c]	35
5. Products of metabolism	Ascorbic acid	HPLC + UV/ECD	37, 38
	Uric acid	HPLC + UV/ECD	37, 38
	glucose	Automated analysis	3, 39
	lactate	Automated analysis	3, 39
	pyruvate	Automated analysis	3, 39
6. Indexes of lipid peroxidation	Malonaldehyde	HPLC + UV	37, 38
7. Potential endogenous neurotoxins (excluding EAA)	Kynurenic acid	HPLC + FD	40
	Quinone formation	HPLC + ECD	41
8. Drugs pharmacokinetic sampling[d]	Drugs include:		
	L-DOPA	HPLC + ECD	e.g., 26, 42
	Acetominophen	HPLC + ECD	
	Phenytoin	HPLC + UV	
	Salicylic acid	HPLC + UV	
	Warfarin	HPLC + UV	
	Chloramphenical	HPLC + UV	
	Morphine	HPLC + UV	
	Cocaine	HPLC + UV	
9. Ions (Na^+, K^+, Mg^{2+}, Ca^{2+}, H^+)			

[a] This list is not comprehensive but provides an indication of the potential range of substances that can be detected in the extracellular space providing suitable analytical techniques are available.
[b] ECD = Electrochemical detection; FD = Fluorescence detection; RIA = Radioimmunoassay; UV = Ultraviolet detection.
[c] IRIA kits available but radioreceptor assay will provide a cheaper alternative.[1]
[d] The use of microdialysis to study central and peripheral pharmacokinetics is increasing.[26,42] Conversely the microdialysis probe can be used to deliver drugs directly to the brain or other site of action.

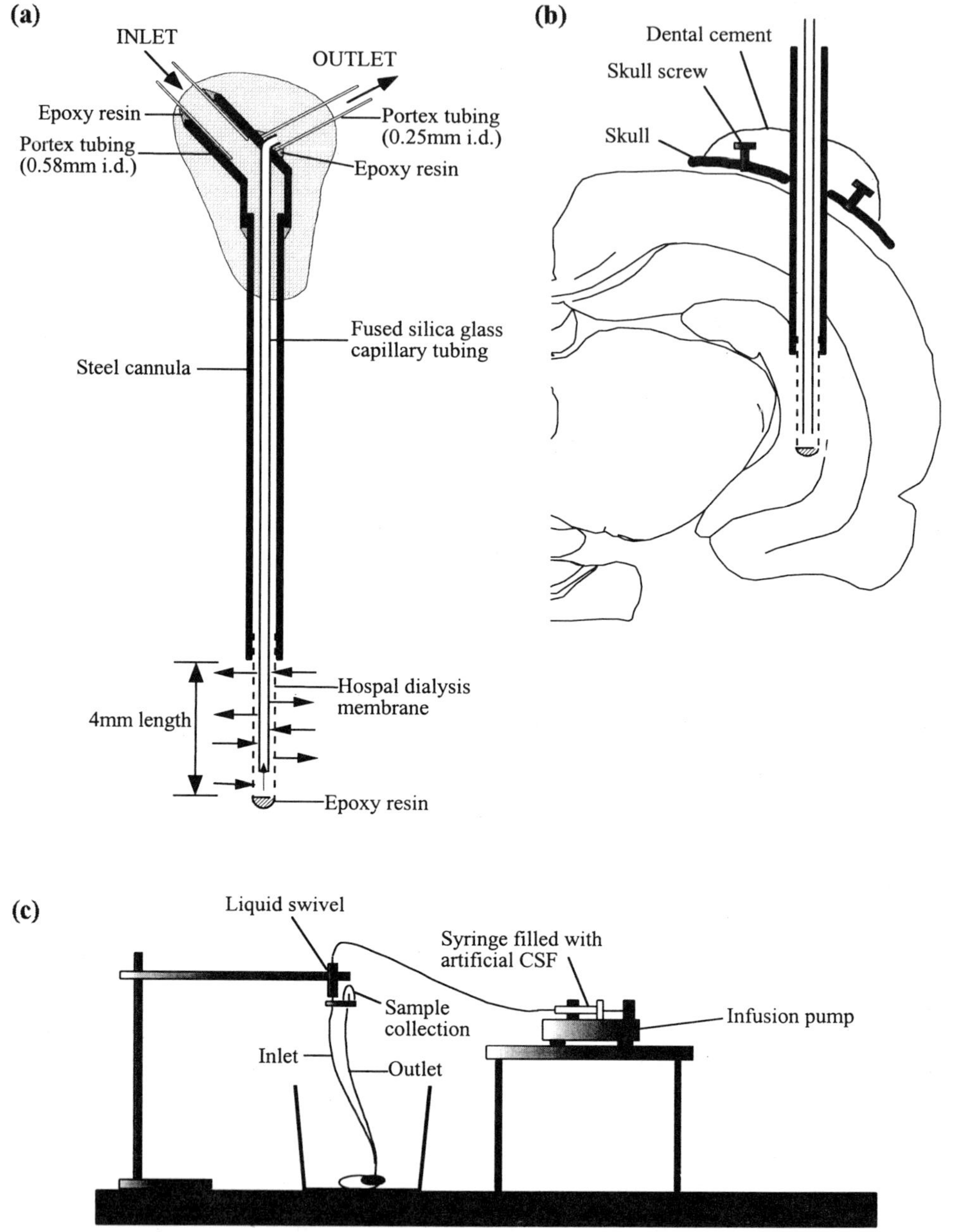

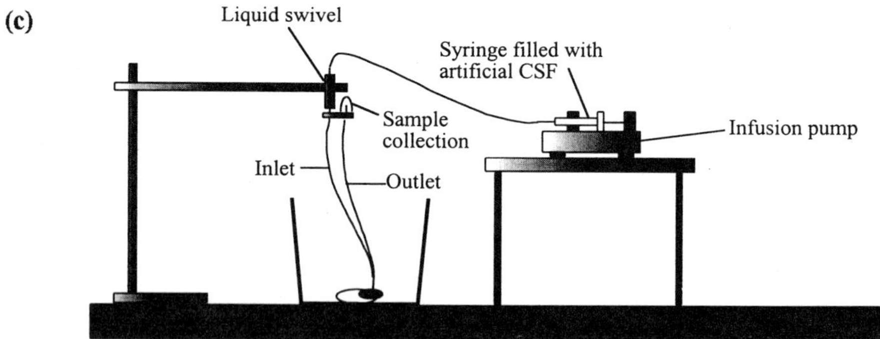

Figure 15.1. Microdialysis probe: (A) basic construction of a concentric microdialysis probe for use in the brain; (B) position of a microdialysis probe implanted into the rat hippocampus (the length of expose dialysis membrane is 4 mm); (C) general arrangement of the equipment during perfusion of a microdialysis probe implanted into a brain region of the rat. (Diagrams kindly provided by Dr Helen Rowley).

in which the methodology might be improved in the future. The methods and their application have been extensively reviewed elsewhere and detailed technical information is available on *in vivo* microdialysis,[1,3,14,15] voltammetry,[6,7,16,17] and antibody probes.[18,19]

15.2. IN VIVO MICRODIALYSIS

15.2.1. The Microdialysis Probe

·Microdialysis can be defined as an *in vivo* bioanalytical sampling technique for the continuous monitoring of chemical events occurring in a living body. When a molecule from the blood enters or leaves a cell, it must traverse the extracellular space. This space is filled with a fluid and makes up about a fifth of the total tissue volume. However, this crucial area of chemical communication between cells (which includes neurotransmission between neurones in the brain) is inaccessible experimentally by conventional methods of bioanalytical chemistry. Microdialysis is based on sampling the levels of endogenous substances within the extracellular space using a microdialysis probe.

The microdialysis probe (Fig. 15.1A) mimics the function of a blood vessel. The essential component of the probe is the dialysis fiber and, in particular, the nature of the membrane from which the fiber is made. The probes are usually concentric, having an outer inward perfusion cannula and an inner collection cannula, with the dialysis membrane attached to the end of the probe and sealed at its base to form a closed perfusion system. The probe is constantly perfused with a physiological solution (artificial cerebrospinal fluid) at a low flow rate (usually less than 2 μL/min). Once the probe is implanted into the tissue (Fig. 15.1B), endogenous substances are filtered by diffusion out of the extracellular fluid into the perfusion medium. Conversely, the probe can be used to infuse exogenous drugs or neurotoxins locally into specific brain areas for periods of up to several days. Samples are collected and then analyzed using any sensitive analytical method. A great advantage of this technique is that the dialysis samples are very clean (high-molecular weight compounds are excluded by the membrane) and can be analyzed directly without further sample preparation.

Microdialysis probes can be readily implanted into specific brain regions under anesthesia using stereotaxic techniques. The probe is then connected to a syringe pump, which delivers the perfusion medium (normally artificial cerebrospinal fluid) to the probe via a flexible connecting tube. Perfusates can be collected either under anesthesia or, more importantly, in freely moving animals (Fig. 15.1C) and injected directly into an HPLC system. Flexible microdialysis probes have been developed to measure extracellular levels of substances in peripheral tissues and blood. Furthermore, dialysis probes have been used successfully to monitor changes in neurotransmitter release in human brain[4,20] and catecholamines in blood. The advantage of the microdialysis approach is that continuous measurement can be made of compounds in the extracellular space during ongoing physiology and behavior. The extracellular levels of neurotransmitters in the brain are very low (*n* moles) and so the technique is dependent upon the availability of very sensitive assay methodology (Table 15.1). The disadvantages include the relatively large size (approx. 250 μm diameter) of the probes and the time required between each sample (usually 5–20 min) so that sufficient levels of the substance of interested can collect to be measured.

15.2.2.   Factors Affecting in Vivo Microdialysis Measurements

Table 15.1 indicates the range of compounds that can be measured at present using *in vivo* microdialysis. While the ability to use the microdialysis approach to measure any particular substance in the extracellular fluid is highly dependent on the availability of suitably sensitive assays other factors also are important. These include the recovery characteristics of the dialysis membrane used in the production of the probe, the composition of the perfusate, and the effect of implanting the probe on the viability of the surrounding tissue.[21,23]

The amount of a particular substance collected by the probe will depend upon its relative recovery across the membrane. This recovery is defined as the ratio between the level of the substance in the perfusate and that in the solution outside the probe. The recovery is dependent on various factors, including the physico-chemical properties of the membrane being used, the size and design of the probe, the flow rate of the perfusate with recovery inversely related to flow rate, the tonicity and composition of the perfusion medium (see below), the nature and structure of the compound being dialyzed, temperature, and finally the need to consider if any interaction might occur between the membrane and the compound. Thus, for any individual substance it is necessary to optimize conditions to obtain sufficient recovery to be able to measure levels in the perfusate (Table 15.2).

It is essential to state the composition of the perfusate as it should have a similar ionic composition to brain extracellular fluid. This is achieved by using artificial cerebrospinal fluid. Changes from normal pH (7.4) or calcium (1.2 μM) values will influence basal levels of extracellular neurotransmitters; e.g., increases in Ca^{2+} and lowered pH both produce increases in basal extracellular amine levels that are not

Table 15.2.   In Vitro Recovery Values for 5-HT, 5-HIAA, GABA, ACh, and CAMP Using the Microdialysis Probes Shown in Fig. 1

5-HT	15.1 $\pm$ 1.9
5-HIAA	12.9 $\pm$ 2.5
ACh	15.0 $\pm$ 2
GABA	36.0 $\pm$ 1.5
cAMP	24.8 $\pm$ 2.4

[a] Data are given as the mean $\pm$ SEM of six rats. The recovery experiments were performed at 1 μL/min. Amines and their metabolites show very similar values whereas amino acid recovery values are somewhat higher.

physiological. Another important factor is the effect of the probe on the brain microenvironment after implantation because of the invasive nature of the microdialysis technique. Despite some degree of brain trauma with insertion of a microdialysis probe, it has been found that the blood–brain barrier is back intact shortly after implantation (30 min–2 h). Changes in local cerebral glucose metabolism and blood flow are sensitive indicators of tissue reaction caused by implantation. Despite consistent changes in both local cerebral glucose metabolism and blood flow during the first 3 h after implantation, these are not apparent at 24 h, indicating that brain metabolism has returned to normal. Cellular reactions (glial cell growth) to implantation of the dialysis fiber does limit the length of time the dialysis probe can be used *in vivo* as recovery declines with time.[21] An experiment to study the long-term application of microdialysis for the measurement of 5-HT concluded that the method was most useful between 24 and 72 h after implantation.

Microdialysis has been widely used to measure changes in extracellular neurotransmitter levels in response to pharmacological, physiological, and behavioral manipulations (see reviews listed for details). Figure 15.2 gives an example of the changes observed in the excitatory amino acid neurotransmitter glutamate and the inhibitory neurotransmitter GABA following experimental epilepsy in the rat.[2,24] In this situation there is an increase in both GABA and glutamate immediately postictally, followed by a prolonged decrease in GABA but a substantial increase in glutamate. The later increase in glutamate may have pathological significance as excessive extracellular levels of this excitatory amino acid are associated with neuronal damage.

15.3. VOLTAMMETRIC BIOSENSORS

Of the different methods currently employed to study the function of neurotransmitter systems, voltammetry arguably provides the most physiologically viable approach. Since Adams[5,43] first applied the principles of voltammetry to the measurement of electroactive neurotransmitters and their metabolites methodological and technical improvements have considerably strengthened and advanced this technique. The basic premise is that electroactive neurotransmitters can be oxidized (at a given potential) at an electrode surface (usually carbon) with the subsequent production of a measurable current, the amount of which is proportional to the concentration of electroactive material. Electrodes are then used *in vivo*, implanted into the CNS, or *in vitro* in tissue slice preparations to measure either basal or evoked neurotransmitter release. The two main elements in this technique that continue to evolve and provide methodological diversity are electrode fabrication and application of the oxidizing potential or waveform. These have produced distinct voltammetric methods aimed at improving different aspects of the technique, such as sensitivity, selectivity, and temporal and spatial resolution.

The different voltammetric methods and electrode configurations commonly used are described below.

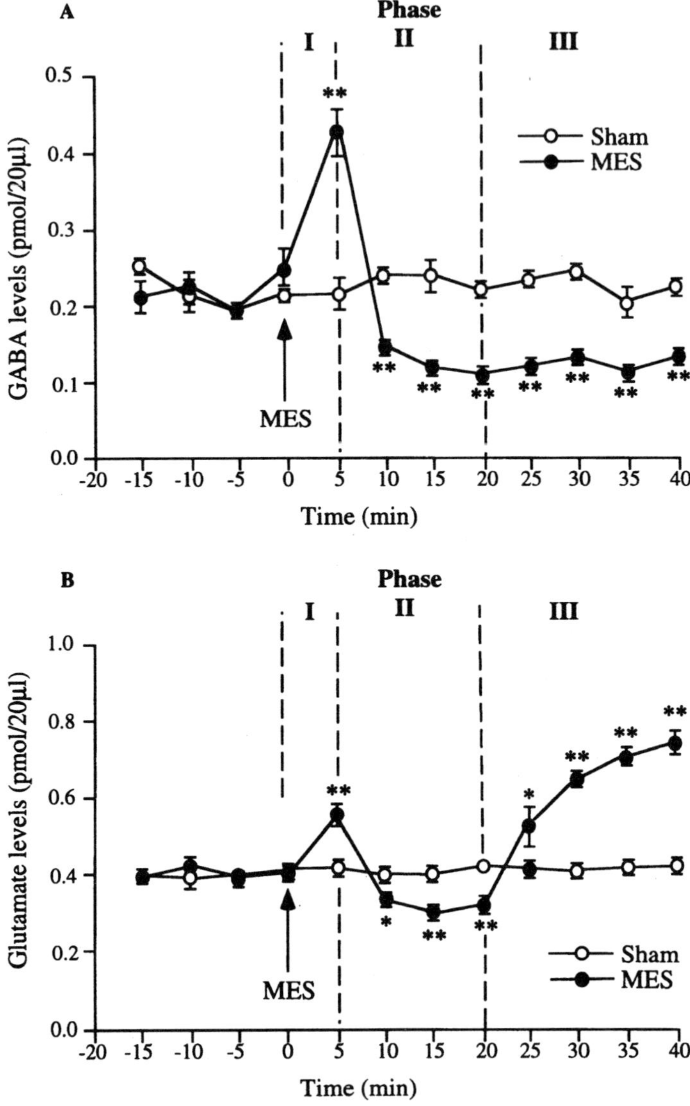

Figure 15.2. Example of an experiment using a microdialysis probe implanted into the ventral hippocampus to monitor changes in extracellular GABA (A) and glutamate (B) following experimental epilepsy (MES) in the rat. Note that there are three phases of MES-induced changes in extracellular GABA and glutamate dialysate levels in the ventral hippocampus of a freely moving rat. Administration of MES is indicated by vertical arrow. Each data point represents mean $\pm$ SEM ($n = 6$). GABA levels were increased by $102 \pm 16\%$ in phase I then reduced by $53 \pm 6\%$ during phase II with the reduction being sustained during phase III. Glutamate levels increased by $26 \pm 6\%$ in phase I, reduced by $16 \pm 2\%$ in phase II, prior to a maximal increase in phase III of $53 \pm 8\%$. *$P < 0.05$, **$P < 0.01$ significantly different from sham control, according to ANOVA with *post hoc* Dunnett's *t*-test. (Figure adapted from data published by Rowley *et al* 1995[24]).

15.3.1. Voltammetric Measurements

15.3.1.1. Differential Pulse

Differential pulse voltammetry (DPV) and the evolved differential normal pulse voltammetry consist of a series of measurement pulses (Fig. 15.3a) applied through a working electrode (usually carbon fiber). Pulses are superimposed on a potential that is "swept" between user-defined limits. Electroactive material is then detected by differential current measurements made just before and at the end of the pulse.

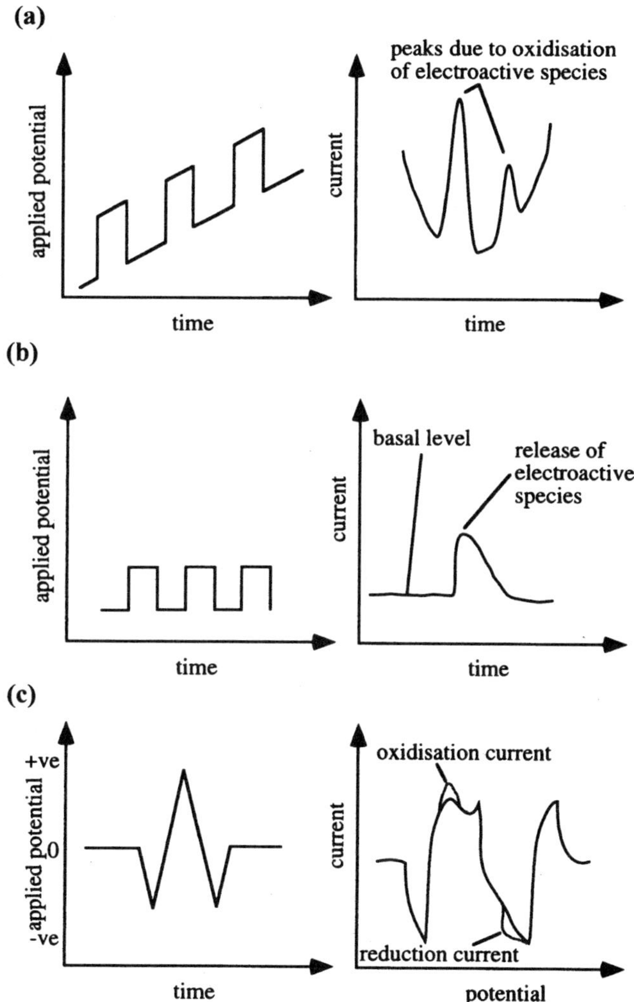

Figure 15.3. Applied waveforms for the different measurement techniques used for *in vivo* voltammetry: (A) differential pulse voltammetry, (B) chronoamperometry, (C) fast cyclic.

The use of pulses and differential measurement avoids problems with interference from capacitive currents that originate as a consequence of the electrode/media bilayer interface. Variations of DPV, such as differential normal pulse voltammetry, involve the use of different pulse configurations and prepulses, which help improve sensitivity and selectivity. The advantage of this method is the ability to examine a range of potentials between the limits of the sweep, making it possible to detect the presence of several electroactive species simultaneously. An associated disadvantage, however, is the time taken to complete a sweep, which can be of the order of minutes, giving this method relatively poor temporal resolution.

15.3.1.2. Chronoamperometry

Chronoamperometry (CA) and differential pulse amperometry use a constant applied potential on which pulses are superimposed (Fig. 15.3b) rather than a potential sweep method. This allows detection of concentration variations of compounds oxidizing within a narrow potential interval. While this method cannot discriminate compounds oxidizing at different potentials it does provide greatly improved temporal resolution. Without the need for a potential sweep pulses can be delivered and measurements made at rates up to 25 per second, providing millisecond resolution times.

15.3.1.3. Fast Cyclic Voltammetry

Fast cyclic voltammetry (FCV) utilizes the current produced by diphasic or triphasic anodic and cathodic potential sweeps. These are delivered over a few milliseconds (Fig. 15.3c) and repeated many times a second to produce an FCV scan. Electroactive compounds are detected as an increase in current above the background levels produced by the impedance of the electrode/electrolyte interface.[44] This occurs across a potential range with a peak in the middle, the height of which is proportional to the concentration of electroactive compound. There are several advantages associated with this method besides the temporal resolution that can be achieved (up to 100 Hz scan rate). Because the applied potential waveform has positive and negative voltage components, electroactive compounds undergo oxidation followed by reduction. This means that there is no neurotransmitter depletion and no buildup of oxidation products around the electrode that can poison the surface, reducing sensitivity, and be toxic to the surrounding neural tissue.

15.3.2. Electrodes

Most voltammetric techniques utilize a three-electrode system comprising a working electrode, a silver/silver chloride reference electrode, and a metal auxiliary electrode. Carbon still remains the most widely used material for construction of the working electrode. It provides the relatively inert, stable, conductive surface conditions for oxidation to take place. The first carbon-based electrodes developed for neurochemical studies consisted of a carbon-paste mixture inserted into Teflon tubing.[43] Used *in vivo* these produced a signal attributable to a combination of ascorbic acid, dopamine, DOPAC, uric acid, 5-HT, and 5-HIAA. Several improvements on this original design have been made, including graphite epoxy[45-47] and glassy carbon electrodes.[48,49]

Carbon fiber electrodes, first used by Gonon and co-workers,[50] are made from pyrolytic strands of carbon sealed in a pulled glass capillary tube (Fig. 15.4). These are by far the most widely used and developed electrodes owing to their comparative ease of use and physical characteristics. While other electrode designs, such as the carbon–stearate electrode (for review see Broderick[51]), are routinely used by some groups,[17] these will not be reviewed here.

15.3.3. Detectable Compounds

Simple carbon-based electrodes can detect most electroactive compounds within a given potential. To do this with the required selectivity and sensitivity depend on an electrode's surface characteristics. Electrodes used in chemically complex environments, such as the brain, must be able to distinguish the compound of interest from other electroactive species with similar oxidation potentials. For example, 5-HT measured with DPV oxidizes at $+240-260\,\mathrm{mV}$. This overlaps with the oxidation range for its primary metabolite, 5-HIAA, which occurs between $260-320\,\mathrm{mV}$. DPV and CA at untreated carbon fiber electrodes cannot resolve such similar oxidation

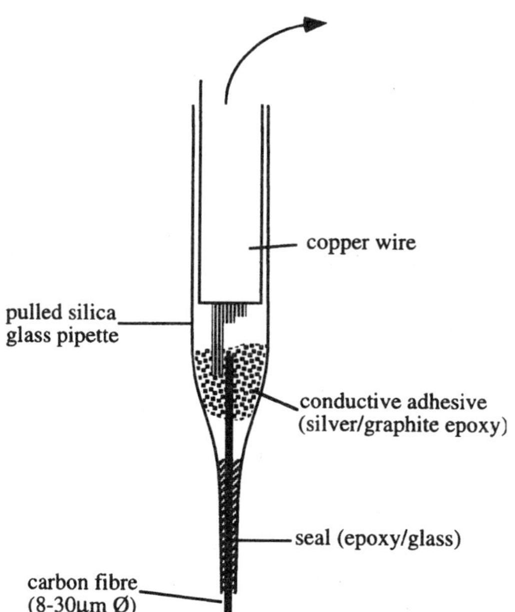

Figure 15.4. Typical carbon fiber voltammetry electrode (see, e.g., Stamford *et al.*[6] and Crespi[93] for details).

potentials without the use of complex mathematical curve-stripping techniques[52] or surface modification using an appropriate pretreatment. FCV overcomes this problem through its inherent fast scanning and sampling rates, which, e.g., allow stimulated 5-HT release to be selectively measured *in vitro* but not basal concentrations.[6] A combination of voltammetric methodology and pretreatment regimes including electrical and chemical surface modifications has enabled the development of several neurotransmitter selective electrodes (Table 15.3).

15.3.3.1. Electrical Pretreatment

As untreated carbon fiber electrodes used with DPV or CA are unable to detect the oxidation of specific electroactive compounds selectively, various electrical pretreatments have been used in order to improve electrode selectivity and sensitivity. These treatments modify the carbon surface and hence its electrical mass transport characteristics, improving the reversibility of the oxidation reactions, which in turn allows signals from different oxidized compounds to be better resolved. For example, a 70-Hz triangular wave (0–2.9 V for 20 s) followed by −0.8 V DC (5 s) and +1.5 V DC (5 s) applied to the electrode increases the selectivity for dopamine to ascorbic acid to 1000:1. Conversely, it is possible to decrease electrode sensitivity to certain molecules such as ascorbate, making the detection of dopamine possible. By using the appropriate sequence of electrical treatments, electrode selectivity and sensitivity can be optimized for a given electroactive compound (for review see Stamford et al.[6] and Crespie et al.[8]).

15.3.3.2. Chemical Treatment

While electrical pretreatment can help resolve closely associated oxidation peaks, compounds with closer oxidation potentials cannot be satisfactorily discriminated without further modifying the electrode surface characteristics. 5-HT and dopamine oxidize at potentials close to those of their major metabolites and often occur in much greater concentrations (100- to 1000-fold), producing a combined oxidation current. However, as the metabolites carry a negative charge they can be excluded from the carbon surface with a layer of perfluorosulfonated ion exchange resin (Nafion) which repels anions but is selectively permeable to cations. Coating electrically pretreated electrodes with several layers of Nafion enables an acceptable selectivity to be achieved for indole- and catecholamine transmitters.[8,53]

Nitric oxide (NO) is an oxidizable neurotransmitter that has recently attracted a great deal of attention. Its detection is made particularly difficult by a short half-life. It is claimed that FCV detects NO specifically owing to fast scanning and sampling.[54] In DPV and CA use is made of chemically treated electrodes coated with polymeric porphyrin (Nickel (II) tetrakis(3-methoxy-4-hydroxy-phenyl) porphyrin), which sensitizes the carbon to NO by catalyzing the oxidation process. The nitrite (NO_2^-) breakdown product of N, which itself is electroactive at a similar potential, is excluded using a Nafion coating over the porphyrin layer.[9]

Chemical pretreatment to modify electrode specificity has more recently been advanced by the use of immobilized enzymes attached to the electrode surface. These

Table 15.3. Measurement of Extracellular Neurotransmitters and Neuromodulators Using Voltammetry

Neurotransmitter	Detection method (oxidizable potential)	Interference	Electrode treatment	Preparation
5-HT	DDP (240–260 mV)[6] CA (240–350 mV) FCV (580–610 mV)[6,66]	5-HIAA, ascorbic acid uric acid None	Electrical: applied waveforms[6,8] Chemical: nafion coating[8] None	Whole brain[8] Brain slices[66]
Dopamine	DPV (80–90 V)[6] CA (240–350 mV)[71] FCV (500–600 mV)[67,68]	DOPAC, HVA, ascorbic acid, uric acid Ca^{2+}, H^{+6}	Electrical: applied waveforms[8] Chemical: nafion coating[8,53] None	Whole brain[69] Brain slice,[67,68] Whole brain,[62]
Melatonin	DPV (570 mV)[72]	5.HIAA	Electrical: applied waveforms[72]	Whole brain (anesthetized)
Nitric oxide	DPV (650 mV)[73] CA (630–750 mV)[9] FCV (1300 mV)[54]	NO_3^-, NO_2^- Dopamine	Chemical: nickel (II) tetrakis (3-methoxy-4-hydroxy-phenyl) porphyrin,[58] mafion[9] None	Isolated tissues,[59] cell culture,[60] whole brain[63,73] Brain slices
Glutamate	CA (700 mV)[10,11]	Ascorbate	Immobilized glutamate oxidase[10,11]	Whole brain[10,11] (anesthetized)
Noradrenaline	DPV (+0.1 V)[74] CA (180 mV)[74] FCV (685 mV)[64]	Ascorbic acid None	Electrical treatment[57] None	Isolated vessels[56] Whole brain[74] Brain slice[64]

biosensors rely on the ability of specific oxidase enzymes to convert electrically inert compounds with the liberation of an electroactive species, usually peroxide. Electrodes selective for glutamate use immobilized glutamate oxidase to produce peroxide molecules, which can be detected at a carbon or platinum electrode surface.[10,11] With a similar approach, electrodes specific for glucose,[12,55] glycerol, ascorbate, and acetyl choline have been reported. While this approach appears promising its true value has not yet been realized.

15.3.4. Application

To a large extent the biological system of interest influences the choice of voltammetric technique. Isolated tissue preparations, e.g., do not have the same chemical complexity as *in vivo* environments and relevant data can be obtained with less rigorous electrode specificity. DPV, CA, and FCV have all been used to measure transmitter levels in isolated arteries and cell culture preparations in association with less stringently selective electrodes.[56–60] Perfused brain slice preparations require a voltammetric approach that demands a greater degree of selectivity and temporal resolution. FCV fulfills these criteria and has emerged as the method of choice for monitoring 5-HT, DA, and NA in slice work.[19,61,64–68]

In vivo measurements made in whole brain are arguably the most technically demanding. The variety of electroactive species dictates the need for optimal selectivity. Additionally, physiological and neurological integrity should be conserved to ensure meaningful results. All three voltammetric techniques have been used to investigate whole-brain environments (see Table 15.1) and each has its relative merits. DPV offers the benefits of maximum sensitivity and specificity and the ability to monitor several electroactive compounds simultaneously. This is, however, at the expense of temporal resolution; scans typically take 1–3 min plus interscan intervals of 1–5 min.[8,63] CA is also able to detect basal neurotransmitter levels with some of the newer systems making as many as 25 measurements per second.[10,11,69,74] FCV has been used less extensively *in vivo*.[62] While FCV offers near "real-time" resolution its inability to measure basal neurotransmitter levels makes it a less popular technique for *in vivo* studies. Voltammetry in conscious, freely moving animals presents the greatest challenge. Aside from the issue of maximum selectivity and sensitivity, freely moving preparations are constrained by physical and technical factors. Chronic studies over days or weeks face the additional problem of electrode degradation. High concentrations of oxidation products can "poison" the electrode surface, which can also be deposited with glial cells, all of which can severely impair electrode function. Examples of conscious voltammetric studies are becoming more frequent with advances in system technology, although they tend to be conducted by groups with expert knowledge.[70–73]

15.4. ANTIBODY MICROPROBES

The antibody microprobe technique[75] is a method for the determination of neuropeptide release sites *in vivo*. The method was devised for peptides by Duggan and Hendry in the mid-1980s and to date has been used almost exclusively by their groups to investigate the distribution and release of several neuropeptides in cat spinal cord, e.g., substance P.[76,77] CGRP and galanin,[78] neurokinin A,[79,80] and neuropeptide Y.[81] The methodology has also been applied to the detection of neuropeptide release in rat brainstem (substance P[82]) and forebrain (β-endorphin,[83] thyrotrophin releasing hormone[84]).

The concept is quite simple; antibody microprobes are similar to microelectrodes, but longer and with a more parallel geometry. They are coated with antibodies specific for a particular peptide before implantation into the central nervous system (CNS). Endogenous peptide in the extracellular fluid binds to the antibodies exposed at the surface of the probe. Subsequent incubation of the probe in a specific tracer *in vitro* and exposure to x-ray film can reveal the site at which endogenous peptide is bound to the probe. Lighter sections of the image suggest that less binding of tracer has occurred, which is considered to be the result of endogenous peptide occupying the antibody binding sites (see Fig. 15.5).

While the concept is simple the practical use of antibody microprobes is less so. The difficulties encountered during fabrication, use, and analysis probably accounts for the relatively small number of laboratories that have attempted to utilize this technology, although a simpler methodology is available for at least one neuropeptide.[19] Furthermore they do provide an opportunity for the detection of neuropeptides in a field with few alternatives especially when sampling from discrete areas of the spinal cord, e.g., lamina II.

15.4.1. Probe Design

Antibody microprobes have been developed using techniques from electrophysiology, radioimmunoassay, and autoradiography. Glass microelectrodes are coated with aminosilane to form a granular finish with a large surface area. Protein G or protein A (depending upon the species used to raise the antiserum) is covalently bound to the amine groups, and these proteins specifically bind the Fc portion of IgG antibodies. The antibodies bind peptide *in vivo* where they lie adjacent to sites of release. The system does not provide a quantitative measure of neuropeptide release, but an indication of sensitivity can be obtained from simple *in vitro* tests and qualitative comparisons can be made between preparations.

The manufacture and use of antibody microprobes is not a trivial procedure. The original protocol for the production of microprobes is both time-consuming and technically demanding, requiring the development of specific skills. Various items of special equipment and facilities are also required. These comprise a glass micropipette puller, a facility to decant silane under argon, an oven to bake the aminosilane coat to the probes, radioactive handling facilities, x-ray film processing and a computerized image analysis system.

It is not possible here to give more than a brief outline of the procedure for manufacture, use, and analysis of microprobes (see Duggan et al.[15,85] and Waterfall[19] for details). Fabrication can be divided into three steps; (1) forming glass micropipettes, (2) coating with aminosilane and, (3) attaching the biological coating.

It is usually convenient to produce microprobes in batches of 25. The glass

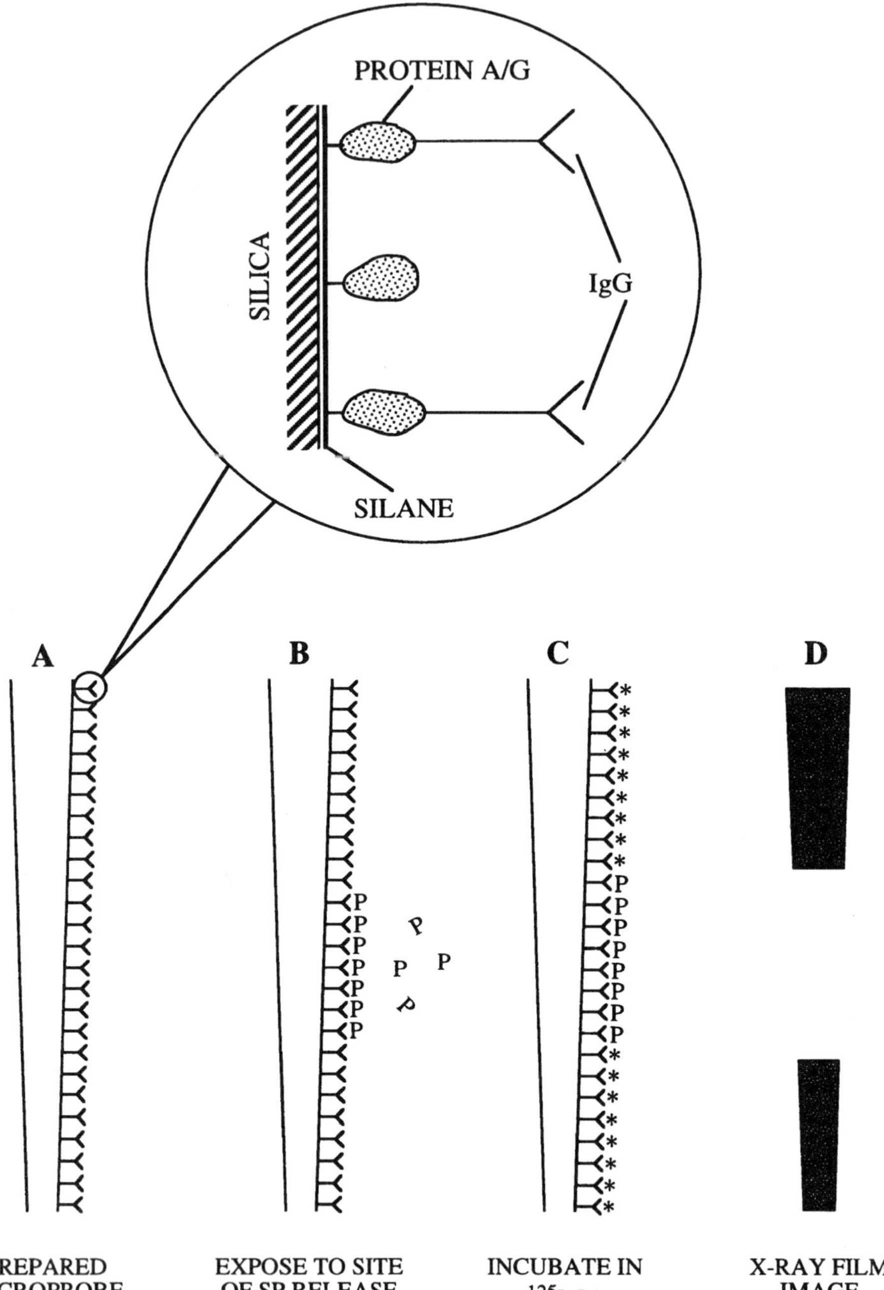

Figure 15.5. Schematic diagram depicting the intended binding of specific IgG antibodies to the surface of glass microelectrodes via protein A or G and aminosilane. The series of microprobes A, B, C, D, depicts the binding of peptide *in vivo* and the formation of an image on x-ray film after incubation in specific tracer. The lower portion of the figure is modified from Duggan and Hendry[94] and the enlargement modified from Jonsson *et al.*[95]

Table 15.4. Comparison of Approximate Times Required to Manufacture a Single Batch of 25 Microprobes Using Alternative Methodologies

Step	Time for original method (h) (hydrated glass/silane/toluene[85])	Time for epoxy charcoal coating over (h) (Tungsten wire[19])
Prepare glass/wire	33	10
Prepare silane, etc.	12	1
Base coat	68	2
Biological coat	23	12
Total time (h)	136	24

micropipettes are cleaned several times and incubated in a solution of aminosilane in toluene at 4°C. If the first coating is not stable then the procedure is repeated until the coating is satisfactory. Two days before experiment day protein A or G is covalently bound to the aminosilane using glutaraldehyde and on the eve of the experiment microprobes are incubated in antibody. Depending on the approach adopted, a batch of probes can take between 24 and 136 h to prepare (Table 15.4).

15.4.2. Experimental Use

Antibody microprobes are invasive devices for use in anesthetized preparations. Individual microprobes remain *in vivo* for several minutes before removal and individual incubation in tracer. Before the final microprobe is removed from a specific location, dye is ejected from the tip to correlate positional information from the probe images with histological sections. Following tracer labeling, the glass micropipettes are snapped off their shafts, glued to cards and apposed to x-ray film for up to 21 days (depending upon the film type and the desired quality of the image). Thus the outcome of an individual microprobe implantation is an autoradiograph film image. Groups of these microprobe images must be pooled for final analysis.

If antibody microprobes have detected release then a lighter area will appear on an image relative to adjacent sections of the probe. Images are scanned using computerized densitometry and the resulting data (units are 0–255 gray scales) are used to construct a graph of image density along the length of the probe (Fig. 15.6). With further analysis (means, standard errors, and *t*-tests) similar sections of microprobes can be compared to determine significant differences.

In practice there may be several factors to consider before attempting any measurements in the CNS using antibody microprobes. First is the availability of an antibody and radioactive tracer. Many antibodies are available commercially but the selectivity and sensitivity of each has to be considered in detail, and it is sensible to characterize the radioimmunoassay for a specific peptide before using it with antibody microprobes. Secondly, only those areas of the CNS with the highest density of peptide-containing nerve terminals should initially be considered, e.g.,

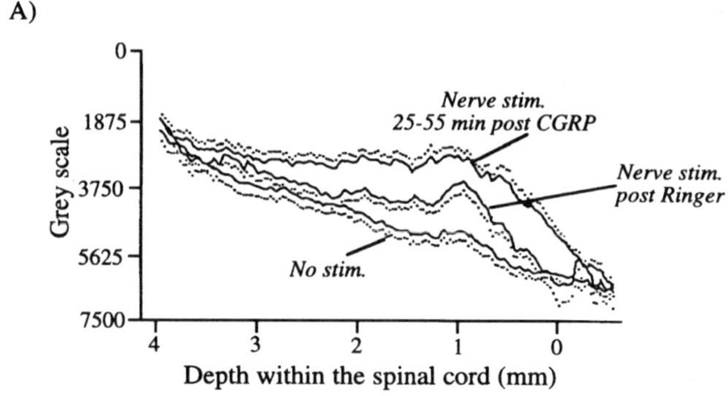

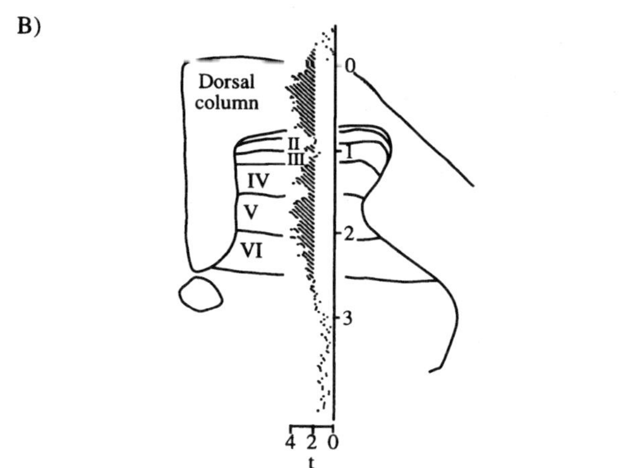

Figure 15.6. The figure (adapted from Schaible et al.[96] with permission) illustrates two methods of expressing *in vivo* data: (A) mean microprobe traces with standard errors of groups of microprobes produced by the original aminosilane method and coated with antibodies specific for substance P. The top trace is derived from a group of microprobes following local microinjection or neuropeptide CGRP and peripheral nerve stimulation. It is compared with controls when either vehicle was accompanied by stimulation or neither simulation nor injection occurred. The probe tip is to the left; (B) illustrates the final stage in the processing with the results of an unpaired *t*-test superimposed over a representation of the anatomy of the implantation site in transverse section.

compare the reported density of the peptide to be investigated with that of substance P in the dorsal horn. Thirdly, the experimental stimulus should be carefully considered to maximize release.

Studies using the technique have revealed new information about the release of neuropeptides, especially in the spinal cord of the cat. The earliest studies using the technique concentrated upon the peptide substance P. The basal and stimulated release of this peptide was observed using antibody microprobes in the lumbar dorsal horn of cats by two groups.[75–77,86] The elevation of extracellular peptide was centered around the substantia gelatinosa of the dorsal gray matter and occurred without an overt stimulus, although the considerable surgical preparation prior to the implantation of microprobes might have been responsible for these basal measurements. Further release was initiated by electrical stimulation of un-myelinated primary afferent nerves (specifically not myelinated nerves) by noxious mechanical and thermal stimuli applied to the ipsilateral hindlimb and following application of capsaicin to the ipsilateral tibial nerve. Such release was not affected by supraspinal input in these anesthetized preparations.

Other antibody microprobe studies have revealed the release of several peptides in the substantia gelatinosa of cat dorsal horn. Basal and stimulated release (in response to noxious thermal stimuli) of somatostatin was observed by Morton and Hutchison,[87] defining the substantia gelatinosa as the site of release. Similarly basal calcitonin gene-relate peptide (CGRP) and galanin were monitored in the same model and CGRP could be elevated by electrical stimuli and noxious thermal and mechanical stimuli.[78,88] Neurokinin A (NKA, a member of the tachykinin family sharing the same C terminal sequence of amino acids) was released by noxious thermal and mechanical stimuli — by electrical stimulation of the tibial nerve and ventral root.[89,79] The stimulation revealed a diffuse distribution in cat dorsal horn. This diffuse distribution and apparent persistence in the tissue was seen as evidence for the peptide as a neuromodulatory substance (as opposed to a fast-acting neurotransmitter substance). Similar patterns were also apparent after inflammation of the knee joint in a model of acute arthritis. In the same cat model elevated dynorphin A was monitored in lamina 1, 5, and 6 of the dorsal horn following electrical stimulation of unmyelinated primary afferents.[90]

Until recently the antibody microprobe technique had been confined to cat lumbar spinal cord but has now been extended to a range of neuropeptides in rat brain and spinal cord. β-Endorphin release has been observed in the periaqueductal gray of rats in response to noxious mechanical stimulation,[83] TRH appeared to be released in the forebrain of rats following amphetamine challenge,[84] spontaneous release of galanin was detected in rat spinal cord,[91] neurokinin A release in rat substantia nigra was dependent upon intact dopaminergic input,[92] dynorphin-A was detected in rat dorsal horn following peripheral stimulation,[93] and substance P was released in the brainstem following noxious stimulation of the nasal mucosa.[82]

Clearly there is considerable evidence to promote the view that this technique is suitable for the location of neurotransmitter release sites in the central nervous system in fine detail for any peptides for which a selective binding protein capable of producing a sensitive assay is available. The first of these studies by Duggan and Hendry demonstrated the release of substance P in the dorsal horn, defining a new level of spatial discrimination for biochemical sampling. This feature, the ability to precisely localize neuropeptide release, is the principal advantage in the overall strategy of this method and the use of fine-tipped microprobes probably provides the best level of resolution outside the synapse. While voltammetric probes can provide similar resolution and theoretically offer measurement of some electroactive peptides, e.g., somatostatin,[94] this is the only currently available method for peptides.

Table 15.5.   Features of Antibody Microprobes

Advantages	Disadvantages
Neuropeptides	Lengthy preparation and analysis time
Fine discrimination of sites of release	Not real-time and noncontinuous
Specific	Invasive
Potential for wide range of molecules	Anesthetized preparations

The microprobe system can be regarded as a qualitative rather than a quantitative measure of release, more suitable to the localization of release than the determination of the magnitude of change, although some quantitative data have been presented from *in vitro* experiments.[78,90] One of the difficulties in the further development of the strategy for the use of microprobes is the considerable practical difficulties that may arise from the application to conscious *in vivo* preparations. Applying other methods to the conscious animal model is not without difficulty but essentially perfusion and telemetry methods involve a single implantation. The antibody microprobe system would require the continual implantation and removal of microprobes throughout the preparation and would have to be accompanied by the repeated use of light anesthesia, which is unacceptable in most experimental paradigms (Table 15.5).

In conclusion, the antibody microprobe technique was developed to measure the release of neuropeptides. Few (if any) other methods enable the location of release with the same level of spatial discrimination. Although implementation of the methodology requires considerable resources it can provide information that would otherwise be unavailable. The technique could be extended to other tissues apart from the CNS and used to detect nonpeptides where a suitable binding molecule can be fixed to a glass micropipette or similar structure.

15.5.   CONCLUSIONS AND THE FUTURE

In vivo biosensors are making an increasing impact on neuroscience research, and their future development will concentrate on certain specific features. The need is for high selectivity combined with high sensitivity and a fast response time. Changes in neurotransmitter release are often very rapid in response to physiological or behavioral events and most of the current generation of probes do not have sufficiently fast response times to measure these changes in "real-time." In the future, probes will be required that have the speed of detection available with the fast voltammetric electrodes combined with the selectivity of the much slower microdialysis approach. Such biosensors would have wide range of application in basic and clinical biomedical research, particularly if the probes could be used together with appropriate image collection technology, such as functional magnetic resonance imaging (MRI).

REFERENCES

1. Cadogan AK, Marsden CA Measurement of neurotransmitters and their associated second messengers *in vivo*. In: *Neurochemistry. A Practical Approach*. Turner AJ, Bachelard HS eds. 2nd Ed. Oxford: IRL Press, 1997, pp 175–199.

2. Rowley HL, Martin, KF, Marsden CA. Decreased GABA release following tonic-clonic seizures is associated with an increase in extracellular glutamate in rat hippocampus *in vivo*. *Neuroscience* 1995;68:415–422.

3. Ungerstedt V. Microdialysis: principles and applications for studies in animals and man. *J Int Med* 1991;230:365–373.

4. During MJ, Spencer DD. Extracellular hippocampal glutamate and spontaneous seizure in the concious human brain. *Lancet* 1993;341:1607–1610.

5. Adams RN, Marsden CA. Electrochemical detection methods for monoamine measurements *in vitro* and *in vivo*. In: Iversen LL, Iversen SD, Suyer SH, eds. *Handbook of Psychopharmacology*, New York: Plenum Press 1982, vol. 15, pp 1–74.

6. Stamford JA, Crespi F, Marsden CA. *In vivo* voltammetric methods for monitoring monoamine release and metabolism. In: Stamford JA, ed. *Monitoring Neuronal Activity: A Practical Approach*. Oxford: IRL Press, 1992, pp 113–145.

7. Marsden CA, Joseph MH, Kruk ZL et al. In vivo voltammetry—present electrodes and methods. *Neuroscience* 1988;25,389–400.

8. Crespi F, Martin KF, Marsden CA. Measurement of extracellular basal level of 5-HT *in vivo* using Nafion coated carbon fibre electrodes combined with differential pulse voltammetry. *Neuroscience* 1988;27,885–896.

9. Malinski T, Taha Z. Nitric oxide release from a single cell measured in situ by a porphyrinic-based microsensor. *Nature* 1992;358,676–678.

10. Hu Y, Mitchell K, Albahadily F, et al. Direct measurement of glutamate release in the brain using a dual enzyme based electrochemical sensor. *Brain Res* 1994;659:117–125.

11. Cooper JM, Foreman PL, Glidle L, et al. Glutamate oxidase enzyme electrodes: microsensors neurotransmitter determination using electrochemically polymerised permselective films. *J Electro Anal Chem* 1995;338:143–149.

12. Lowry JP, Fillenz M. Simultaneous real-time monitoring of glucose and oxygen *in vivo*. In: Gonzalez-Mora JL, Borges R, Mas M, eds. *Monitoring Molecules in Neuroscience*. Santa Cruz de Tenerife: La Laguna University Press, 1996; pp 50–51.

13. Brazell MP, Marsden CA. Intracerebral infection of ascorbate oxidase: effect on *in vivo* electrochemical recordings. *Brain Res* 1982;249:167–172.

14. Sharp T, Zetterström T. *In vivo* measurement of monoamine neurotransmitter release using brain microdialysis. In: Stamford JA, ed. *Monitoring Neuronal Activity. A Practical Approach*, Oxford: IRL Press, 1992, pp 147–179.

15. Westerink BHC. Brain microdialysis and its application for the study of animal behaviour. *Behav Brain Res* 1995;70:103–124.

16. Blaha CD, Liu D, Phillips AG. Improved electrochemical properties of stearate-graphite electrodes after albumin and phospholipid treatments. *Biosens Bioelectron* 1996;11:63–79.

17. Kruk ZL, O'Connor JJ. Fast electrochemical studies in isolated tissues. *Trends Pharm Sci* 1995;16:145–149.

18. Duggan, AW. Antibody Microbes. In: Stamford JA, ed. *Monitoring Neuronal Activity: A Practical Approach*, Oxford: IRL Press 1992, pp 181–201.

19. Waterfall AH, Clarke RW, Bennett GW. Novel methods for the preparation of antibody microprobes. *J Neurosci Meth* 1994;55:41–45.

20. Hillerd L, Persson L, Ponten U, et al. Neurometabolic monitoring of the ischemic human brain using microdialysis. *Acta Neurochir (Wien)* 1990;102:91–97.

21. Benveniste, H. Brain microdialysis. *J Neurochem* 1989;52:1667–1679.

22. Benveniste H, Diemer NH. Cellular reactions to implantation of a microdialysis tube in the rat hippocampus. *Acta Neuropathol (Berlin)* 1987;74:234–238.

23. Obrenovich TP, Zilkha E, Vrenjak J. Intracerebal microdialysis: electrophysiological evidence of a critical pitfall. *J Neurochem* 1995;64:1884–1887.

24. Rowley HL, Marsden CA, Martin KF. Differential *in vivo* effects of phenytoin and sodium valproate on generalised seizure-induced changes in extracellular GABA and glutamate release. *Eur J Pharm* 1995;294,541–546.

25. Marsden CA, Aspley SA, Rowley HL. Neurotransmitters: determination in physiological samples using electrochemical detection. In: *Encyclopedia of Analytical Science*, New York: Academic Press 1995, volume 6, pp 3295–3303.

26. Deleu D, Sarre S, Ebinger G, Michotte Y. Simultaneous monitoring of levodopa, dopamine and their metabolites in skeletal muscle and subcutaneous tissue in different pharmacological conditions using microdialysis. *J Pharm Biomed Anal* 1993;11:377–385.

27. Rowley HL, Martin KF, Marsden CA. Determination of *in vivo* amino acid neurotransmitters by high-performance liquid chromatography with *o*-phthalaldehyde-sulphite derivatisation. *J Neurosc Meth* 1995;57:93–99.

28. Maidment NT, Siddall BJ, Rudolph VR. et al. Dual determination of extracellular cholecystokinin and neutotensin fragments in rat forebrain: microdialysis combined with a sequential multiple antigen radioiommunoassay. *Neuroscience* 1991;45:81–93.

29. Becker C, Benoliel JJ, Hamon M. Stimulation of 5HT$_{1A}$ receptors prevents the restraint stress—and yohimbine-induced increase in the extracellular levels of cholecystokinin (CCK) in the frontal cortex. In: Gonzalez-Mora JL, Borges R, Mas M, eds. *Monitoring Molecules in Neuroscience* Santa Cruz de Tenerife: La Laguna University Press, 1996, pp 201–202.

30. Maidment NT, Brumbaugh DR, Rudolph VD, et al. Microdialysis of extracellular endogenous opioid peptides from rat brain *in vivo*. *Neuroscience* 1989;33:549–557.

31. Gruber S, Nomikos GG, Stevensson TH, et al. Effects of *d*-amphetamine and dopamine receptor antagonists on calcitonin gene-related peptide levels in ventral striatum. In: Gonzalez-Mora JL, Borges R, Mas M, eds. *Monitoring Molecules in Neuroscience*. Santa Cruz de Tenerife: La Laguna University Press, 1996, pp 205–206.

32. Mertes PM, Beck B, Jaboin Y, et al. Microdialysis in the estimation of interstitial myocardial neuropeptide release. *Regul Pept* 1993;49:81–90.

33. Orosco M, Gerozissis K, Rouch C, et al. Feeding-related immunoreactive insulin changes in the PVN-VMH revealed by microdialysis. *Brain Res* 1995;671:149–158.

34. Kendrick KM, Ohkura S, Guevara R, et al. Nitric oxide and the neural basis of olfactory memory. In: Gonzalez-Mora JL, Borges R, Mas M, eds. *Monitoring Molecules in Neuroscience*. Santa Cruz de Tenerife: La Laguna University Press, 1996, pp 217–218.

35. Vallebuona F, Raiteri M. Monitoring cyclic GMP during cerebellar microdialysis in freely-moving rats as an index of nitric oxide synthase activity. *Neuroscience* 1993;57:577–585.

36. Van Wylen DL, Willis J, Sodhi J, et al. Cardiac microdialysis to estimate interstitial adenosine and coronary blood flow. *Am J Physiol* 1990;258:H1642–H1649.

37. Waterfall AH, Singh G, Fry JR, et al. Detection of the lipid peroxidation product malonaldehyde in rat brain *in vivo*. *Neurosci Lett* 1995;200:69–72.

38. Waterfall AH, Singh G, Fry JR, et al. Acute acidosis elevates malonaldehyde in rat brain *in vivo*. *Brain Res* 1996;712:102–106.

39. Camilli F, Hallström A, Darvelid M, et al. Microdialysis in injured human brain. In: Gonzalez-Mora JL, Borges R, Mas M, eds. *Monitoring Molecules in Neuroscience*. Santa Cruz de Tenerife: La Laguna University Press, 1996, pp 395–396.

40. Wu H-Q, Schwarcz R. Endogenous kynurenic acid in the normal and dysfunctional rat brain *in vivo*. In: Gonzalez-Mora JL, Borges R, Mas M, eds. *Monitoring Molecules in Neuroscience*. Santa Cruz de Tenerife: La Laguna University Press, 1996, p 362.

41. Pravda M, Bogaert L, Sarre S, et al. On-line *in vivo* monitoring of extracellular quinones in rat brain. In: Gonzalez-Mora JL, Borges R, Mas M, eds. *Monitoring Molecules in Neuroscience*. Santa Cruz de Tenerife: La Laguna University Press, 1996, pp 371–372.

42. Scott DO, Sorenson LR, Steele KL, et al. *In vivo* microdialysis sampling for pharmacokinetic investigations. *Pharm Res* 1991;8:389–392.

43. Kissinger PT, Hart JB, Adams RN. Voltammetry in brain tissue: a new neurophysiological approach. *Brain Res* 1973;55:209–213.

44. Stamford JA. Fast cyclic voltammetry-measuring transmitter release in real-time. *J Neurosci Meth* 1990;34:67–72.

45. Conti JC, Strope E, Adams RN, et al. Voltammetry in brain tissue chronic recording of stimulated dopamine and 5-hydroxy tryptamine release. *Life Sci* 1978;23:2705–2716.

46. Cheng HY, Shenk, J, Huff, R, et al. In vivo electrochemistry-behaviour of microelectrodes in brain tissue. *J Electroanal Chem* 1979;100:23–31.

47. Wang J. Epoxy based microelectrodes for voltammetric measurements. *Anal Chem* 1981;53:2280–2283.

48. Mos J, Broxterman HJ, Van Bennekem WP. *In vivo* voltammetric investigations into the action of HA 966 on central dopaminergic neurones. *Brain Res* 1981;207;465–470.

49. Gonon F, Cespuglio R, Ponchon JL, et al. Mesure electrochimique cintinue de la liberation de DA realisé *in vivo* dans le neostriaum du rat. *C R Acad Sci Paris* 1978;286:1203–1206.

50. Ponchon JL, Cespuglio R, Gonon F, et al. Normal pulse polarography with carbon fibre electrodes for *in vitro* and *in vivo* determination of catecholamines. *Anal Chem* 1979;51:1483–1489.

51. Broderick PA. State-of-the-art microelectrodes for *in vivo* voltammetry. *Electroanalysis* 1990;2:241–251.

52. Gonzalez Mora JL, Guadalupe T, Fumero B, et al. Mathematical resolution of mixed *in vivo* voltammetry signals: models, equipment, assessment by simultaneous microdialysis sampling. *J Neurosci Meth* 1991;39:231–244.

53. Crespi F, Mobius C. *In vivo* selective monitoring of basal levels of cerebral dopamine using voltammetry with Nafion modified (NA-CRO) carbon fibre microelectrodes. *J Neurosci Meth* 1992;42:149–161.

54. Iravani MM, Kruk ZL. Electrochemical detection of nitric oxide using fast cyclic voltammetry. *J Physiol* 1993;496:48P.

55. Moussy F, Harrison DJ, O'Brien DW, et al. Performance of subcutaneously implanted needle type glucose sensors employing a novel trilayer. *Anal Chem* 1993;65:2072–2077.

56. Gonon F, Msghina M, Stjärne L. Kinetics of noradrenaline released by sympathetic nerves. *Neuroscience* 1993;56:535–538.

57. Gonon F, Bao JX, Msghina M, et al. Fast and local electrochemical monitoring of noradrenaline release from sympathetic terminals in isolated rat tail artery. *J Neurochem* 1993;60:1251–1257.

58. Malinski T, Taha Z, Grunfeld S, et al. Measurements of nitric oxide in biological materials using a porphyrinic microsensor. *Anal Chim Acta* 1993;279:135–140.

59. Malinski T, Mesaros S, Patton SR. Direct measurement of nitric oxide in the cardiovascular system. *Physiol Res* 1996;45:279–284

60. Blatter LA, Taha Z, Mesaros S, et al. Simultaneous measurements of Ca^{2+} and nitric oxide in bradykinin-stimulated vascular endothelial cells. *Cir Res* 1995;76:922–924.

61. Tonner C, Stamford JA. 'Real time' measurement of dopamine release in an *in vitro* model of neostriatal ischaemia. *J Neurosci Meth* 1996;67:133–140.

62. Garris PA, Christensen JRC, Rebec GV, et al. Real-time measurement of electrically evoked extracellular dopamine in the striatum of freely moving rats. *J Neurochem* 1997;68:152–161.

63. Desvignes C, Robert F, Vachette C, et al. Monitoring nitric oxide (NO) in rat locus coeruleus: differential effects of NO synthase inhibitors. *Neuroreport* 1997;8:1321–1325.

64. Stamford JA, Palij P. Real-time monitoring of endogenous noradrenaline release in rat brain slices using fast cyclic voltammetry: 1. Characterisation of evoked noradrenaline efflux and uptake from nerve terminals in the bed nucleus of stria terminalis, pars ventralis. *Brain Res* 1992;587:137–146.

65. Jones SR, Mickelson GE, Collins LB, et al. Interference by pH and Ca^{2+} ions during measurements of catecholamine release in slices of rat amygdala with fast-scan cyclic voltammetry. *J Neurosci Meth* 1994;52:1–10.

66. O'Connor J, Kruk Z. Fast cyclic voltammetry can be used to measure stimulated endogenous 5-HT release in untreated brain slices. *J Neurosci Meth* 1991;38:25–34.

67. Palij P, Bull DR, Sheehan MJ, et al. Presynaptic regulation of dopamine release in corpus striatum monitored *in vivo* in real time by fast cyclic voltammetry. *Brain Res.* 1990;509:172–174.

68. Bull DR, Palij P, Sheehan MJ, et al. Application of fast cyclic voltammetry to measurement of electrically evoked dopamine overflow from brain slices *in vitro*. *J Neurosci Meth* 1990;32:37–44.

69. Friedemann M, Gerhardt GA. *In vivo* electrical studies of the effects of N-methyl-D-aspartate (NMDA) on dopamine nerve terminals in the neostriatum of the anesthetized rat. *Proc West Pharmacol Soc* 1989;32:143–147.

70. Friedemann MN, Gerhardt GA. Regional effects of ageing on dopaminergic function in the Fischer 344 rat. *Neurobiol Aging* 1992;13:325–332.

71. Crespi F, Ratti E, Trist DG. Melatonin: a hormone monitorable *in vivo* by voltammetry? *Analyst* 1994;119:2193–2197

72. Burlet S, Cespuglio R. Voltammetric detection of nitric oxide (NO) in the rat brain: its variations throughout the sleep–wake cycle. *Neurosci Letts* 1997;226:131–135.

73. Renner KJ, Pazos L, Adams RN. *In vivo* voltammetric evidence for the detection of norepinephrine release in the thalamus of freely moving rats. *Brain Res* 1992;577:49–56.

74. Duggan AW, Hendry, IA. Laminar localisation of the sites of release of immunoreactive substance-P in the dorsal horn with antibody coated micro-electrodes. *Neurosci Lett* 1986;68:134–140.

75. Duggan AW, Morton CK, Zhao ZQ, et al. Noxious heating of the skin releases immunoreactive substance P in the sub-gelatinosa of the cat: a study with antibody microprobes. *Brain Res* 1987;403:345–349.

76. Zhao Z-Q, Yang HQ, Zhang KM, et al. Release and depletion of substance P by capsaicin in substantia gelatinosa studied with the antibody microprobe technique and immunohistochemistry. *Neuropeptides* 1992;23:161–167.

77. Morton CR, Hutchison WD. Release of sensory neuropeptides in the spinal cord: studies with CGRP and galanin. *Neuroscience* 1989;31:807–815.

78. Hope PJ, Lang CW, Duggan AW. Persistence of immunoreactive neurokinins in the dorsal horn of barbiturate anesthetized and spinal cats, following release by tibial nerve stimulation. *Neurosci Lett* 1990;118:25–28.

79. Hope PJ, Jarrott B, Schaible H-G, et al. Release and spread of immunoreactive neurokinin-A in the spinal cord in a model of acute arthritis. *Brain Res* 1990;533:292–299.

80. Mark MA, Jarrott B, Colvin LA, et al. The release of immunoreactive neuropeptide Y in the spinal cord of the anaesthetized rat and cat. *Brain Res* 1997;754:195–203.

81. Schaible H-G, Ebersberger A, Peppel P, et al. Release of immunoreactive substance P in the trigeminal brainstem nuclear complex evoked by chemical stimulation of the nasal mucosa and the dura mater encephali: a study with antibody microprobes. *Neuroscience* 1997;76:273–284.

82. Duggan AW, Hope PJ, Lang CW, et al. Noxious mechanical stimulation of the hind paws of the anaesthetized rat fails to elicit release of immunoreactive beta-endorphin in the periaqueductal grey matter. *Neurosci Lett* 1993;149:205–208.

83. Waterfall AH, Clarke RW, Bennett GW. Detection of thyrotrophin releasing hormone in rat brain *in vivo* using novel antibody microprobes. *Neurosci Lett* 1993;151:97–100.

84. Duggan AW, Hendry I, Green JL, et al. The preparation and use of antibody microprobes. *J Neurosci Meth* 1988;23:241–247.

85. Duggan AW, Hendry IA, Morton CR, et al. Cutaneous stimuli releasing immunoreactive substance P in the dorsal horn of the cat. *Brain Res* 1988;451:261–273.

86. Morton CR, Hutchison WD, Hendry IA, et al. Somatostatin: evidence for a role in thermal nociception. *Brain Res* 1989;488:89–96.

87. Morton CR, Hutchison WD. Morphine does not reduce the intraspinal release of CGRP in the cat. *Neurosci Lett* 1990;117:319–324.

88. Duggan AW, Hope PJ, Jarrott B, et al. Release, spread and persistence of immunoreactive neurokinin A in the dorsal horn of the cat following noxious cutaneous stimulation studies with antibody microprobes. *Neuroscience* 1990;35:195–202.

89. Hutchison WD, Morton CR, Terenius L. Dynorphin A: *in vivo* release in the spinal cord of the cat. *Brain Res* 1990;532:299–306.

90. Hope PJ, Lang CW, Grubb BD, et al. Release of immunoreactive galanin in the spinal cord of rats with ankle inflammation: studies with antibody microprobes. *Neuroscience* 1994;60:801–807.

91. Furmidge LJ, Duggan AW, Arbuthnott GW. *In vivo* detection of immunoreactive neurokinin A release within rat substantia nigra and its dependency on a dopaminergic input. *Brain Res* 1995;679:241–248.

92. Riley RC, Zhao ZQ, Duggan AW. Spinal release of immunoreactive dynorphin A(1-8) with the development of peripheral inflammation in the rat. *Brain Res* 1996;710:131–142.

93. Crespi F. *In vivo* voltammetry with micro-biosensors for analysis of neurotransmitter release and metabolism. *J Neurosci Meth* 1990;34:53–65.

94. Duggan AE, Hendry I. Immobilised antibodies: a newer approach to measurement of released neuropeptides. *News Physiol Sci* 1988;3:44–46.
95. Jonsson U, Malmqyvist M, Olofsson G, et al. Surface immobilization techniques in combination with ellipsometry. *Meth Enzym* 1988;137:381–388.
96. Schaible H-G, Hope PJ, Lang CW, et al. Calcitonin gene-related peptide causes intraspinal spreading of substance P released by peripheral stimulation. *Eur J Neurosci* 1992;4:750–757.

16

Biosensors for Agrochemicals

Rosie B. Wong

16.1. INTRODUCTION

Population growth and a reduction in farming acreage over the last few decades has brought about an increasing emphasis on the use of agrochemicals in modern agricultural practice to improve yield for cost-effective production of food and fiber. Although regulatory requirements for registration of agrochemical products have restricted or banned the use of many proven toxic substances and the industry is striving to produce safer and lower-use-rate chemicals, the environmental impact of some long-lasting compounds in soil and water and bioaccumulation of some compounds in the ecosystem are still of concern. Many agencies worldwide have mandated monitoring of groundwater, drinking water, and food commodities for chemical residues, and awareness of the impact of these residues on daily life has been heightened.

The traditional methods of analysis for chemical residues have relied on gas chromatography (GC) or high-performance liquid chromatography (HPLC) carried out in a laboratory environment by trained personnel. This approach has proved to be inadequate under the current monitoring requirements because it is costly and labor-intensive and only a small fraction of the required monitoring samples can be processed. When immunoassays came on the scene, their lower costs and more rapid turnaround times resulted in their acceptance in some areas of food and water monitoring. However, immunoassay is not a panacea for all agrochemical monitoring needs nor will the ever-inventive scientists stop from improving upon this solution. Biosensors, marrying as they do the properties of biological materials such as enzymes, antibodies, organelles, and even cells with those of electrochemical, photochemical, and physical phenomena of molecular interactions promise to bring monitoring technology to an even higher plane, e.g., multianalyte, continuous, on-site, or remote monitoring.

Rosie B. Wong ● Agriculture Research Division, American Cyanamide Company, Princeton, New Jersey 08543-0400.

Biosensors and Their Applications, edited by Yang and Ngo, Kluwer Academic/Plenum Publishers, New York, 1999.

283

Recent biosensor development activity and instrumentation for environmental monitoring have been extensively reviewed.[1-5] This chapter, however, is a survey of biosensors from the perspective of plant pesticide categories. Plant pesticides can be divided into three major functional groups: insecticides, herbicides, and fungicides. Insecticides include organophosphates, organochlorines, carbamates, pyrethroids, and pyrroles; fungicides are various thio-containing compounds and morpholine classes; herbicides include the largest classes of organic chemicals such as triazines, sulfonylureas, and imidazolinones. The major biological materials used in biosensors for agrochemical monitoring are enzymes, antibodies, lipid bilayer membranes, photosynthetic reaction centers, and whole cells. An attempt will be made here to compare sensor approaches with respect to application and detection levels. To conclude, some thoughts on the direction of future biosensor development will be presented.

16.2.   INSECTICIDES

16.2.1.   Organophosphates, Carbamates, and Organochlorines

The organophosphorus compounds include six different substitutions on the phosphate moiety as depicted in Table 1. Carbamate insecticides are esters of carbamic acids, which include among others aldicarb, carbofuran, carbaryl, and methomyl. Chlorinated organic insecticides comprise a large family of chlorinated compounds such as aldrin, chlordane, DDT, endosulfen, heptachlor, and lindane among many others. The common thread in all these compounds is their ability to inhibit cholinesterase enzymes.

16.2.1.1.   Enzyme Sensors

Acetylcholinesterase and butylcholinesterase enzymes are by far the most often used enzymes for agrochemical sensing owing perhaps to their commercial availability and stability. By monitoring the formation of acid during the hydrolysis of the substrates acetylcholine or butylcholine, quantitation of the enzyme activity can be simply achieved potentiometrically. Detection can be achieved by ion selective electrodes[6-9] or ion-selective field-effect transistors.[10-12] The sensitivity can approach 10^{-10}–10^{-9} M for paraoxon, malathion, and parathion-methyl.[13] An amperometric sensing mode that makes use of the substrate acetylthiocholine was described by Skládal and Mascini.[14] The thiocholine produced is electrochemically oxidized at $300\,mV$ using a cobalt phthalocyanine-modified electrode. With this system, sensitivities for paraoxon and heptenophos are 1.5 and 8.4 μg/L, respectively, and the test can be done in 3 min. Another means of amperometric sensing is by using acetylcholinesterase coupled with choline oxidase enzyme.[15-17] Measurement of the consumption of oxygen or the production of hydrogen peroxide, a product of oxidase enzyme is the indirect indicator of choline concentration, thus cholinesterase activity. Since cholinesterase enzyme inhibition by the insecticides is irreversible, most often the sensors cannot be reused. Martorell et al.,[18] however, reported a

Table 16.1. Six Leading Types of Organophosphorous Insecticides

Name	Generic formula	Example
Phosphate	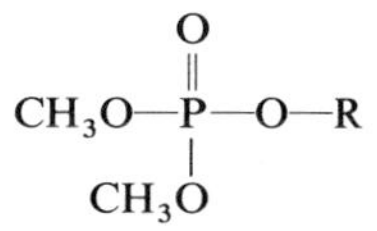 $\underset{\displaystyle CH_3O}{\overset{\displaystyle O}{CH_3O-\overset{\|}{P}-O-R}}$	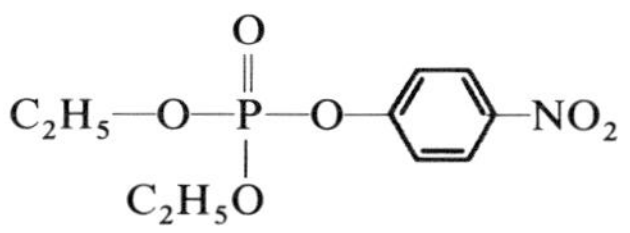 $\underset{\displaystyle C_2H_5O}{\overset{\displaystyle O}{C_2H_5-O-\overset{\|}{P}-O-}}\!\!\!\bigcirc\!\!\!-NO_2$ Paraoxon
Phosphorothioate	$\underset{\displaystyle C_2H_5O}{\overset{\displaystyle S}{C_2H_5O-\overset{\|}{P}-O-R}}$	$\underset{\displaystyle C_2H_5O}{\overset{\displaystyle S}{C_2H_5-O-\overset{\|}{P}-O-}}\!\!\!\bigcirc\!\!\!-NO_2$ Parathion
Phosphorothiolate	$\underset{\displaystyle C_2H_5O}{\overset{\displaystyle O}{C_2H_5O-\overset{\|}{P}-S-R}}$	$Br-\bigcirc-O-\overset{\displaystyle O}{\underset{\displaystyle OC_2H_5}{\overset{\|}{P}}}-S-CH_2CH_2CH_3$, Cl Profenofos
Phosphorodithioate	$\underset{\displaystyle C_2H_5O}{\overset{\displaystyle S}{C_2H_5O-\overset{\|}{P}-S-R}}$	$\underset{\displaystyle C_2H_3O}{\overset{\displaystyle S}{C_2H_5-O-\overset{\|}{P}-S-CH_2-S-C_2H_5}}$ Phorate
Phosphonate	$\underset{\displaystyle CH_3O}{\overset{\displaystyle O}{CH_3O-\overset{\|}{P}-R}}$	$\underset{\displaystyle CH_3O}{\overset{\displaystyle O}{CH_3O-\overset{\|}{P}-\overset{\overset{\displaystyle OH}{\|}}{CH}-CCl_3}}$ Trichlorfon
Phosphoroamidate	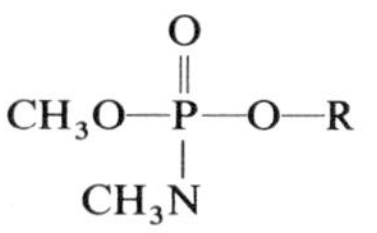 $\underset{\displaystyle CH_3N}{\overset{\displaystyle O}{CH_3O-\overset{\|}{P}-O-R}}$	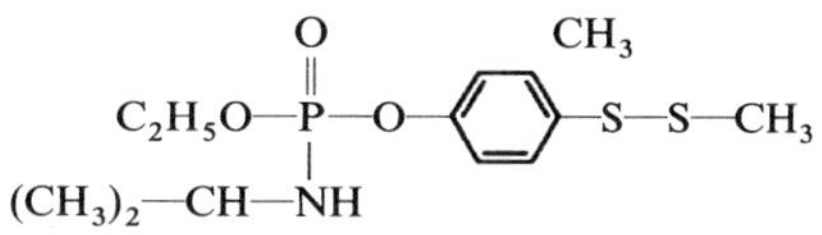 $\underset{\displaystyle (CH_3)_2-CH-NH}{\overset{\displaystyle O}{C_2H_5O-\overset{\|}{P}-O-}}\!\!\!\bigcirc\!\!\!-\underset{\displaystyle}{\overset{\displaystyle CH_3}{S-S-CH_3}}$ Fenamiphos

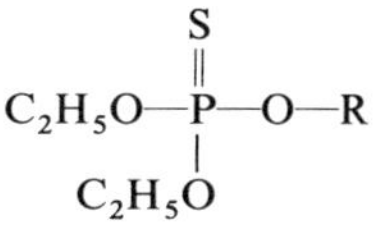

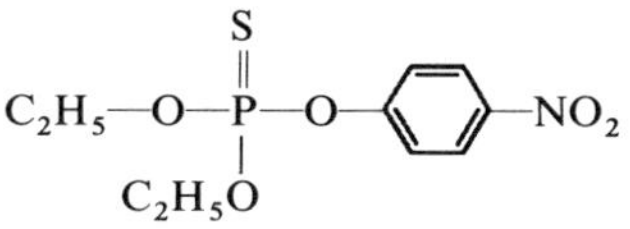

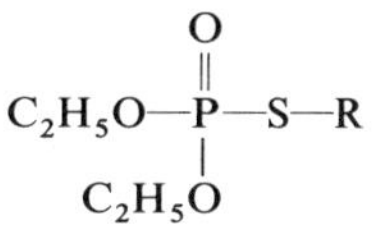

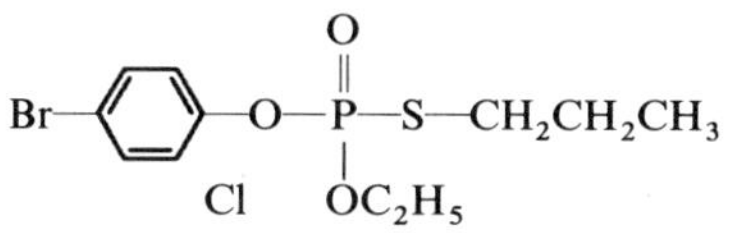

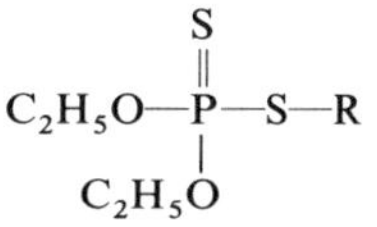

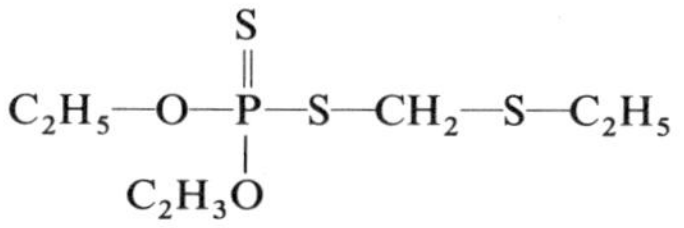

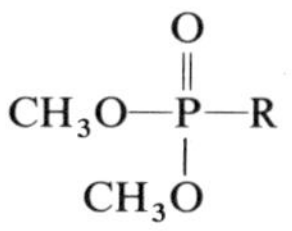

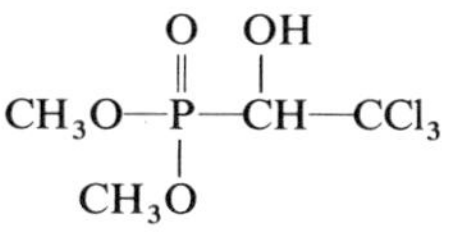

polishable graphite–epoxy biocomposite that can be easily polished for reuse. A major drawback in most of these sensor reports is the lack of test data on matrices containing real-world samples. Kumaran and Morita[19] demonstrated a one-use polyamide membrane immobilized butylcholinesterase electrode system that successfully quantifies diazinon and fenitrothion in soil extracts. The level of quantitation for this system ranges between 10 and 500 ppb.

The presence of organic solvent in aqueous solutions frequently causes enzyme denaturation and loss of enzymatic activity. However, Mionetto et al.[20] demonstrated the stability of immobilized acetylcholinesterase in many alkane and nonpolar solvents such as hexane and *n*-octane. Iwuoha and Smyth[21] showed that organic phase horseradish peroxidase enzyme electrodes can detect the presence of thiocarbamate metabolites. It is postulated that an acetylcholinesterase organic phase enzyme electrode biosensor might be useful for detecting low levels of organophosphate or carbamate insecticides in water or food substances if the compounds can be first partitioned and concentrated in the organic solvent.

Rogers et al.[22] attached fluorescein-labeled acetylcholinesterase onto optical fibers. Excitation of the fluorescein dye within the evanescent zone of the fiber causes fluorescence, and this fluorescence signal is captured by internal reflection through the fiber. When acetylcholine is hydrolyzed by the fiber-bound cholinesterase enzyme, a proton is generated, causing fluorescein fluorescence to be quenched. In the presence of inhibitors such as paraoxon, bendiocarb, methyomyl, dichlorphos, and parathion, cholinesterase activity is inhibited, thus reducing proton formation, which in turn reduces the quenching effect. Pesticide concentration can thus be related to the fluorescein signal. This test can be done in 1 min. However, the sensitivity only ranges from 10^{-4} to 10^{-8} M, which is perhaps not sufficient for agrochemical monitoring. The same laboratory also reported a light addressable potentiometric sensor (LAPS) using biotinylated cholinesterase.[23] The enzyme is captured on a strapavidin membrane and a potentiometric measurement is taken upon hydrolysis of the substrate to form acetic acid. The important features of the system are its multiple-sampling capability and its reusability, but its sensitivity is not impressive; with the best being 10 nM for paraoxon and bendiocarb.

Other enzyme systems applicable to organophosphates are: (1) inhibition of alkaline phosphatase hydrolysis of 1,2-dioxetane phenyl phosphate compound, which generates *in situ* chemiluminescence upon hydrolysis[24]; paraoxon and methyl parathion can be quantified to a level of 50 ppb using this system. (2) Hydrolysis of the phosphate insecticides by phosphotriesterase from bacterium, *Flavobactrium sp*[25]; however, the sensitivity is only in the μM range. (3) Inhibition of lipase by organophosphorus pesticides have been reported by Wei et al.[26] using surface acoustic wave measurement; the detection limit is 58 ng/mL.

16.2.1.2.   Immunosensors

Owing to the general success of enzyme sensing for organophosphate and carbamate insecticides, very little work has been devoted to immunosensing of these compounds. Anis et al.[27] demonstrated a fiber-optic immunosensor for parathion that can detect 0.3 ppb using a quartz fiber coated with casein–parathion conjugate and antibody bound to the surface is quantified with a fluorescein-labeled second antibody. The fluorescein fluorescence signal is captured through the evanescent waveguide. Enzyme immunoassays for other organophosphates have been reported, so it will not be long before more immunosensors for this class become available.

16.2.1.3.   Bilayer Lipid Membrane Electrochemical Sensors

In 1994 Nikolelis and Kull[28] observed that insecticides such as monocrotofos, carbofuran, and methylparathion interact with bilayer lipid membranes made of phosphatidyl choline and dipalmitylphosphatidic acid, producing a transient current signal with a duration of seconds. The signal generation is directly related to the concentration of the insecticides. Dose-vs.-signal plots are reported in the μM to mM range.

16.2.2. Pyrethroids Immunosensors

Pyrethroids are synthetic compounds produced to duplicate or improve on the biological activity of the active component of the pyrethrum plant, a botanical insecticide. This class includes allethrin, permethrin, cypermethrin, and fluvalinate, to name just a few. Although pyrethroids are used in many areas of agriculture and household insect control, there has been only one report (Northrup et al.[29]) describing an attempt to use a fiber-optic immunosensor to detect a pyrethroid.

16.3. HERBICIDES

16.3.1. Triazines

Triazines are compounds based on a symmetrical triazine structure with substitution sites on each of the carbon atoms. Compounds from this class most often used as herbicides are atrazine, propazine, simazine, prometon, and terbutryn. Atrazine has been used historically in many agricultural practices and much groundwater has been contaminated because of it. Monitoring efforts for atrazine are becoming routine in many countries so most of the development work in sensor technology has focused on atrazine.

16.3.1.1. Enzyme Sensors

Atrazine has been reported to inhibit tyrosinase[30–31] and acetylcholinesterase.[32] An amperometric detector was used for the tyrosinase system allowing a 2 ppm detection. The multiresidue monitoring system reported by Cowell et al.[32] includes atrazine inhibition of acetylcholinesterase, but no detection level was given.

16.3.1.2. Immunosensors

Perhaps the lack of suitable enzyme systems useful for sensitive triazine monitoring inspired the development of an immunosensor for this class of chemicals. Both monoclonal and polyclonal antibodies raised against numerous triazine analogues, ranging from very compound-specific to class-specific, have been prepared. Detection methods also cover the entire spectrum of sensing technologies. With the indirect approaches, labels such as enzymes and fluorophore are most commonly used as the signal generator. Concern for groundwater contamination by the triazines especially on the European continent has motivated the development of automated systems.

Bier et al.[33] immobilized a derivative of s-triazine, aminohexylatrazine onto hard-clad silica fiber and used fluorescein-conjugated monoclonal antibody to bind to the fiber. Quantitation of the fluorescence coupled onto the fiber is accomplished via the evanescent waveguide. The limit of detection for terbutryn approaches 0.1 ng/L.

Several flow injection systems such as the membrane system developed by Krämer and Schmid[34,35] and the capillary assay employing amperometric detection of alkaline phosphatase[36] all demonstrate sensitivity of detection within the European Drinking Water Act limit of 0.1 ppb. Owing to the nature of the flow systems, the analyte–antibody complex is established in a very short time frame and the matrix effect is usually minimal. Thus no pretreatment of the sample is necessary. Other systems such as the Eu(III) chelate W8044-Eu fluorescence labeled anti-triazine antibody used by Wortberg et al.[37] strives for high throughput in anticipation for the large numbers of samples to be monitored.

Many direct optical measurement methods that record the binding events of antibody to the analyte or vice versa have been used for the triazines. Guilbault et al.[38] reported a piezoelectric immunosensor using a polyclonal sheep antibody layered onto gold electrodes of 10-MHz piezoelectric crystals precoated with protein A. Determinations of atrazine in drinking water can be carried out eight or nine times with drying of the crystal between runs. The sensitivity was at 0.03 ppb. Steegborn and Skládal[39] were able to coat a silanized gold electrode of piezoelectric quartz crystal with albumin-conjugated atrazine via glutaraldehyde cross-linking and took antibody-binding measurements in a flow cell system. In the presence of atrazine, the rate of frequency decrease due to decrease in antibody bound or the relative change in frequency after a 10 min binding period provides an indicator of the quantity of atrazine. Decreases in relative binding of 5 and 30% are recorded with 0.1 and 1.0 μg/L atrazine, respectively. The crystal can be regenerated with a 5-min flow of 100 mM NaOH with no drift in signal.

A grating coupler waveguide system with antigen coated on the surface was constructed by Bier et al.[40] Assays are performed by preincubating the triazine-containing sample with antibody for 1 h before introducing it into the flow cell. This grating coupler waveguide is an automatic flow-through system, regenerable by treatment with protease. A sensitivity of 15 μg/L is reported. Owing to the lengthy incubation time required, a matrix effect is observed, specifically with bovine serum albumin. A surface plasmon resonance-based BIAcore system has been demonstrated by Minunni and Mascini[41] with a detection limit of 0.05 μg/L. A surface transverse wave acoustic sensor reported by Tom-Moy et al.[42] detects 0.06 μg/L atrazine in 3 min. This system immobilizes biotinylated atrazine to the device surface through an avidin bridge. Direct measurement of antibody attachment is accomplished by the surface transverse acoustic wave device. The surface is refreshed with acid and the reproducibility of the system within nine cycles is 7%.

16.3.1.3. Photosystem Reaction Center Sensors

Atrazine is known to inhibit photosystem II (PSII) in various light and dark cycles. Ways to isolate and store functional PS II for sensor use has proved to be the major hurdle in sensor development. Thylakoid membranes isolated from spinach leaves entrapped in polyvinylalcohol as reaction centers has been demonstrated by Rouillon et al.[43] Photoproduction of H_2O_2 by this photosystem is determined electrochemically and reduction of such a signal is correlated with the concentration of atrazine. Extended storage stability of spinach thylakoids has been demonstrated by Brewster et al.[44] By using calcium alginate gel, activity can be retained for 1 week at 4°C or for 6 months by rapid freezing. The sensitivity of these systems is at best 10 ppb. The better stability of bacterial photoreaction centers allows for more successful sensor development.

Herbicide detection based on the interference of photobleaching of the bacteriochlorophyll dimer has been successfully demonstrated. This process is monitored by measuring the decrease in absorption at 860 nm. A purple bacteria *Rhodobacter-sphaeroides* reaction center for atrazine sensing has been reported by Jockers et al.[45] The soluble reaction center is stable for 6 months but the sensitivity for atrazine is only at the 600-ppb level. A microalga, *Chlorella vulgaris*, biosensor employing a carbon-cathode oxygen electrode for monitoring the evolution of oxygen is used for detecting atrazine,[46] and the system is stable for more than 24 h of continuous use, but the detection limit is only around 100 ppb. The fact that PSII sensor sensitivity is intimately governed by the inhibition constant of the herbicides and that atrazine is a strong inhibitor of plant photosynthesis but a weak inhibitor of bacterial photosynthesis presents a dilemma for sensor development. Progress is being made to modify the bacterial system genetically to improve sensitivity to atrazine inhibition and provide useful and stable material for sensors.

16.3.1.4. Bilayer Lipid Membrane Electrochemical Sensors

Nikolelis and Siontorou[47] developed a filter-supported bilayer lipid membrane of egg phosphatidylcholine and dipalmitoylphosphatidic acid for sensing the triazine family of compounds. The time delay between introduction of the herbicide to the bilayer lipid membrane and the transient signal response was specific for each of the triazine compounds studied. Thus, a mixture of the simazine, atrazine, and propazine compounds can be monitored simultaneously, with detection limits in the tens of ppbs.

16.3.2. Phenylacetic Acids

The herbicides 2,4-dichlorophenoxyacetic acid (2,4-D) and 2,4 5-trichlorophenoxyacetic acid (2,4,5-T) are used extensively for field crop, turf and non-crop-land applications. The potential health hazard associated with these compounds, which are now known to be carcinogenic, has made monitoring of the environment and drinking water desirable. Since no specific enzyme inhibition system is available for these compounds, immunochemical approaches are the only means of developing sensors. Radioimmunoassays and enzyme-linked immunosorbent assays have been reported since the early 1980s.[48-50] Eremin[51] reported on using a fluorescence polarization technique to detect 100 ppb of 2,4-D. Although this system has the benefit of not requiring reagent separation and the test is completed in 7 min, the sensitivity may not be adequate for monitoring purposes. A competitive ELISA flow system using peroxidase enzyme-linked antibody, which monitors the enzyme activity either by a light addressable potentiometric sensor or a redox potential sensor with a sensitivity of 50 ppb was developed by Piras et al.[52] Khomutov et al.[53] immobilized 2,4-D specific antibody on a porous photoactivated cellulose matrix and was able to measure the pH change of the peroxidase enzyme action on the substrate mixture of ascorbic acid, *o*-phenyleneamine and H_2O_2. The sensitivity of this system is reported at 1 ppb. Further development in automation along this line of immunoelectrodes was shown by Dzantiev et al.[54] By immobilizing the antibody on a graphite electrode and using 5-aminosalicyclic acid and H_2O_2 as a substrate, the assay is accomplished in 12 min and can be repeated 60 times. However, the sensitivity seemed to be compromised to a level of 40–50 ppb.

Two very sensitive systems have been reported by Whittmann et al.[55] A fiber-optic evanescent waveguide immunosensor using a fluorescein-labeled second antibody as the signal generator has a sensitivity of 0.2 ppb. Preincubation of the analyte with the antibody is needed before it flows through the optical-fiber sensor. Regeneration of the sensor can be accomplished by proteinase K digestion of the surface-attached protein. In the same report, a flow injection system using antibody bound to porous microglass beads through a biotin–avidin bridge was used. This antibody column forms the basis for a competitive sequential saturation assay in which the analyte and corresponding enzyme-linked analyte compete for a limited number of antibody sites. A fluorescent substrate is employed to quantify the bound enzyme. The system can be regenerated, and the cycling time for each assay is 15 min. A 0.03-ppb detection limit has been reported for this system. The water matrix was tested in both systems and found to be compatible.

Improvements in sensitivity, throughput, and ease of assay are addressed by Dzgoev et al.[56] in their microformat approach. Using a microscope slide containing four rows of twenty 2-mm-square wells they conducted ELISA with an alkaline phosphatase chemiluminescence substrate with a slow scan CCD camera to monitor the light signal. A detection limit of 6 pg/mL was achieved with precision at less than 10% for samples below 1 ng/mL.

16.3.3. Substituted Ureas and Sulfonylureas

Diuron, linuron, and difenoxuron are a few of the herbicides containing the common urea moiety with substitution on the two nitrogen atoms. The sulfonylureas such as bensulfuron-methyl, chlorimuron-ethyl, and chlorsulfuron share a common structure of sulfone substitution on one nitrogen atom of the urea molecule with further substitutions extending from the other nitrogen and the sulfur atoms. Both classes of herbicides have low toxicity. The fact that the sulfonyl ureas are applied at very low rates reduces the environmental impact; however, monitoring of these herbicides is needed to address carryover phytotoxic issues on follow crops, which may be sensitive to these herbicides.

16.3.3.1. Enzyme Sensors

The sulfonylureas are potent inhibitors of the enzyme acetolactate synthetase (ALS),[57] a key enzyme in the biosynthesis of the branched-chain amino acids valine,

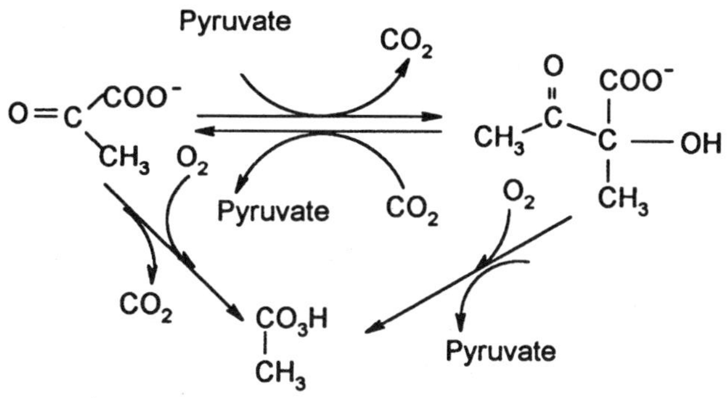

Scheme 1. Acetolactate and oxygenase activities catalyzed by ALS.

leucine, and isoleucine in plants. Marty et al.[58] first demonstrated the feasibility of preparing polymerized gel using polyvinyl alcohol bearing styrylpyridinium groups (PVA-SbQ) in the presence of the ALS enzyme. This membrane retains some enzyme activity for a short period of time. Since ALS is synthesized in small amounts in plant material, the enzyme supply is an issue. Seki et al.[59] used an *E. coli* enzyme to assemble sensors that has the same sensitivity to the herbicides. By harnessing the oxygenase reaction catalyzed by the ALS enzyme as shown in Scheme 1, the sensor can monitor oxygen consumption as a record of enzyme activity. Unfortunately, this oxygenase activity monitoring approach compromises the sensitivity for herbicide detection. The concentrations of sulfometuron methyl, thifensulfuron methyl, or chlorsulfuron necessary for a detectable reduction of oxygen consumption ranges between 2.5 and 10 μM.

16.3.3.2. Immunosensors

Antibodies to urea herbicides are available, so immunosensors have been developed as well. A fluorescence polarization sensing system was reported by Mel'nichenko et al.[60] When a marker fluorescent molecule and the antiserum to methabenzthiauron are mixed, a polarized fluorescent signal is produced. In the presence of unlabeled methabenzthiauron, fluorescence polarization is reduced, indicating the displacement of marker molecules from the antibody by the analyte. This signal reduction is directly correlated to the concentration of the analyte and the signal can be monitored without separation of the displaced marker. The concentrations of the analyte tested are 0.05 to 5 μg/mL, which is perhaps too high for environmental monitoring.

16.3.3.3. Photosynthetic Reaction Center and Whole-Cell Sensors

The photobleaching phenomenon of the reaction center of *R. sphaeroides*[45] and inhibition of immobilized thylakoids[43] described for the triazine herbicides have also been used for diuron. The detection limit for diuron in the thylakoid system is 10^{-8} M, better than that of atrazine. Other whole-cell systems using microalga *C.*

vulgaris[46] with a carbon-cathode electrode can, as noted before, also detect iso-proturon at 100 μg/mL in 30 min.

16.3.4. Imidazolinones

Members of the class of imidazolinone herbicides share the central feature of a substituted imidazolinone ring linked to a nicotinic acid, benzoic acid, or other ring compound, as shown in Fig. 16.1. This class includes imazapyr, imazethapyr, imazaquin, imazamox, imazampic, and imazabenzmethyl. The imidazolinones inhibit the branched-chain amino acid synthesis enzyme acetohydroxy acid synthetase (AHAS)[61] or ALS. As with the sulfonylureas, they are nontoxic to mammals and fish and used in small quantities are of very little environmental concern. However, soil residues of these compounds may present a problem for sensitive rotational crops, which is the major reason for sensor development.

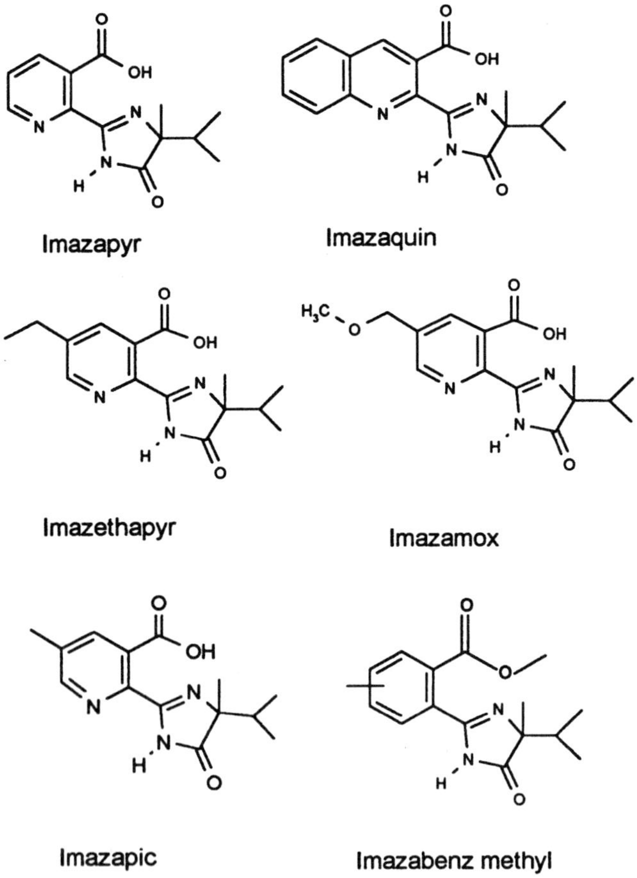

Figure 16.1. Imidazolinone class herbicides.

16.3.4.1. Enzyme Sensor

The ALS enzyme sensor reported by Seki et al.[59] is the only enzyme sensor tested for the imidazolinones. The detection level is in the $2-10\,\mu M$ concentration.

16.3.4.2. Immunosensors

Both polyclonal and monoclonal antibodies to the imidazolinones have been produced.[62] These antibodies bind the entire class of imidazolinone compounds with varying affinities. A fiber-optics sensor using an evanescent waveguide has been demonstrated.[63] This system employs antibody physically adhered to the surface of quartz fibers that is inserted into a flow cell. Fluorescein-labeled analyte is the signal generator. Two protocols are possible: (1) the label is introduced simultaneously with the imazethapyr for direct competition with the antibody sites on the fiber; (2) the fiber is preloaded with fluorescein-labeled analyte and then the label and free imazethapyr mixture are introduced to displace the bound signal, with the latter leading to higher sensitivity of detection. A similar dose response can be obtained with buffered water extracts of soil matrix or a buffer solution.

A capillary-column flow injection liposome system[64] uses immobilized mono-clonal antibody on the surface of a capillary tube. The tube is preloaded with fluorescent dye-filled liposomes composed of imidazolinone previously conjugated in phospholipid molecules. The dose response of imazethapyr is established by intro-ducing the analyte solutions to displace the bound liposome and then introducing a lysis solution to quantify the bound liposomes remaining on the column. The fluorescence generated by the lysed liposomes is monitored downstream. When a fluorescein-labeled analyte is used instead of the dye-filled liposome, the sensitivity of detection is about 1000-fold less, thus demonstrating the power of liposome amplification. The same fluorescein-labeled analyte and antibody is used in a polarization fluorescence system, with a resulting detection level of 100 ppb.[65]

Table 16.2 shows a comparison of the same analyte and antibody in four sensing systems with respect to sensitivity, time required for assay, precision, matrix effect,

Table 16.2. Comparison of Imazethapyr Immunosensors

	Evanescent waveguide[63]	Flow injection capillary column	Liposome-amplified flow injection capillary column[64]	Fluorescence polarization
Detection limit	0.3 ppb	100 ppb	0.1 ppb	10 ppb
Cycle time	10–15 min	10–15 min	10–15 min	7 min (one use)
Precision	<10% SD	<10% SD	11–28% SD	<6% SD
Reusability	Yes	Yes	Yes	No
Matrix effect	Compatible to soil extract	Not determined	Water matrix reduced sensitivity 10-fold	Not determined
Automation	Not developed	Yes	Yes	Yes

and automation. The liposome flow injection system provides the most sensitive detection level and the best possibilities for automation. However, since the liposomes have to be prebound onto the antibody in the capillary column, nonspecific binding of matrix to the lipophilic liposomes causes interference and lowers precision. As demonstrated in a surface water test, the matrix reduces the sensitivity 10-fold in this liposome amplified assay. The fluorescein-conjugated analyte direct competition system is the least sensitive, although matrix effect may not be a problem, as was observed in fiber-optic waveguide flow systems. A fiber optic waveguide is perhaps a more efficient mode of capturing fluorescence signal compared with straight fluorometry as in the capillary column system.

16.4. FUNGICIDES

16.4.1. Dithiocarbamate Enzyme Sensor

The number of fungicide-monitoring reports is negligible compared with those for herbicides. Marty et al.[58] demonstrated fungicide dithiocarbamate sensing using aldehyde dehydrogenase inhibition. By using propionic aldehyde substrate for the aldehyde dehydrogenase enzyme, the cofactor NAD–NADH is coupled with a diaphorase catalyzing the oxidation of $K_3Fe(CN)_6$, which generates a low potential current (Scheme 2). The decrease in the current reflects the enzyme inhibition by the fungicide. No sensitivity level was established.

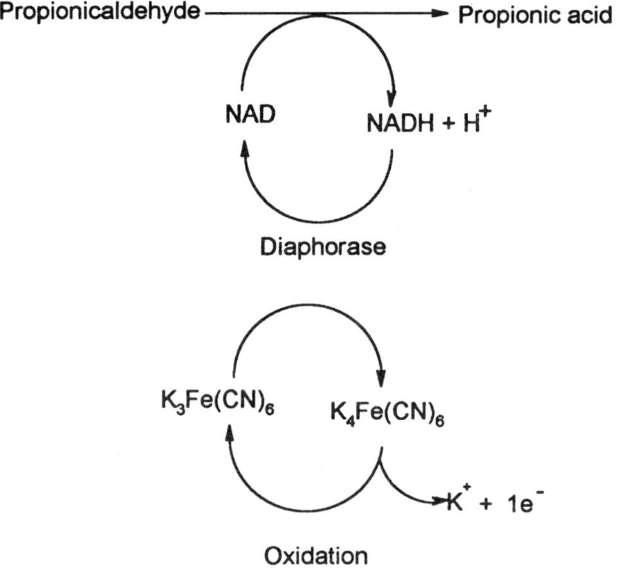

Scheme 2. Generation of current by two enzyme coupled reaction.

16.4.2. Benzimidazole Immunosensor

An antibody to thiabendazole immobilized to silicon, silicon dioxide, and silicon nitride has been reported by Flounders et al.[66] Using thiabendazole-HRP conjugate, they showed that surface-bound antibody can bind the analyte and can be repeatedly cycled. Ellipsometry measurements of the HRP-thiabendazole conjugate is the goal in a competition immunosensing. However no data are available.

16.5. FUTURE TRENDS

Sensor development for agrochemical monitoring has experienced tremendous growth during the past 10 years. As more antibodies to specific chemicals or classes of compounds become available, more immunosensors will be developed. Modifications made in antibody attachment such as sol-gel entrapment,[67] or molecular mutations of engineered single-chain Fv antibodies to render them more solvent tolerant, with handles for easy attachment to sensor surfaces and other properties not attainable in the native antibody molecules can now be realized. The technology introduced by Dickinson et al.[68] for depositing multiple antibodies on a small sensor surface and that of fluorescence resonance energy transfer (FRET) reported by Schultz et al[69] will allow multiresidue as well as remote sensing. With neural network and pattern recognition methods,[70] quantitative multiresidue analysis can become a reality. Microformat imaging ELISA, as reported by Dzgoev et al.,[56] using thick-film patterned glass slides containing 80 wells. Projections of 984 wells in a microtiter-sized footprint combined with chemiluminescence detection using a slow-scan CCD camera can lead to a 4000-fold improvement in sensitivity over the colorimetric assay, and a 10-fold increase in throughput. Even without specific antibodies, one can envision preparing imprints of chemicals by the method of molecular imprinting polymers.[71] These imprinted polymers may possess solvent tolerance and specificity not easily obtainable in immunoglobulins.

In the realm of enzyme sensing, production of a sufficient quantity of stable enzyme may be a major hurdle. Here again, biotechnology techniques may be employed. Gene cloning can solve the enzyme supply needs while gene mutations may be able to produce stable and sensitive enzymes useful in sensor formating. As catalytic antibody techniques improve, this approach may be another source of biomolecules for molecular recognition in which the product of enzyme catalysis can be monitored. An example of such a sensor was reported by Blackburn et al.[72] Photosynthetic reaction centers for herbicide monitoring may become popular as a general warning monitor because of the nonspecific nature of the system. However, the levels of detection at the moment are not sufficiently sensitive to be practical. Improvement in detection sensitivity as well as reaction center stability should be emphasized.

There is no doubt that new and improved sensors will be available for agrochemical monitoring. Many reports cited in this chapter did not address the matrix issue. Thus, it remains to be determined if such techniques can be useful monitoring tools in the real world. It is thus imperative that the matrix effect of environmental samples or food extracts be addressed along with sensor development so that application needs can be met without delay.

REFERENCES

1. Marco M, Barcelo D. Environmental application of analytical biosensors. *Meas Sci Technol* 1996;7:1547–1562.
2. Kramer P M. Biosensors for measuring pesticide residues in the environment: past, present and future. *AOAC Intl* 1996;79:1245–1252.
3. Reshetilov AN. Models of biosensors based on principles of potentiometric and amperometric transducers: use in medicine, biotechnology and environmental monitoring. *Appl Biochem Microbio* 1996;32:72–85.
4. Marty J, Garcia RR. Biosensors: potential in pesticide detection. *Trends Anal Chem* 1995;14:329–333.
5. Trojanowicz M, Hitchman M. Determination of pesticides using electrochemical biosensors. *Trends Anal Chem* 1996;15:38–45.
6. Kumaran S, Tran-Minh C. Insecticide determination with enzyme electrodes using different enzyme immobilization techniques. *Electroanalysis* 1992;4:949–954.
7. Imato T, Ishibashi N. Potentiometric butyrylcholine sensor for organophosphate pesticides. *Biosens Bioelectron* 1995;10:435–441.
8. Palleschi G, Bernabei M, Cremisini C, et al. Determination of organophosphorous insecticides with a choline electrochemical biosensor. *Sens Act* 1992;B7:513–517.
9. El Yamani H, Tran-Minh C, Abdul MA, et al. Automated system for pesticide detection. *Sen Act* 1988;8:193–198
10. Dumschat C, Müller H, Stein K, et al . Pesticide-sensitive ISFET based on enzyme inhibition. *Anal Chim Acta* 1991;252:7–9.
11. Nyamsi Hendju AM, Jaffrezic-Renault N, Martelet P, et al. Sensitive detection of pesticides using a differential ISFET-based system with immobilized cholinesterases. *Anal Chim Acta* 1993;281:3–11.
12. Marty JL, Sode K, Karube I. Biosensor for detection of organophosphate and carbamate insecticides. *Electroanalysis* 1992;4:249–252.
13. Tran-Minh C, Pandey PC, Kumaran S. Studies on acetylcholine sensor and its analytical application based on the inhibition of cholinesterase. *Biosens Bioelectron* 1990;5:461–471.
14. Skládal P, Mascini M. Sensitive detection of pesticides using amperometric sensors based on cobalt phthalocyanine-modified composite electrodes and immobilized cholinesterases. *Biosens Bioelectron* 1992;7:335–343.
15. Bernabei M, Chiavarini S, Cremsini C, et al. Anticholinesterase activity measurement by a choline biosensor: application in water analysis. *Biosens Bioelectron* 1993;8:265–271.
16. Navera EN, Suzuki K, Yokohama E, et al. Micro-choline sensor for acetylcholinesterase determination *Anal Chim Acta* 1993;281:673–679.
17. Cagnini A, Palchetti I, Lionti I, et al. Disposable ruthenized screen-printed biosensors for pesticides monitoring. *Sens Act B* 1995;24/25:85–89.
18. Martorell D, Cespedes F, Martinez-Fabregas E, et al. Amperometric determination of pesticides using biosensor based on a polishable graphite-epoxy biocomposite. *Anal Chim Acta* 1994;290:343–348.
19. Kumaran S, Morita M. Application of a cholinesterase biosensor to screen for organophosphorous pesticides extracted from soil. *Talanta* 1995;42:649–655.
20. Mionetto N, Marty J, Karube I. Acetylcholinesterase in organic solvents for the detection of pesticides: biosensor application. *Biosen Bioelectron* 1994;9:463–470.
21. Iwuoha EI, Smyth MR. Organic phase enzyme electrodes: kinetics and analytical applications. *Biosens Bioelectron* 1997;12:53–75.
22. Rogers KR, Cao CJ, Valdes JL, et al. Acetylcholinesterase fiber-optic biosensor of anticholinesterases. *Fund Appl Toxicol* 1991;16:810–820.

23. Fernando JC, Rogers KR, Anis NA, et al. Rapid detection of anticholinesterase insecticides by areusable light addressable potentiometric biosensor. *J. Ag Food Chem* 1993;41:511–516.

24. Ayyagari MS, Kamtekar S, Pande R, et al. Chemiluminescence-based inhibition kinetics of alkaline phosphatase in the development of a pesticide biosensor. *Biotech Prog* 1995;11:699–703.

25. Sode K, Togo H, Yamazaki T, et al. Development of a noval conductometric determination of organophosphate insecticide using phosphotriesterase. *Denki Kagaku* 1996;12:1234–1238.

26. Wei W, Cai Q, Wang R. et al. Determination of organophosphorus pesticides based on the inhibition of lipase with a surface acoustic wave sensor system. *Intl J Environ Anal Chem* 1996;64:279–287.

27. Anis NA, Wright J, Rogers, KR, et al. A fiber-optic immunosensor for detecting parathion. *Anal Lett* 1992;25:627–635.

28. Nikolelis DP, Krull UJ. Direct electrochemical sensing of insecticides by bilayer lipid membrane. *Anal Chim Acta* 1994;288:187–192.

29. Northrup MA, Stanker LH, Vanderlaan M, et al, Development and characterization of a fibre optic immuno-sensor in: Theophanides T, ed. *Spectroscopy of Inorganic Bioactivators: Theory and Applications — Chemistry, Physics, Biology, and Medicine.* Dordrecht: Kluwer Academic, 1989: 229–241.

30. McArdle FA, Persaud KC, Development of an enzyme-based biosensor for atrazine detection. *Analyst* 1993;118:419–423.

31. Besombes JL, Cosnier S, Labbe P, et al. A biosensor as warning device for the detection of cyanide, chlorophenyls, atrazine and carbamate pesticides. *Anal Chim Acta* 1995;311:255–263.

32. Cowell DC, Dowman AA, Ashcroft T, et al. The detection and identification of metal and organic pollutants in potable water using enzyme based sensors. *Biosensors 94*, Third World Congress on Biosensors June 1–3, 1994, New Orleans USA.

33. Bier FF, Stöcklein W, Böcher M, et al. Use of a fibre optic immunosensor for the detection of pesticides. *Sens Act B* 1992;7:509–512.

34. Krämer P, Schmid R. Flow-injection immunoanalysis (FIIA) — a new immunoassay format for the determination of pesticides in water. *Biosens Bioelectron* 1991;6:239–243.

35. Dietrich M, Kramer PM. Continuous immunochemical determination of pesticides via flow injection immunoanalysis using monoclonal antibodies against terbutryn immobilized to solid supports. *Food Agr Immunol* 1995;7:203–220.

36. Jiang TB, Halsall HB, Heineman WR, et al. Capillary enzyme immunoassay with electrochemical detection for the determination of atrazine in water. *J Ag. Food Chem* 1995;43:1098–1104.

37. Wortberg M, Middendorf C, Katerkamp A, et al. Flow-injection immunosensor for triazine herbicides using Eu(III) chelate label fluorescence detection. *Anal Chim Acta* 1994;289:177–186.

38. Guilbault GG, Hock B, Schmid RA. A piezoelectric immunosensor for atrazine in drinking water. *Biosens Bioelectron* 1992;7:411–419.

39. Steegborn C, Skládal P. Construction and characterization of the direct piezoelectric immunosensor for atrazine operating in solution. *Biosens Bioelectron* 1997;12:19–27.

40. Bier FF, Jockers R Schmid RD. Integrated optical immunosensor for s-triazine determination: regeneration, calibration and limitations. *Analyst* 1994;119:437–441.

41. Minunni M, Mascini M. Detection of pesticide in drinking water using real-time biospecific interaction analysis (BIA). *Anal Lett* 1993;26:1441–1460.

42. Tom-Moy M, Baer RL, Spirasolomon D, et al. Atrazine measurements using surface transverse wave devices. *Anal Chem* 1995;67:1510–1516.

43. Rouillon R, Sole M, Carpentier R, et al. Immobilization of thylakoids in polyvinylalcohol for the detection of herbicides. *Sens Act B* 1995;26/27:477–479.

44. Brewster JD, Lightfield AR, Bermel OL. Storage and immobilization of photosystem II reaction centers used in an assay for herbicides. *Anal Chem* 1995;67:1296–1299.

45. Jockers R, Bier F, Schmid RD. Herbicide biosensor based on photobleaching of the reaction centre of *Rhodobacter-sphaeroides*. *Anal Chim Acta* 1993;274:185–190.

46. Pandard P, Rawson DM. An amperometric algal biosensor for herbicide detection employing a carbon-cathode oxygen electrode. *Environ Toxicol Water Qual* 1993;8:323–333.

47. Nikolelis DP, Siontorou CC. Flow injection monitoring and analysis of mixtures of simazine, atrazine and propazine using filter-supported bilayer lipid membrane (BLMs). *Electroanalysis* 1996;8:907–912.

48. Rinder DF, Fleeker JR. A radioimmunoassay to screen for 2,4-dichlorophenoxyacetic acid and 2,4,5-trichlorophenoxyacetic acid in surface water. *Bull Environ Contam Toxicol* 1981;26:375–380.

49. Knopp D, Nuhn P, Dobberkau HJ. Radioimmunoassay for 2,4-dichlorophenoxyacetic acid. *Arch Toxicol* 1985;58:27–32.

50. Fleeker JR Two enzyme immunoassay to screen for 2,4-dichlorophenoxyacetic acid in water. *J Assoc Anal Chem* 1987;70:874–878.

51. Eremin SA. Polarization fluoroimmonoassay for rapid specific detection of pesticides. In: Nelson JC, Karu, AE, Wong RB, eds. *Immunoanalysis of Agrochemicals: Emerging Technologies* ACS Symposium Series 586, Washington DC: ACS, 1995;223–234.

52. Piras L, Adami M Fenu S, et al. Immunoenzymatic application of a redox potential biosensor. *Anal Chim Acta* 1996;335:127–135.

53. Khomutov SM, Zherdev AD, Dzantiev BB, et al. Immunodetection of herbicide 2,4-dichlorophenoxyacetic acid by field-effect transistor-based biosensors. *Anal Lett* 1994;27:2983–2995.

54. Dzantiev BB, Zherdev AV, Yulaev MF, et al. Electrochemical immunosensors for determination of the pesticides 2,4-dichlorophenoxyacetic acid and 2,4,5-trichlorophenoxyacetic acids. *Biosens Bioelectron* 1996;11:127–185.

55. Whittmann C, Bier FF, Eremin SA, et al. Quantitative analysis of 2,4-dichlorophenoxyacetic acid in water samples by two immunosensing methods. *J Ag Food Chem* 1996;44:343–350.

56. Dzgoev A, Mecklenburg M, Larsson P-O, et al. Microformat imaging ELISA for pesticide determination. *Anal Chem* 1996;68:3364–3369.

57. Chaleff RS, Mauvais CJ. Acetolactate synthetase is the site of action of two sulfonylurea herbicides in higher plants. *Science* 1984;224:1443–1444.

58. Marty JL, Mionetto N, Noguer F, et al. Enzyme sensors for the detection of pesticides. *Biosens Bioelectron* 1993;8:273–280.

59. Seki A, Ortega F, Marty JL. Enzyme sensor for the detection of herbicides inhibiting acetolactate synthetase. *Anal Lett* 1996;29:1259–1271.

60. Mel'nichenko OA, Eremin SA, Egorov SA. Single-reagent immunoassay for the determination of methabenzthiazuron. *J Anal Chem* 1996;51:512–515.

61. Shaner DL, Anderson PC, Stidham MA. Imidazolinone: potent inhibitors of acetohydroxyacid synthetase. *Plant Physiol* 1984;76:545–550.

62. Wong RB, Ahmed ZH. Development of an enzyme-linked immunosorbent assay for imazaquin herbicide. *J Ag Food Chem* 1992;40:811–816.

63. Wong RB, Anis N, Eldefrawi ME, Reusable fiber-optic-based immunosensor for rapid detection of imazethapyr herbicide. *Anal Chim. Acta* 1993;270:141–147.

64. Lee M, Durst RA. Determination of imazethapyr using capillary-column FILIA (flow-injection liposome immunoanalysis). *J Ag Food Chem* 1996;44:4032–4036.

65. Wong RB, Reeves SG, Anis N, et al. Rapid reusable immunosensing of imidazolinone herbicides. Pacific Chem '95 Congress 1995 Abstract Agro 143, Honolulu, HI USA. Dec 17–22.

66. Flounders AW, Brandon DL, Bates AH. Immobilization of thiabendazole-specific monoclonal antibodies to silicon substrates via aqueous silanization. *Appl Biochem Biotech* 1995;50:265–283.

67. Turnansky A, Avnir D. Sol-gel entrapment of monoclonal anti-atrazine antibodies. *J Sol-Gel Sci Tech* 1996;7:135–143.

68. Dickinson TA, White J, Kauer JS, et al. A chemical-detecting system based on a cross-reactive optical sensor array. *Nature* 1996;382:697–700.

69. Schultz SL, Agayn V, Walt DR. Continual immunosensor monitoring of herbicide in discharge streams. 213th ACS National Meeting, San Francisco, CA, Abst. # 43. April, 1997

70. Yatsenko V. Determining the characteristics of water pollutants by neural sensors and pattern recognition methods. *J Chromatog* 1996;A722:233–243.

71. Anderson LI, Nicholls IA, Mosback K. Antibody mimic obtained by non-covalent molecular imprinting. In: Nelson JC, Karu, AE, Wong RB, eds. *Immunoanalysis of Agrochemicals: Emerging Technologies* ACS Symposium Series 586; Washington DC: ACS, 1995: 89–97.

72. Blackburn GF, Talley DB, Booth PM, et al. Potentiometric biosensor employing catalytic antibodies as the molecular recognition element. *Anal Chem* 1990;62:2211–2216.

17

Thick-Film Biosensors

C.A. Galán-Vidal, J. Muñoz, C. Domínguez, and S. Alegret

17.1. INTRODUCTION

The use of manual production techniques in the field of biosensors yields devices of high cost and poor reproducibility. While techniques are not readily adaptable to mass fabrication, building sensors by hand is easy and allows rapid development of prototypes involving a very modest infrastructure. Consequently, the fabrication of sensing devices by hand is highly recommended for early developmental stages, but its practical application is highly limited.

The development and the application of solid-state biosensors is rising owing to their small dimensions and mechanical robustness. They are also produced at low cost with higher reproducibility because well-established planar mass fabrication techniques are employed.[1]

Two different methodologies for the mass production of planar devices can be distinguished: thin-film and thick-film techniques. Thin-film technology is based on a series of depositions and etching procedures involving microlithographic processes. These operations can deposit layers 1 to 10 μm thick. The materials forming these layers are applied on a substrate (silicon wafers, for instance) using microelectronic techniques such as sputtering or chemical vapor deposition (CVD). This technology is highly suited for the development of miniaturized solid-state devices. It has the advantage of showing high reproducibility but requires sophisticated and expensive infrastructure.[2] As a consequence, research in this field can be prohibitively expensive and thus makes sense only when success is highly likely. Thick-film technology can be seen as an intermediate step between thin-film and hand-made methodologies as it permits the mass production of reproducible biosensors using modest facilities.

C.A. Galán-Vidal • Centro de Investigaciones Químicas, Universidad Autónoma del Estado de Hidalgo, 42076 Pachuca, Hidalgo, Mexico.   J. Muñoz and C. Dominguez • Departmento de Microsistemas y Technología de Silicio, Instituto de Microelectrónica de Barcelona, Centre Nacionale de Microelectronica, 08193 Bellaterra, Barcelona, Spain.   S. Alegret • Sensor and Biosensor Group, Department de Química, Universitat Autònoma de Barcelona, 08193 Bellaterra, Barcelona, Spain.

Biosensors and Their Applications, edited by Yang and Ngo, Kluwer Academic/Plenum Publishers, New York, 1999.

This technology has a great potential and some of its products are already on the market.[3]

17.2.   THICK-FILM TECHNOLOGY

Thick-film technology is based on the successive and selective deposition of layers made of different materials (pastes or inks). These products are laid on a supporting material or substrate in layers from 10 to 50 μm thick (Fig. 17.1).[2,4,5] The use of this technology offers significant advantages, such as design flexibility, a wide choice of materials, reduced expense in production facilities, and low unit costs. Additionally, the process can be automated, the devices can be miniaturized, and technology transfer (from thick-film to other technologies and vice versa) is relatively easy.

The application of thick-film technology permits the mass production of planar solid-state biosensors. Devices so obtained show good reproducibility and high electrochemical activity owing to the large surface area that results from their microporous structure. There is a wide range of techniques for the deposition of materials,[6] including ink-jet printing, casting, dipping, spray-coating, dropping, spinning, fluidized-bed coating, brushing, and roller coating. However, most devices reported to date have been produced using screen-printing techniques.

Although different piezoelectric and thermometric transducers have been produced with thick-film technology, it is by far the electrochemical devices that have been showing the most interesting results. These transducers include conductometric, potentiometric, and, especially, amperometric devices.[5] One of the more significant advantages of this group of transducers is based on the fact that the required instrumentation is inexpensive.

Amperometric sensors and biosensors have been reported in three configurations: with only the working electrode, with the reference and working electrodes,

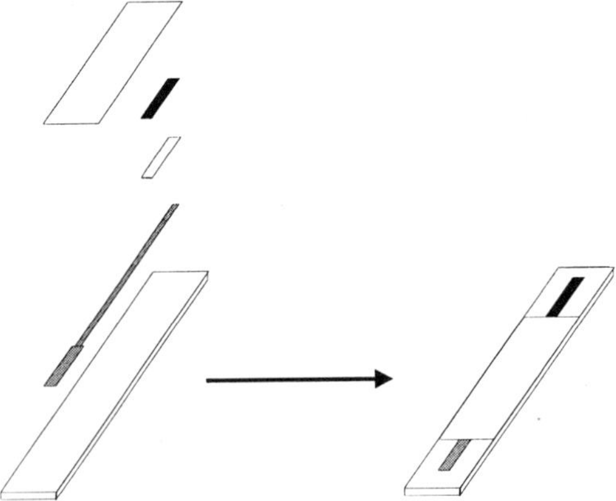

Figure 17.1.   Sequential deposition of materials using thick-film technology.

and with the reference, working, and auxiliary electrodes. The characterization of these devices has been done with electrochemical techniques, but several researchers have started to use differential thermal analysis, thermogravimetric analysis, x-ray diffraction and, with increasing frequency, scanning electron microscopy.

17.2.1. Screen-Printing Technique

The screen-printing process depends on the materials to be deposited and the intended application of the device being produced. Thus, a single general description of the technique cannot be proposed. Nevertheless, the operations described below can be categorized.[5]

A conductive or dielectric paste is deposited on a substrate by forcing it through a mask made of sealed mesh with holes at selected points. The paste is passed through the mask by pressing it with a squeegee, which causes the material to be deposited on the substrate following the pattern of the mask (Fig. 17.2 and 17.3). After printing, the paste solvents are left to evaporate and the paste is then cured. These stages are repeated for each of the layers, which implies a different set of masks for each design. The final layer is generally the one containing the sensing material. However, in some cases it is the encapsulant that is deposited last, with a window being left for the contact between the sensing area and the analyte. Its versatility means that thick-film technology employs widely varying procedures for the fabrication of sensors, which depend on features such as the materials used, particular fabrication techniques, and integration with other technologies.

17.2.2. Materials

The performance of chemical sensors constructed with thick-film technology is highly dependent on the materials employed. Furthermore, the construction process and the operation of the devices have an effect on these materials. Three types of materials are used: the substrate the pastes, and the sensing material.

The substrate is the supporting material, where the functional elements of the sensor are printed. The properties required of this material are a function of the type of tranducer, the fabrication process, and the medium in which the sensor will be applied. In general, the substrate has to be robust, low-cost, compatible with the rest of the components of the sensor (adhesion, thermal expansion coefficient, etc.) and it has to be inert.[4] Alumina ceramics and PVC are two of the substrate materials that are used frequently.[5] Other materials include enameled steel, polycarbonate, glass fiber, and cardboard covered with acrylic paint.

The pastes are formulations usually made with polymer or ceramic materials. They are deposited sequentially on the substrate forming the sensor's functional structures after a drying and curing process. A wide range of these pastes can be obtained commercially with different physical and chemical properties (e.g., viscosity, conductivity, thermal resistance, and water resistance). Usually, these pastes are modified to adapt them to a particular fabrication process. Sometimes special formulations have to be custom-made following indications in the literature.[5,7,8]

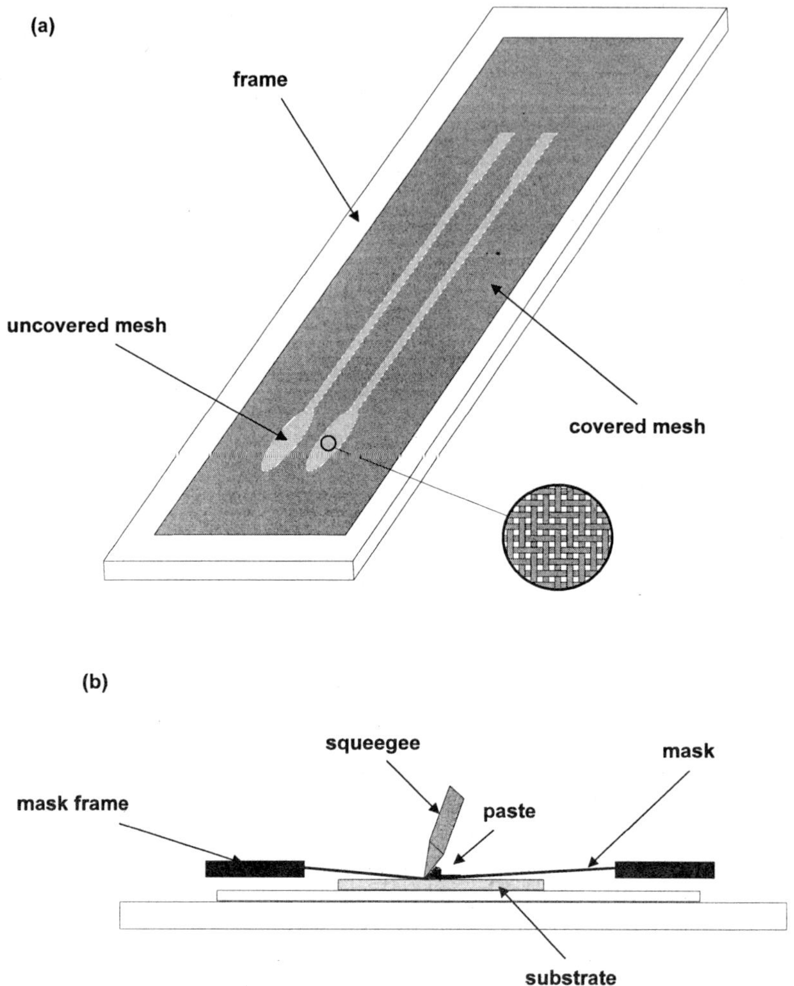

Figure 17.2. Mask and screen-printing process: (a) The mask is formed with a stainless steel, nylon, or polyester mesh stretched on a metallic frame. The mesh is covered with a photosensitive emulsion that is opened selectively using a photolithographic technique. (b) Cross section of a screen-printing process.

Pastes are based on a mixture of several components, such as binding agents (e.g., ground glass and resins), solvents or carriers (e.g., terpinol, ethylenglycol, and cyclohexanone) and other additives that confer certain functional features on the paste. If a conductive paste is needed, a metal powder such as gold, platinum, palladium, or silver is added. Powdered graphite is also used for this purpose because of its reduced cost and biocompatibility.[9–12] On the other hand, if dielectric pastes are required (for encapsulation, for instance) metal oxides or insulating binding agents such as alumina or silica have to be used.

In addition to the thermally cured pastes there is a growing use of photolithography and photocurable formulations in the fabrication of sensors using thick-film

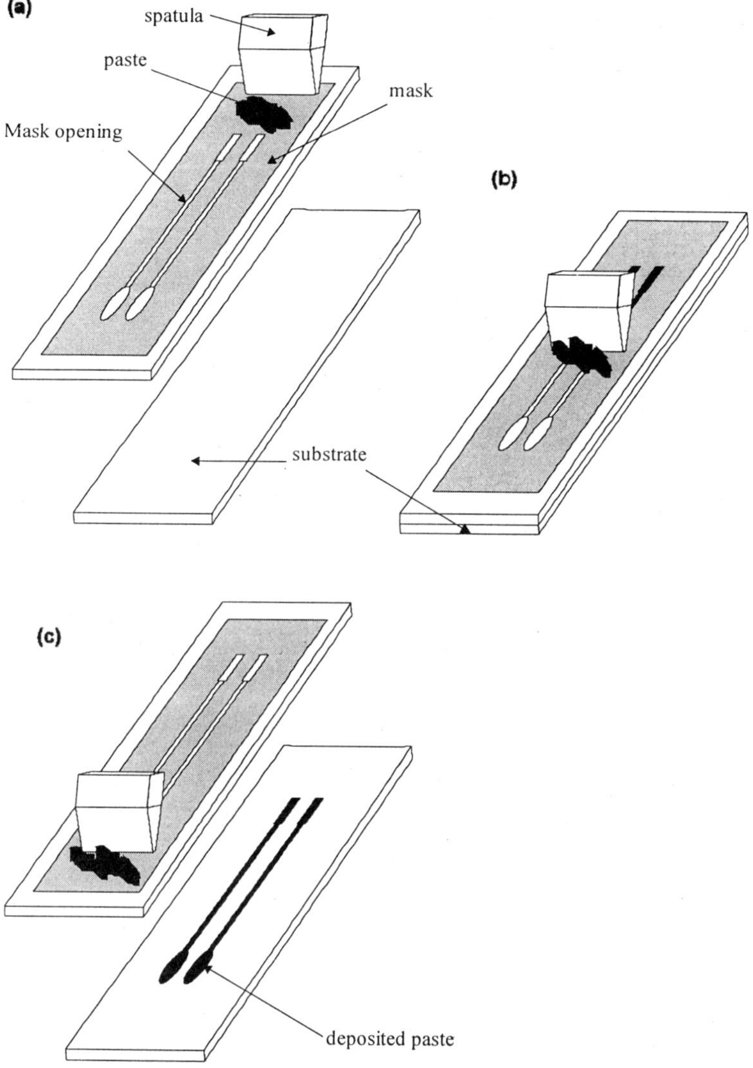

Figure 17.3. Screen-printing process used in the fabrication of sensors and biosensors using thick-film technology: (a) mask and substrate preparation prior to the printing, (b) printing process, (c) paste deposited on the substrate after printing.

technology. This has brought about substantial technological advances such as a shortening of the fabrication time and an increase in resolution.[13] The biological sensing material may be added to the bulk of the paste, thus producing biocomposites. It can also be immobilized on the surface of the transducer by, e.g., adsorption, cross-linking, or covalent binding. In both instances, the sensing materials are applied in the last stages of fabrication so that they are not exposed to the high temperatures that are needed during the initial thermal curing processes, thus preserving their sensitivity and stability.[14] Finally, in order to reduce the influence of possible interference, either to protect the sensing element or to increase the linear range, coatings of semipermeable membranes (e.g., Nafion or cellulose acetate) can be applied to the surface of the biosensor. This modification is dependent on the particular application of the biosensor.

17.2.3. Trends

The most important current efforts in thick-film technology are directed toward simplifying the fabrication process by reducing the number of processing steps. This concern has led to the development of new transducer materials suitable for screen-printing, such as biocomposites.[7,8,10,14,16−18,20−24] On the other hand, the optimization of promising new inks is generating more interest in the ink-jet techniques, which can increase both resolution of the planar pattern deposition and reproducibility of the finished devices.[18] For instance, thick-film biosensors fabricated by screen-printing are showing good reproducibility [lower than 5% (relative standard deviation)] among biosensors from the same batch (Table 17.1).

Although the technology described is an attractive choice for the construction of chemical sensors and biosensors, it is not free of problems and shortcomings. As noted earlier, it requires complicated curing cycles at high temperatures for lengthy periods of time, poor reproducibility can result from the evaporation of the paste solvents during printing; and the technology has a resolution limit of 100 μm. The possible unavailability of a given material for a particular device may also be an obstacle. However, some of these limitations may be overcome by designing disposable devices in cases where long lifetimes or good signal reproducibility for long periods are not required. In such cases, the construction process of the device has to be highly reproducible.

Table 17.1. Biosensor Fabrication Reproducibility between Same Batch Devices

Analyte	Reproducibility % (RSD)[a]	References
Glucose	5, 10, 11, 11.3, 17	14, 29, 23, 32, 36
Lactate	12.8, 6.1, 8.7	27, 31, 33
Creatinine	15–20	35
Uric acid	9.2	30
Phenols	5, 15.3	20, 28
Pesticides and herbicides	6.2, 8, 15	21, 26, 34

[a]Relative standard deviation.

Currently, the convergence of difference technologies is permitting the optimization of integrated transducers in microsystems and expanding the range of applications for thick-film biosensors.

17.3. APPLICATIONS

Most thick-film biosensors have been developed for biomedical applications, as can be seen in Table 17.2. Among the group of reported analytes, glucose is the one most studied because, owing to its chemical stability, glucose oxidase is the enzyme used as a model in work connected with new technological advances. However, the glucose biosensor also presents problems in certain applications.

Table 17.2. Some of the Analytes Determined Using Thick-Film Biosensors

Receptor	Transducer	Analyte	References
Glucose oxidase (GOD)	Amperometric	Glucose	7–10, 14, 16–18, 22, 23, 25, 29, 32, 36–48
Glucose dehydrogenase (GDH)	Amperometric	Glucose	38, 47, 49
Lactate oxidase (LOD)	Amperometric	Lactate	13, 22, 41, 42, 52, 53
Lactate dehydrogenase (LDH)	Amperometric	Lactate	27, 33, 50, 51
GOD + HRP	Amperometric	Glucose	15
GOD + invertase (INV)	Amperometric	Sucrose	14, 37, 40, 48
GOD + INV + mutarotase (MUT)	Amperometric	Sucrose	14
β-galactosidase + GOD	Amperometric	Lactose	14, 31
Acetylcholinesterase AChE	Amperometric	Pesticides	54, 69
Acetylcholinesterase AChE	Potentiometric	Pesticides	55
Tyrosinase	Amperometric	Phenols, pesticides	20, 21, 28, 56
Choline oxidase (ChO)	Amperometric	Pesticides	26, 59
Salicylate hydroxylase (SH)	Amperometric	Salicylate	57
Ascorbate oxidase (AAO)	Amperometric	Ascorbic acid	37
Alcohol oxidase (AOD)	Amperometric	Ethanol	38
Alcohol dehydrogenase (ADH)	Amperometric	Ethanol	38, 24, 44, 65
Urease	Conductimetric	Urea	58
Urease	Potentiometric	Urea	41
Putrescine oxidase	Amperometric	Biogenic amines	10
Malate dehydrogenase (MDH)	Amperometric	Malate	50
L-Amino acid oxidase	Amperometric	L-Phenylalanine	22
Horseradish peroxidase (HRP)	Amperometric	Peroxides	23
Glutathione peroxidase	Amperometric	Reduced glutathione	60
Salidase	Potentiometric	Bound sialic acid (tumor marker)	61
Uric acid oxidase	Amperometric	Uric acid	30
Cholesterol oxidase	Amperometric	Cholesterol	62
3-Hydroxybutyrate Dehydrogenase	Amperometric	Ketone bodies	63
Aryl acylamidase	Amperometric	Acetaminophen	64
Butyrylcholinesterase	Amperometric	Pesticides	67

Table 17.3.   Analysis of Real Samples by Thick-Film Biosensors

Analyte	Real sample	References
Glucose	Fruit juices	10, 14, 37, 38, 48
Glucose	Fermentation broth	36
Glucose	Whole blood	15
Glucose	Wine	10
Glucose	Blood serum	40, 41
Lactose	Dairy products	14, 31
Sucrose	Fruit juices	48
Lactate	Dairy products	50, 53
Lactate	Fermentation broth	13
Free fatty acids	Milk	68
Ethanol	Wine	38
Ethanol	Gin	65
Ketone bodies	Whole blood	63
Urea	Blood serum	41
Uric acid	Urine	30
Creatine and creatinine	Blood serum	35
Reduced glutathione	Whole blood	60
Bound sialic acid (tumor marker)	Serum samples	61
Acetaminophen	Whole blood	64
Ascorbic acid	Fruit juices	48
Hydrazine compounds	Drinking and river water	56
Pesticides	Drinking and brook water	69
Phenols	Drinking and groundwater	20
Phenols	Industrial waste waters	28
Pesticides and herbicides	River water	21

Recent work has shown that thick-film biosensors are used increasingly in fields such as pharmaceutics, agriculture, environment, and the food industry (see Table 17.2). In many cases these applications are linked to industrial process control and to the monitoring of toxic, corrosive, or explosive substances. However, applications are still limited and greater knowledge and further technological development is needed.

Enzymes are the most common sensing element used for thick-film biosensors as their specificities are compatible with electrochemical transduction, but biosensors incorporating antibodies,[12,34] DNA,[66] and different species of bacteria[68] have been also constructed.

Some applications in real samples have been carried out (Table 17.3) as a validation of the biosensors that have been developed. In this case, it is necessary to emphasize that most of the determinations have been done in biological fluids, food samples, different types of water, and fermentators.

In summary, the biosensors constructed by thick-film technology are in an early stage of development. Nevertheless, in a short period of time they have been applied successfully in several industrial fields and have even been commercialized. The development of new materials, immobilization techniques, and the availability of new biological sensing elements may permit the fabrication of a wider variety of thick-film biosensors.

REFERENCES

1. Madou MJ, Morrison SR. *Chemical Sensing with Solid State Devices.* San Diego: Academic Press, 1989.
2. Lambrechts M, Sansen W, *Biosensors: Microchemical Devices,* Bristol: IOP Publishing, 1992.
3. Alvarez-Icaza M, Bilitewski U, Mass production of biosensors. *Anal Chem* 1993;65:525A–533A.
4. Prudenziati M, ed. *Thick Film Sensors* Amsterdam: Elsevier, 1994.
5. Galán-Vidal CA, Muñoz J, Domínguez C, et al. Chemical sensors, biosensors and thick-film technology. *Trends Anal Chem* 1995;14:225–231.
6. Licari JJ, Hugues LA, *Handbook of Polymer Coatings for Electronics,* New Jersey: Noyes Publications, 1990.
7. Cardosi MF, Birch SW, Screen printed glucose electrodes based on platinized carbon particles and glucose oxidase. *Anal Chim Acta* 1993;276:69–74.
8. Nagata R, Yokoyama K, Clark SA, et al. A glucose biosensor fabricated by the screen printing technique. *Biosensen Bioelectron* 1995;10:261–267.
9. Atanasov P, Kaisheva A, Gamburzev S, et al. Metallocene-mediated amperometric biosensors for one-shot glucose determination. *Sens Act B* 1992;8:59–64.
10. Bilitewski U, Chemnitius GC, Rüger P, et al. Miniaturized disposable biosensors. *Sens Act B* 1992;7:351–355.
11. Gilmartin MAT, Hart JP, Development of amperometric sensors for uric acid based on chemically modified graphite-epoxy resin and screen-printed electrodes containing cobalt phthalocyanine. *Analyst* 1994;119:243–252.
12. Weetall HH, Hotaling T, A simple, inexpensive, disposable electrochemical sensor for clinical and immuno-assay. *Biosensors* 1987/88;3:57–63.
13. Rohm I, Genrich M, Collier W, et al. Development of ultraviolet-polymerizable enzyme pastes: bioprocess applications of screen printed L-lactate sensors. *Analyst* 1996;121:877–881.
14. Bilitewski U, Jäger A, Rüger P, et al. Enzyme electrodes for the determination of carbohydrates in food. *Sens Act B* 1993;15/16:113–118.
15. Marcinkeviciene J, Kulys J, Bienzyme strip-type glucose sensor. *Biosens Bioelectron* 1993,8:209–212.
16. Koopal CGJ, Bos AACM, Nolte JM, Third-generation glucose biosensor incorporated in a conducting printing ink. *Sens Act B* 1994;18/19:166–170.
17. Nagata R, Clark SA, Yokoyama K, et al. Amperometric glucose biosensor manufactured by a printing technique. *Anal Chim Acta* 1995;304:157–164.
18. Newman JD, White SF, Tothill IE, et al. Catalytic materials, membranes, and fabrication technologies suitable for the construction of amperometric biosensors. *Anal Chem* 1995;67:4594–4599.
19. Nagata R, Yokoyama K, Durliat R, et al. An enzyme-containing ink for screen-printed glucose sensors. *Electroanalysis* 1995;7:1027–1031.
20. Wang J, Chen Q, Microfabricated phenol biosensors based on screen printing of tyrosinase containing carbon ink. *Anal Lett* 1995;28:1131–1142.
21. Wang J, Nascimento VB, Kane SA, et al. Screen-printed tyrosinase-containing electrodes for biosensing of enzyme inhibitors. *Talanta* 1996;43:1903–1907.
22. Sampath S, Lev O, Inert metal-modified, composite ceramic-carbon, amperometric biosensors: renewable, controlled reactive layer. *Anal Chem* 1996;68:2015–2021.
23. Wang J, Pamidi PVA, Park DS, Screen-printable sol-gel enzyme-containing carbon inks. *Anal Chem* 1996;68:2705–2708.
24. Park J, Yee H, Kim S, Amperometric biosensor for vapor determination of ethanol vapor. *Biosens Bioelectron* 1995;10:587–594.
25. Varlan AR, Suls J, Jacobs P, Sansen W, A new technique of enzyme entrapment for planar biosensors, *Biosensens Bioelectron* 1995;10:xv–xix.
26. Cagnini A, Palchetti I, Mascini M, et al. Ruthenized screen-printed choline oxidase-based biosensors for measurement of anticholinesterase activity. *Mikrochim Acta* 1995;121:155–166.
27. Sprules SD, Hart P, Pittson R, et al. Evaluation of a new disposable screen-printed sensor strip for the measurement of NADH and its modification to produce a lactate biosensor employing microliter volumes. *Electroanalysis* 1996;8:539–543.

28. Kotte H, Gründig B, Vorlop K, et al. Methylphenazonium-modified enzyme sensor based on polymer thick films for subnanomolar detection of phenols. *Anal Chem* 1995;67:65–70.
29. Rohm I, Künnecke W, Bilitewsky U, UV-polymerizable screen-printed enzyme pastes. *Anal Chem* 1995;67:2304–2307.
30. Gilmartin MAT, Hart JP, Novel, reagentless, amperometric biosensor for uric acid based on a chemically modified screen-printed carbon electrode coated with cellulose acetate and uricase. *Analyst* 1994;119:833–840.
31. Jäger A, Bilitewski U, Screen-printed enzyme electrode for the determination of lactose. *Analyst* 1994;119:1251–1255.
32. Gilmartin MAT, Hart JP, Patton DT, Prototype, solid-phase, glucose biosensor. *Analyst* 1995;120:1973–1981.
33. Sprules SD, Hart P, Wring SA, et al. A reagentless, disposable biosensor for lactic acid based on a screen-printed carbon electrode containing meldola's blue and coated with lactate dehydrogenase, NAD$^+$ and cellulose acetate. *Anal Chim Acta* 1995;304:17–24.
34. Kaláb T, Skládal P. A disposable amperometric immunosensor for 2,4-dichlorophenoxyacetic acid. *Anal Chim Acta* 1995;304:361–368.
35. Madaras MB, Popescu IC, Ufer S, et al. Microfabricated amperometric creatine and creatinine biosensors *Anal Chim Acta* 1996;319:335–345.
36. White SF, Tothill YE, Newman JD, et al. Development of a mass-producible glucose biosensor and flow-injection analysis system suitable for on-line monitoring during fermentations. *Anal Chim Acta* 1996;321:165–172.
37. Scholze J, Hampp N, Bräuchle C. Enzymatic hybrid biosensors. *Sens Act B* 1991;4:211–215.
38. Rüger P, Bilitewski U, Schmid RD, Glucose and ethanol biosensors based on thick-film technology. *Sens Act B* 1991;4:267–271.
39. Johnson KW, Reproducible electrodeposition of biomolecules for the fabrication of miniature electro-enzymatic biosensors. *Sens Act B* 1991;5:85–89.
40. Bisenberger M, Bräuchle C, Hampp N. A triple-step potential waveform at enzyme multisensors with thick-film gold electrodes for detection of glucose and sucrose. *Sens Act B* 1995;28:181–189.
41. Silber A, Bisenberger M, Bräucle C, et al. Thick-film multichannel biosensors for simultaneous amperometric and potentiometric measurements. *Sens Act B* 1996;30:127–132.
42. Wang J, Chen Q, Enzyme microelectrode array strips for glucose and lactate. *Anal Chem* 1994;66:1007–1011.
43. Wang J, Chen Q, Screen-printed glucose strip based on palladium-dispersed carbon ink. *Analyst* 1994;119:1849–1851.
44. Wang J, Chen Q, Pedrero M, et al. Screen-printed amperometric biosensors for glucose and alcohols based on ruthenium-dispersed carbon inks. *Anal Chim Acta* 1995;300:111–116.
45. van Os PJHJ, Bult A, van Bennekom WP. A glucose sensor, interference free for ascorbic acid. *Anal Chim Acta* 1995;305:18–25.
46. Chen Q, Pamidi PVA, Wang J, et al. β-Cyclodextrin cation exchange polymer membrane for improved second-generation glucose biosensor. *Anal Chim Acta* 1995;306:201–208.
47. Bilitewski U, Rüger P, Schmid RD. Glucose biosensors based on thick-film technology. *Biosen Bioelectron* 1991;6:369–373.
48. Popp J, Silber A, Bräuchle C, et al. Sandwich enzyme membranes for amperometric multi-biosensor applications: improvement of linearity and reduction of chemical cross-talk. *Biosens Bioelectron* 1995;10:243–249.
49. Silber A, Hampp N, Schuhmann W. Poly(methylene blue)-modified thick-film gold electrodes for the electrocatalytic oxidation of NADH and their application in glucose biosensors. *Biosens Bioelectron* 1996;11:215–223.
50. Silber A, Bräuchle C, Hampp N. Dehydrogenase-based thick-film biosensors for lactate and malate. *Sens Act B* 1994;18/19:235–239.
51. Hart AL, Turner APF, Hopcroft D. On the use of screen- and ink-jet printing to produce amperometric enzyme electrodes for lactate. *Biosens Bioelectron* 1996;11:263–270.
52. Hart AL, Cox H, Janssen D. Stabilization of lactate oxidase in screen-printed enzyme electrodes. *Biosens Bioelectron* 1996;11:833–837.

53. Collier WA, Janssen D, Hart AL. Measurement of soluble L-lactate in dairy products using screen-printed sensors in batch mode. *Biosens Bioelectron* 1996;11:1041–1049.
54. Hartley IC, Hart JP. Amperometric measurement of organophosphate pesticides using a screen-printed disposable sensor and biosensor based on cobalt phthalocyanine. *Anal Proc* 1994;31:333–337.
55. Eppselsheim C, Hampp N. Potentiometric thick-film sensors for the determination of the neurotransmitter acetylcholine. *Analyst* 1994;119:2167–2171.
56. Wang J, Chen L. Hydrazine detection using a tyrosinase-based inhibition biosensor. *Anal Chem* 1995;67:3824–3827.
57. Zhou D, Nigam P, Jones J, et al. Production of salicylate hydroxylase from pseudomonas putida UUC-1 and its application in the construction of a biosensor. *J Chem Tech Biotechnol* 1995;64:331–338.
58. Bilitewski U, Drewes W, Schmid RD. Thick film biosensor for urea. *Sens Act B* 1992;7:321–326.
59. Cagnini A, Palchetti I, Lionti I. Disposable ruthenized screen-printed biosensors for pesticide monitoring. *Sens Act B* 1995;24–25:85–89.
60. Wring SA, Hart JP. Chemically modified, screen-printed carbon electrodes. *Analyst* 1992;117:1281–1286.
61. Aubeck R, Eppelsheim C, Bräuchle C. Potentiometric thick-film sensor for the determination of the tumour marker bound sialic acid. *Analysis* 1993;118:1389–1392.
62. Gilmartin MAT, Hart JP. Fabrication and characterization of a screen-printed, disposable, amperometric cholesterol biosensor. *Analyst* 1994;119:2331–2336.
63. Batchelor MJ, Green MJ, Sketch CL. Amperometric assay for the ketone body 3-hydroxybutyrate. *Anal Chim Acta* 1989;221:289–294.
64. Vaughan PA, Scott LDL, McAleer JF, Amperometric biosensor for the rapid determination of acetominophen in whole blood. *Anal Chim Acta* 1991;248:361–365.
65. Sprules SD, Hartley IC, Wedge R, et al. A disposable reagentless screen-printed amperometric biosensor for the measurement of alcohol in beverages. *Anal Chim Acta* 1996;329:215–221.
66. Wang J, Rivas G, Cai X. Accumulation and trace measurements of phenothiazine drugs at DNA-modified electrodes. *Anal Chim Acta* 1996;332:139–144.
67. Kulys J, D'Costa EJ. Printed amperometric sensor based on TCNQ and cholinesterase. *Biosens Bioelectron* 1991;6:109–115.
68. Schmidt A, Standfuss-Gabisch C, Bilitewski U, Microbial biosensor for free fatty acids using and oxygen electrode based on thick-film technology. *Biosens Bioelectron* 1996;11:1139–1145.
69. Günther A, Bilitewski U. Characterisation of inhibitors of acetylcholinesterase by an automated amperometric flow injection system. *Anal Chim Acta* 1995;300:117–125.

18

Alternative Polymer Matrices for Potentiometric Chemical Sensors

Hakhyun Nam and Geun Sig Cha

18.1. INTRODUCTION

Potentiometric chemical and biosensors based on solvent polymeric membranes, typically the highly plasticized poly(vinyl chloride) (PVC) doped with small amounts of electroactive components (e.g., ionophore and/or lipophilic ionic salt), are now routinely utilized in the field of biomedical analysis.[1−4] However, some inherent problems associated with the use of PVC-based membranes,[5] i.e., the weak adhesion of the membrane to solid-state sensor devices, lack of immobilizing functional groups for bioreagents, dissolution of membrane components into the biological fluids, and poor biocompatibility, have led many researchers to search for alternative polymer matrix-based sensing membranes. In this chapter, we will summarize such efforts as they have been reported in the literature, including our own studies.

18.2. SENSOR MEMBRANES FOR ALL-SOLID-STATE ELECTRODES

In the early 1970s, Freiser and Cattral found that simple metal wires (e.g., Cu, Ag, Pt) coated with PVC-based ion-selective membranes (coated wire electrode or CWE) also exhibit potentiometric behavior comparable to that of conventional ion-selective electrodes (i.e., ISEs with an internal reference solution).[6,7] Although no theoretical model has as yet satisfactorily described the electrochemical events occurring at the metal electrode/membrane interface,[6−15] a logical extension of CWEs has led to the development of various types of functional all-solid-state ISEs; the exposed end of a metallic conductor printed on a silicon wafer, alumina plate,

---

Hakhyun Nam and Geun Sig Cha • Department of Chemistry, Kwangwoon University, Seoul 139–701, Korea.

Biosensors and Their Applications, edited by Yang and Ngo, Kluwer Academic/Plenum Publishers, New York, 1999.

311

or other insulating polymer substrates, or the silicon nitride surface of a field effect transistor (FET) gate may be modified with PVC-based sensing membranes to fabricate a miniaturized and integrated array of ISEs on a single chip (see Fig. 18.1).[16-20] However, it has long been recognized that the poor adhesion of the PVC-based membranes to solid-state sensor devices leads to electrolyte shunts around the membrane, rendering the sensor inoperative within a short time. Furthermore, the formation of an acidic aqueous layer — a result of water vapor and various dissolved gases (e.g., CO_2, O_2, and organic acids) permeating through the membrane — at the membrane/metal electrode interface causes polymer membrane-based all-solid-state ISEs to exhibit unstable potential drift.[10-14]

On the commercial side, shrewdly exploiting both the advantages and disadvantages of the polymer membrane-based all-solid-state ISEs, several companies have produced disposable ion-sensor cartridges with chip-based ISE arrays [e.g., Multi-PLY cartridge for the Dimension system (Dade-Behring International Chemistry Systems), i-STAT system (i-STAT Co.) and IRMA system (Diametrics Inc.)].[16,21,22] Nevertheless, the use of poorly adhering membranes on the chip-based electrodes could lower the production yield of operable sensors and reduce their shelf life.

Several methods have been proposed to improve membrane adhesion to solid substrates and solve the problem of signal drift; some examples include the modification of PVC matrix for binding to hydroxyl-bearing surfaces,[23-25] mechan-

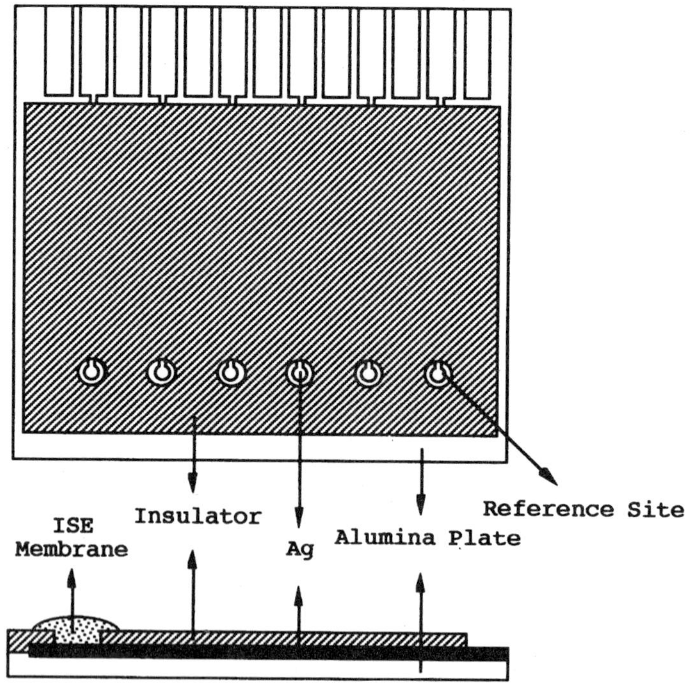

Figure 18.1.   Schematic drawing of a typical array-type all-solid-state ion-sensor.

ical attachment of the membranes,[26] electrode surface treatment with adhesion promoters (e.g., $SiCl_4$), photocuring techniques,[27-31] self-assembled layers,[32] Langmuir–Blodgett films,[33-42] modification of the electrode surface with conducting polymer layers,[43-47] introduction of a chemically bound hydrogel layer,[48] and the use of materials other than PVC-based polymers.[49-69] In many cases, however, these methods, even if they improve membrane adhesion, sensor stability, and lifetime, are hardly practical, as they are too sophisticated to be employed for mass fabrication[70] or result in poor electrochemical performance. Biocompatibility of the sensing membrane is another issue to consider. Protein adsorption and thrombus formation on the surface of sensing membranes and high affinity to lipid-soluble ionic species not only limit the operable lifetime of biomedical sensors but also result in faulty analytical values.[63-68] Thus, solvent polymeric ion-sensing membranes that are easily applicable on any all-solid-state device without compromising its potentiometric performance and yet are highly compatible with biological fluids are a prerequisite in developing microfabricated ion- and biosensor arrays.

18.3. SILICONE RUBBER MATRIX-BASED ISE MEMBRANES

18.3.1. One-Component Room Temperature Vulcanizing-Type Silicone Rubber (RTV-SR) Matrix

Silicone rubber (SR)-based membranes have received some attention from the earliest days of ISE research[71-74]: unlike PVC, SR-based membranes can be formulated without a solvent mediator (or plasticizer) owing to their low glass transition temperature, adhere strongly to most solid surfaces, and exhibit less interference from lipophilic anions present in biological fluids. Among various types of SR matrices, one-component room-temperature vulcanizing-type silicone rubbers (RTV-SRs) are particularly attractive materials for preparing the ISE membranes.[50-60] RTV-SR, commercially available one-part adhesive or sealant, is readily dissolved or dispersed in common organic solvents, and easily cured at room temperature by atmospheric moisture activation. The curing process causes condensation of the silicone prepolymers into a high-molecular-weight rubber while evolving alcohol, acetic acid, oxime, or acetone as a by-product.* However, the use of RTV-SR-based membranes for all-solid-state ion sensors has not been widely studied owing in part to their high electrical resistance,[55] which causes a slow response and a noisy signal.† Furthermore, the electrochemical performance of RTV-SR-based membranes is extremely sensitive to the type of electroactive components, i.e., ionophore, lipophilic additive, and plasticizer, employed.

---

*Our recent study showed that the variations in the potentiometric performance of ISE membranes based on different types of RTV-SR matrices are insignificant.

†Recently, Högg et al.[58] reported that fluorosilicone rubber (DOW Corning 730 RTV)-based membranes exhibit significantly reduced impedance.

18.3.2. RTV-SR Membrane-Based Ion-Selective Electrodes

18.3.2.1. Sodium-Selective Electrodes

As can be seen from the four potentiometric response curves of conventional-type sodium-selective electrodes (Fig. 18.2), each mounted with the plasticizer-free RTV-SR membrane containing a different ionophore, only those based on ETH 2120 and *tert*-butylcalix[4]arene tetraacetic acid tetraethyl ester (calix[4]arene ester) provide a near-Nernstian response to sodium. In many cases, however, the RTV-SR membrane compositions optimized with the conventional-type electrode are not directly applicable to all-solid-state-type electrodes; they often result in noisy and irreproducible potentiometric signals.[55,56]

Figure 18.3 shows the dynamic response curves of several all-solid-state-type sodium-selective ISEs based on RTV-SR membranes containing calix[4]arene ester and various kinds of lipophilic salts, respectively. It clearly demonstrates that the potentiometric performance of RTV-SR membrane-based electrodes varies significantly depending on the type of lipophilic salts incorporated. Only the RTV-SR membrane with potassium tetrakis[3,5-bis(trifluoromethyl)phenyl] borate (KTFPB)[75] exhibit potentiometric behavior comparable to that of the corresponding conven-

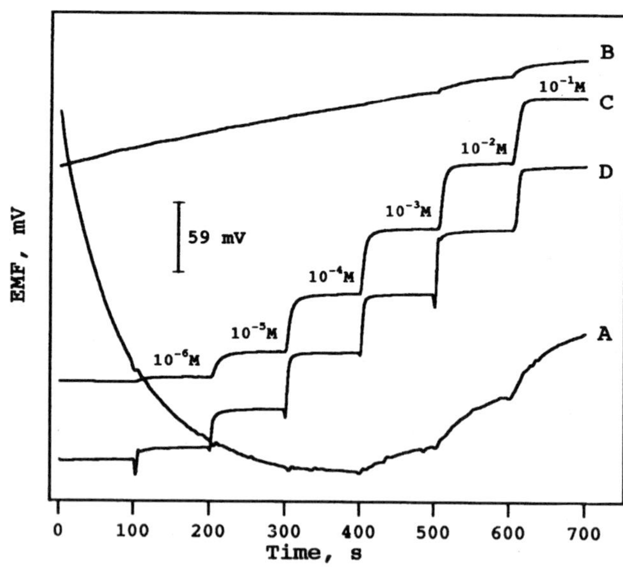

Figure 18.2. Dynamic response curves for conventional-type electrodes mounted with plasticizer-free RTV-SR membranes containing four different sodium-selective neutral carriers to the change in NaCl concentration from 10^{-6} to 10^{-1} M (background electrolyte; 0.05 M tris-HCl, pH 7.2): (A) *N,N'*-dibenzyl-*N,N'*-diphenyl-1,2-phenylenedioxydiacetamide (ETH 157), (B) bis[(12-crown-4)methyl] dodecyl-methylmalonate, (C) *N,N,N',N'*-tetracyclohexyl-1,2-phenylene-dioxydiacetamide (ETH 2120), (D) 4-tert-butyl-calix[4]arene-tetraacetic acid tetraethyl ester (calix[4]arene ester).

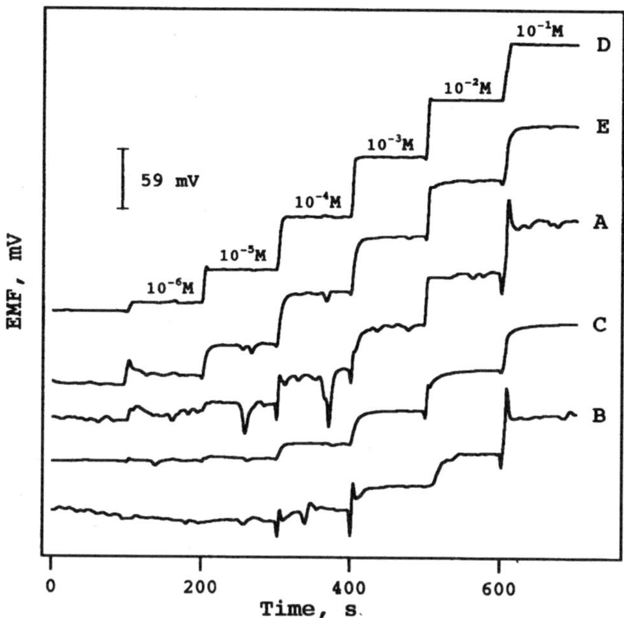

Figure 18.3. Dynamic response curves for all-solid-state electrodes coated with RTV-SR/calix[4]arene ester-based membranes containing different type of lipophilic additives (0.5 wt%): (A) no additive, (B) ETH 500, (C) KTpClPB, (potassium tetrakis[p-chlorophenyl]borate) (D) KTFPB, (E) NaHFPB (sodium tetrakis[3,5-bis(1,1,1,3,3,3-hexafluoro-2-methoxy-2-propyl)]borate).

tional-type electrode. Impedance spectra shown in Fig. 18.4 also indicate that the potentiometric performance of each electrode is closely related to the magnitude of bulk membrane resistance; the incorporation of KTFPB into the calix[4]arene ester-based RTV-SR membrane results in an impedance spectrum similar to that of the corresponding PVC/DOS* membrane. It seems that the highly polar trifluoromethyl groups on phenyls of TFPB⁻ help increase the electrode's compatibility in the polar siloxane matrix medium, effectively delocalize the anion charge centered on borate, and facilitate the dissociation of countercations (potassium), resulting in low membrane resistance. Other advantages of the calix[4]arene ester-based RTV-SR membranes with KTFPB include their flat response to pH change, high signal stability (less than ± 1 mV/h), and extended lifetime (>230 days; see Fig. 18.5).

Kimura et al. also reported that plasticizer-free RTV-SR-based ISE membranes can be prepared if the neutral carriers are highly soluble in the matrix; they were able to prepare the high-performance sodium-selective FET with RTV-SR and specially designed asymmetrical calix[4]arene ionophore.[52]

*bis(2-ethylhexyl) sebacate.

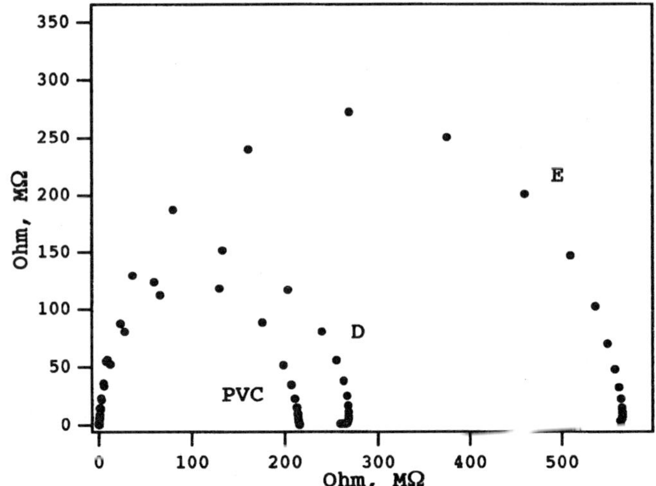

Figure 18.4. Impedance plots for the calix[4]arene ester-based RTV-SR membranes. Curves D and E correspond to the membranes shown in Fig. 18.3.

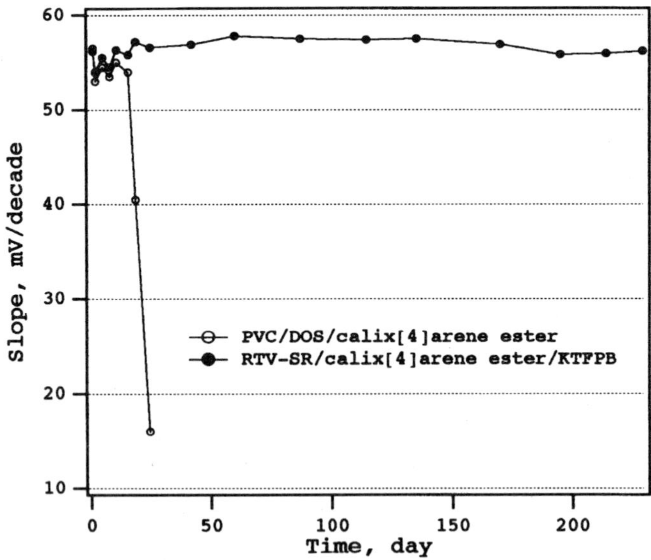

Figure 18.5. Variation of response slopes over time for all-solid-state electrodes with PVC- and RTV-SR-based membranes.

18.3.2.2. Potassium-Selective Electrodes

The potassium-selective RTV-SR membranes based on valinomycin can be applied to both conventional-type and all-solid-state-type electrodes with no extra caution.[50,51,54,71–74] The all-solid-state-type electrodes with valinomycin-based RTV-SR membranes also exhibit virtually the same potentiometric performance as the corresponding conventional-type electrodes in terms of their selectivity, signal stability, and lifetime.

18.3.2.3. Calcium-Selective Electrodes

Development of calcium-selective membranes based on an RTV-SR matrix has been another challenge.[55] Previous experimental data show that the RTV-SR membrane containing calcium-selective neutral carriers, ETH 129* or ETH 1001,[†] exhibits very slow response to step titrations. Addition of common lipophilic salt (KTpClPB) to RTV-SR/ETH 129 membrane rather worsens its potentiometric performance. The sluggish potentiometric responses of the RTV-SR-based membranes has been attributed to the slow ion-exchange kinetics at the membrane/ sample interface owing to the high surface resistance of SR.[71] It was also observed that the calcium-selective neutral carriers tend to crystallize out of the membrane during the curing period.

To solve such problems, a small amount of plasticizer ($\sim$10–22 wt%) was incorporated into the RTV-SR membranes, which lowered the bulk resistance of the membranes substantially (about tenfold decrease) without impairing their adhesive strength. However, the calcium-selective neutral carriers (ETH 129 or ETH 1001) doped in plasticized RTV-SR tend to exhibit a non-Nernstian response with reduced calcium selectivity over other cations. It was found that further addition of ETH 500[‡] ($<$0.5 wt%) to the membranes composed of RTV-SR/DOS/ETH 129 (or ETH 1001) not only reduces their bulk membrane resistance but also corrects their non-Nernstian behavior with surprisingly enhanced calcium selectivity. Figure 18.6 demonstrates the potentiometric performance of all-solid-state-type electrode coated with these membranes. These electrodes also exhibit small potential drift ($<$1 mV/day) with an extended lifetime ($>$9 weeks).

18.3.2.4. Carbonate-Selective Electrodes

Anion-selective membranes can be also formulated with the RTV-SR matrix. The carbonate-selective membranes based on trifluoroacetophenone derivatives [e.g., trifluoroacetyl-p-butylbenzene (TFABB) or trifluoroacetyl-p-decylbenzene (TFADB)] have been used to estimate the CO_2 content in physiological samples, as they offer the advantages of rapid response times and easy fabrication and miniatur-

---

*ETH 129: N,N,N',N'-tetracyclohexyl-3-oxapentanediamide.

[†]ETH 1001: N,N'-[(4R, 5R)-4,5-dimethyl-1,8-dioxo-3,6-dioxaoctamethylene]bis(12-methylaminodode-canoate)

[‡]ETH 500: tetradodecylammonium tetrakis (p-chlorophenyl)borate.[76]

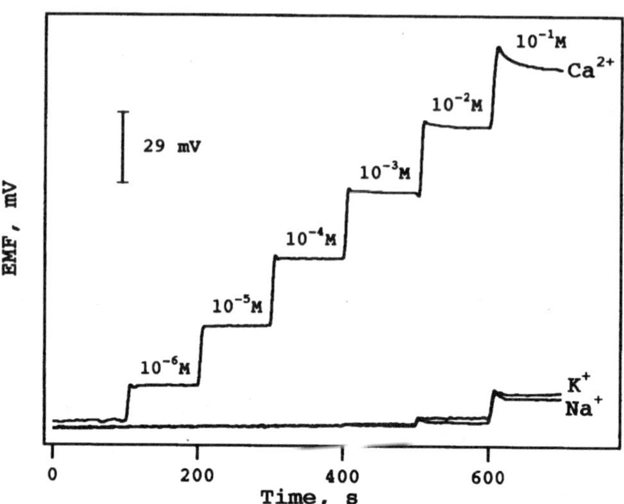

Figure 18.6. Dynamic response curves for the all-solid-state electrode coated with RTV-SR/DOA/ETH 1001/ETH 500 membrane to the change in concentrations of Ca^{2+}, Na^+, and K^+ from 10^{-6} to 10^{-1} M.

ization compared to Severinghaus-type gas sensors.[77–79] However, the use of PVC membrane-based carbonate-selective electrodes has been limited by the serious interference from lipophilic anions, such as salicylate, which are present in increased levels in the blood samples of patients who take aspirin (therapeutic level: 1–3 mM). Recently, the use of asymmetric membranes, which are formed by first casting a thin layer of cellulose triacetate without a carrier, hydrolyzing one side of this thin film with base, and, subsequently, casting a second layer of cellulose triacetate containing TFABB, has overcome the salicylate interference problems in serum carbon dioxide measurements.[79] However, the fabrication process for such membranes is rather cumbersome and difficult to apply for the all-solid-state-type electrode.

As noted earlier, the SR matrix-based membranes exhibit less interference from lipophilic anions. To take advantage of SR matrix-based membranes, the carbonate-selective membranes based on RTV-SR have been prepared after a careful optimization process. It was shown that in terms of the carbonate response, the TFABB- or TFADB-based membranes prepared with appropriately plasticized RTV-SR matrix not only function as well as conventional PVC matrix membranes, but they also exhibit a substantially reduced interfering response to salicyclate. Their analytical performance could be enhanced further by employing a high-pH buffer diluent [e.g., 2-amino-2-methyl-1-propanol (AMP)-H_2SO_4; pH 9.5–10.5]. Figure 18.7 shows a comparison of the potentiometric response of the PVC- and RTV-SR-based carbonate-selective membranes in the absence and presence of salicylate, which clearly demonstrates the enhanced performance and analytical utility of the RTV-SR-based membrane system for CO_2 measurements.

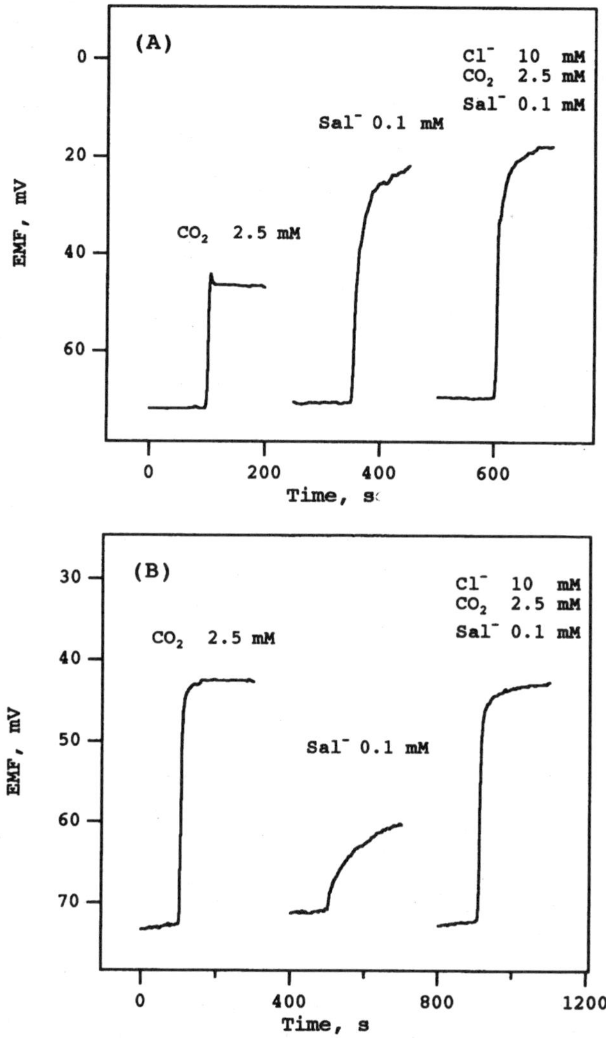

Figure 18.7. Response comparison of the conventional and new CO_2 measurement systems: (A) PVC/DOA/TFADB/TDMACl membrane with 0.2 M Tris-H_2SO_4, pH 8.6, and (B) RTV-SR/DOA/TFADB/TDMACl membrane with 0.2 M AMP-H_2SO_4, pH 9.7.

18.3.2.5. pH Electrodes

In physiology, the pH values of physiological fluids indicate the status of acid–base balance in the body, accounting for the problems associated with the functions of the buffer systems of blood, the respiratory system, and the renal mechanisms. Hence, it is desirable that biosensor cartridges for clinical application be equipped with a pH sensor. Although the pH-FET may be integrated into the cartridge-type sensor array, it is not convenient for use with different types of electrode systems, e.g., the simple metal conductor printed on alumina plate or silicon wafer. There have been some efforts to develop polymer-based pH-selective membranes for such electrode systems.[54]

The potentiometric behavior of RTV-SR membrane doped with H^+-selective ionophore, tridodecylamine (TDDA), is non-Nernstian, exhibiting an inflection point at pH 6.4 due to the drastic change in its response slope from 94 to 40 mV/pH.[80] The cause of this nonideal behavior is not yet fully understood. The calibration plots for the RTV-SR/TDDA membranes are gradually rectified as the content of plasticizer is increased, eventually fitting to the Nernstian-type equation in the pH 3–12 range when it becomes 21 wt% of the total composition. However, the use of polymer membrane-based all-solid-state-type pH electrodes is still problematic because they exhibit a considerable interfering response to dissolved carbon dioxide and other permeable gases. Although the introduction of a buffered hydrogel layer between the pH-sensitive membrane and the metal electrode greatly reduces such interference, it not only complicates the sensor fabrication process but also causes detachment of the outer sensing membrane from the sensor device because of the high osmotic pressure developed at the inner layer. The strong adhesion of RTV-SR membranes to solid substrate can help contain the swollen inner hydrogel layer. For the time being, the application of polymer membrane-based all-solid-state-type pH electrode with no hydrogel layers seem limited.

18.3.2.6. Other Ion-Selective Electrodes

Ammonium-selective RTV-SR membranes based on nonactin can be prepared without plasticizer, and their potentiometric performance is comparable to that of corresponding membranes based on plasticized PVC.[80] Chloride-selective RTV-SR membranes prepared with about 50 wt% tridodecylmethyl ammonium chloride (TDMACl) exhibit reduced interfering response to physiological concentrations of salicylate (~ 1 mM), but with a high detection limit (8.7×10^{-3} M) and small response slope (-24 mV/pCl).[81] In summary, the RTV-SR-based membranes, although a rather laborious process is often required to determine their optimal compositions, can be applied easily on most solid-state sensor devices, and the resultant all-solid-state-type Na^+-, K^+-, NH_4^+-, Ca^{2+}-, and CO_3^{2-}-selective electrodes exhibit virtually the same potentiometric performance, as corresponding conventional ISEs.

18.3.3. A pCO_2 Sensor with the Valinomycin-Based RTV-SR Membrane

An interesting application has been proposed for the valinomycin-based RTV-SR membrane, exploiting its high potentiometric performance, low impedance, insensitivity to pH change, intrinsic adhesive strength, and high CO_2 gas permeability.[57,72] As shown in Fig. 18.8A, a typical ISFET-based differential pCO_2 sensor system, which is a miniaturized version of the Severinghaus-type gas sensor system, requires placing Ag/AgCl reference sites beside each ISFET to complete a measuring

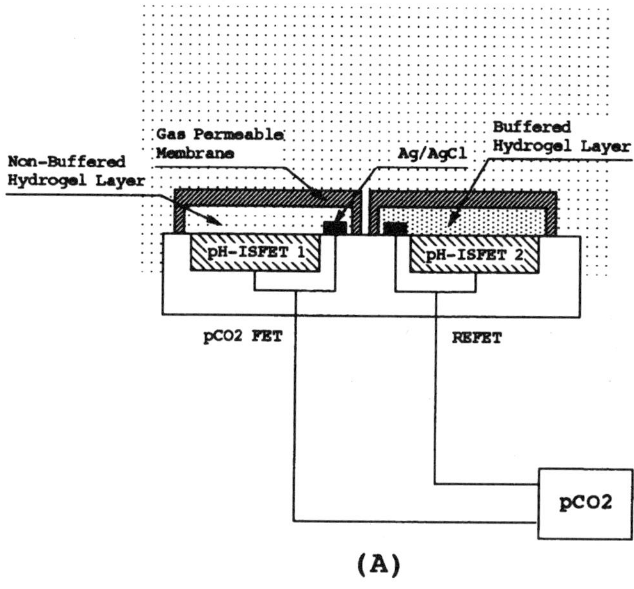

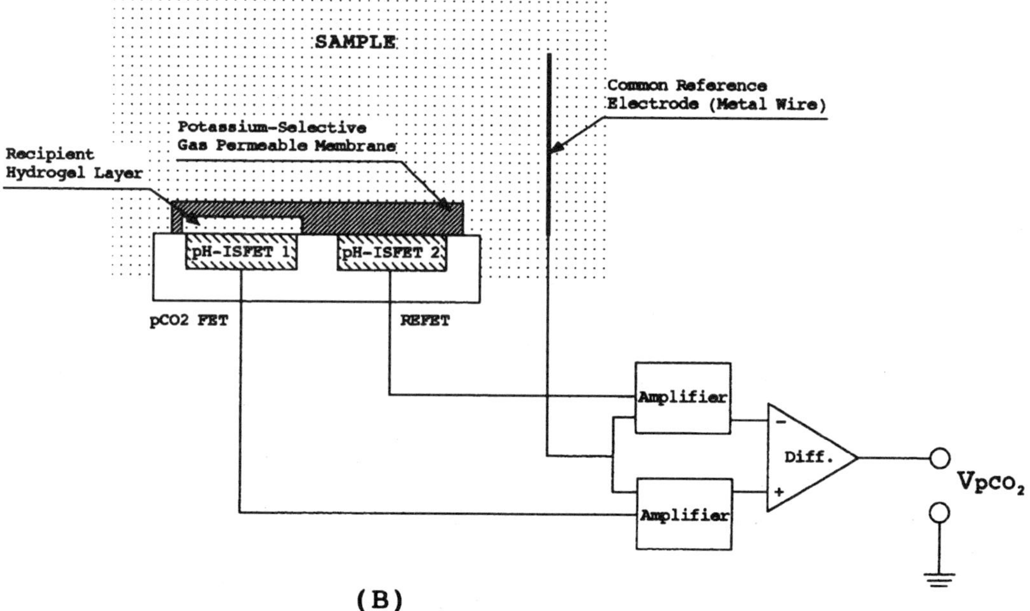

Figure 18.8. Schematic of the Severinghaus-type ISFET-based pCO_2 sensor: (A) conventional-type differential CO_2 measurement setup, (B) differential CO_2 measurement setup employing a low-resistance gas-permeable membrane.

circuit, bypassing the very high electrical resistance of the gas-permeable membrane. With a low-resistance gas-permeable membrane, the inner reference electrode sites can be replaced with one common reference (or any metallic electrode integrated on the chip), which greatly simplifies the construction of all-solid-state-type pCO_2 sensor system. Figure 18.8B is a schematic diagram of ISFET-based differential pCO_2 sensor system based on RTV-SR membrane doped with valinomycin; output signals from the pCO_2-FET and REFET (reference FET) are measured against the common reference, and the difference between these two signals is measured through a differential amplifier. Figure 18.9 clearly illustrates the advantage of the differential pCO_2 sensor system based on the RTV-SR/valinomycin membrane; potential and temperature drifts, and potassium and other ion responses of the two identical ISFETs cancel out, providing an excellent potentiometric signal to changes in pCO_2.

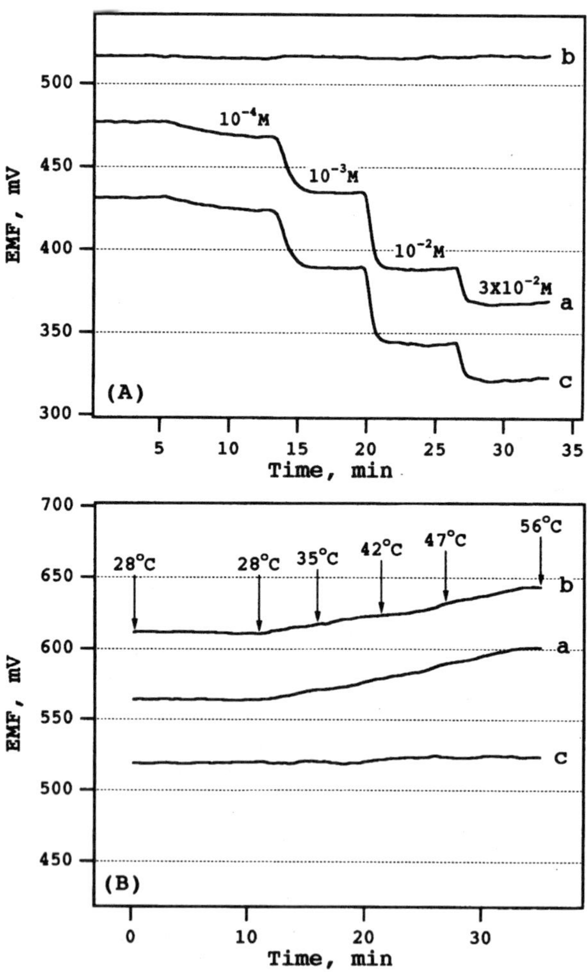

Figure 18.9. Response of the individual (a) pCO_2-FET and (b) REFET and of (c) the differential pCO_2 measurement setup measurement to the change in (A) dissolved pCO_2 concentration and (B) temperature.

18.4. POLYURETHANE-BASED ISE MEMBRANES

18.4.1. Polyurethane Matrix

Polyurethanes (PUs) are a class of polymer consisting of alternating hard- and soft-segment units. The hard segments are commonly based on aromatic or aliphatic diisocyantes and butanediol, whereas the soft segments are typically low-molecular-weight polyesters or polyethers. Figure 18.10 shows the structures of typical PUs; the hard-segment units act like thermally labile cross-link sites through hydrogen bonding and fillers for the rubbery soft-segment matrix (glass transition temperature, $T_g < -60°C$).[82,83] The physical and chemical properties of PUs are easily controlled by the types and relative compositions of the building blocks employed, additives (e.g., catalysts, surfactants, and foaming agents), and industrial processing techniques; a great variety of PUs have been formulated and used as basic materials for manufacturing a wide range of industrial and commercial goods, including implantable biomedical devices. PUs have also been examined as an alternative matrix for plasticized PVC for more than two decades — initially to prepare the ISE membranes with no plasticizer,[84] and recently to design adhesive and biocompatible membrane systems employing some medical-grade PUs such as Tecoflex (Thermedics Inc., Woburn, MA) and Pellethane (DOW, Midland, MI).[61–69]

Since the soft segments in a PU assume the role of a built-in plasticizer, when PUs with appropriate soft/hard-segment compositions (ca. 60:40–80:20 in wt%) are solvent-cast, they yield transparent, elastic, adhesive, and chemically stable sensing membranes even in the absence of any plasticizer.[83] However, in general, the potentiometric performance of plasticizer-free PU membrane is much inferior to that of the plasticized PVC membranes: insufficient solubility of neutral carriers in the PU matrix may be the cause of poor potentiometric performance. Addition of plasticizer and/or lipophilic additive is necessary for the PU-based membranes to result in near-Nernstian response to the primary ion. However, the amount of plasticizer needed could be much less for PU-based membranes than for those based

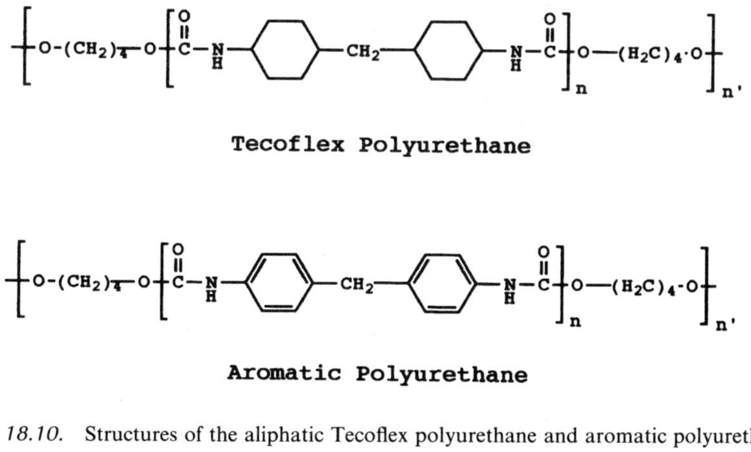

Tecoflex Polyurethane

Aromatic Polyurethane

Figure 18.10. Structures of the aliphatic Tecoflex polyurethane and aromatic polyurethane.

on PVC.[65,67,83] Since small cations (e.g., Li^+, Na^+, and K^+) tend to disrupt the hydrogen-bonded cross-links in PU,[85] resulting in mechanically weak (or waxy) membranes, it is advisable to incorporate a measured amount of lipophilic additive to PU-based membranes. Such effects are more pronounced in the plasticizer-free PU membranes with higher hard-segment composition.

Although PU-based membranes offer several advantages in biomedical applications, their potentiometric properties are strongly dependent on the types and relative compositions of neutral carriers and the other incorporated electroactive components. PU matrices are less inert than PVC matrices, and rather a careful optimization process is necessary to determine the compositions of PU-based ion-selective membranes. It should be remembered that the potentiometric properties of PU-based membranes are determined primarily by the type of PU matrix chosen.* Since PU-based membranes tend to exhibit higher detection limits with reduced response slopes compared to those of PVC-based membranes, PU is often blended with PVC-based polymers to take advantage of both matrix systems.[61-64]

18.4.2.　Ion- and Biosensor Membranes Based on PU-Blended Matrices

Recent interest in PU-based membranes has been engendered primarily by the research activities of Myerhoff's group. Cha et al.[63] found that the ammonium- and potassium-selective membranes prepared with the Tecoflex PU/hydroxylated PVC (PVC/PVC-OH/PVC-COOH)/DOA matrix exhibit electrochemical performance (i.e., slope, detection limit, and selectivity) that is essentially equivalent to that of PVC membranes. This membrane system also exhibits enhanced blood compatibility, significantly retarding blood clotting time on its surface. Treating the various membrane systems with an adhesion promoter, such as $SiCl_4$, they observed that only the PU-containing membranes exhibited drastically improved adhesion to the silicon nitrite-based surface, potentiometric stability, and lifetime.

Liu et al.[64] further developed this membrane system into bioselective electrodes; a thin layer of hydrophilic polyurethane (HPU) containing cross-linking reagent polylysin (PLS) was applied on one side of the PU/hydroxylated PVC/DOA/TDDA (or nonactin) membrane, and enzymes were then immobilized on the surface of the asymmetric HPU layer. They demonstrated that such asymmetric membranes are easily applied on the all-solid-state-type electrodes, resulting in conventional-electrode-like performance. However, we recently noted that the enzyme-containing HPU layer can be prepared with no cross-linking reagent[69]; as shown in Fig. 18.11, the enzyme, urease, can be mixed directly with HPU in a methanol/THF cosolvent and cast on the ammonium-selective PU membrane. Interestingly, this new asymmetric membrane exhibits a better response time, slope, and detection limit to the change in urea concentration than membranes with HPU/PLS-immobilized enzyme layer (see Fig. 18.12).

---

*The blank PU membrane composed of 60 wt% hard segment exhibits low electrical resistance, but hardly responds to most ionic species below 10^{-2} M. An all-solid-state electrode modified with this membrane functions like a reference electrode. On the other hand, blank membranes based on PVC and SR content exhibit cationic responses, indicating the presence of anionic impurities in the matrices.[86]

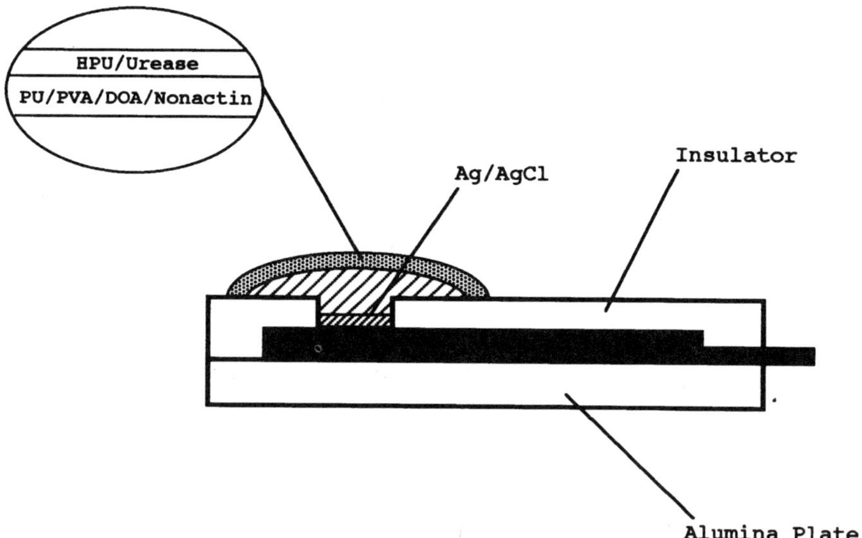

Figure 18.11. Schematic of all-solid-state urea sensor.

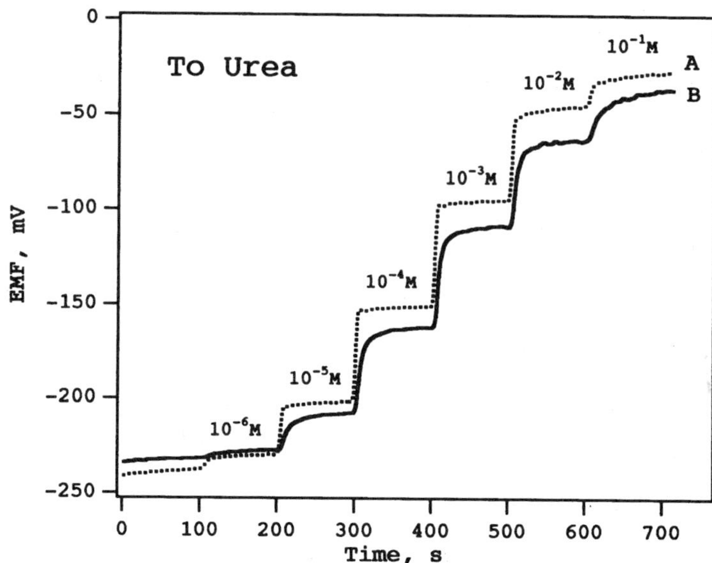

Figure 18.12. Response comparison of the ammonium-selective membrane-based urea sensors with (A) HPU/urease layer and (B) urease cross-linked to HPU/PLS layer.

As discussed above, the soft segments in PU behave like a built-in plasticizer. Thus, it is possible to fabricate plasticizer-free membranes blending the PVC with the PU containing high soft-segment compositions (e.g., $w_{\text{soft}} > 80\,\text{wt}\%$). For example, the valinomycin-doped PU/PVC membranes exhibit fair detection limits ($7 \times 10^{-5}\,\text{M}$) and response slopes ($54\,\text{mV/pK}^+$). However, in general, the potentiometric performance of plasticizer-free PU/PVC membranes is not comparable to that of the plasticized membranes. The use of the PU/PVC matrix should thus be counterweighed by other advantages, such as enhanced adhesion and improved biocompatibility.

18.4.3. Biocompatible ISE Membranes

It has long been recognized that ISE measurements may suffer from interference caused by various types of components in physiological fluids, e.g., a shift in the standard cell potential due to the induction of a membrane asymmetry potential caused by protein adsorption on the surface of the solvent polymeric membranes,[66,87,88] a negative interference caused by lipophilic anionic species,[73] or erroneous potentiometric response caused by thrombus formation.[67,68,89] Furthermore, the membrane components, especially the plasticizer, leached out of an implanted sensor could dissolve into the physiological fluids, eliciting a serious inflammatory reaction in the vicinity of the electrode.[65] Attempts to improve the blood and biomedical compatibility of conventional ISE membranes have been carried out using well-known medical-grade PUs.

Park et al. compared the shifts in the standard cell potentials of various calcium-selective electrodes upon repeated exposure to the protein (bovine serum albumin; BSA)-containing solutions, employing a flow-injection system.[66] It was observed that the baselines of the electrodes with PVC/NPOE/ETH 129 membranes were significantly shifted in the positive direction, while those of the electrodes with Tecoflex PU/NPOE/ETH 129 membranes were within instrumental resolution. In addition, there is a report that the use of Tecoflex membrane with nonpolar plasticizer (e.g., DOS) substantially reduces the interference from hydrophobic anions.[65]

Espadas-Torre and Meyerhoff performed *in vitro* platelet adhesion studies to compare the thrombogenic properties of various polymer matrices that are useful for preparing implantable ion-selective membrane electrodes.[67] While both PVC and Tecoflex PU membranes doped with TDDA or valinomycin were shown to be thrombogenic, they found that incorporation of high-molecular-weight block copolymers of poly(ethylene oxide) (PEO) and poly(propylene oxide) (PPO) into those membranes reduces platelet adhesion. A marked decrease in platelet adhesion was achieved by applying a thin layer of photo-cross-linked PEO on the surface of Tecoflex PU-based membranes without degrading their potentiometric performance. Recently, the same group introduced an ingenious method to develop nonthrombogenic sensors[68]: a thin layer of H^+- or K^+-selective membrane that also contained $4\,\text{wt}\%$ of nitric oxide (NO)-releasing compound, N,N'-dimethyl-hexanediamine nitric oxide adduct (DMHD/N_2O_2) is sandwiched between the two identical ISE membrane layers with no $DMHD/N_2O_2$. The ISE membranes modified in this manner slowly emit NO, a potent platelet anticoaggregation and vasodilator agent, and exhibit remarkably reduced levels of platelet adhesion on their surface, while providing the same potentiometric performance as the unmodified ISE membranes. However, the potential toxicity of this approach needs further investigation.

The problem of toxicity for implantable solvent polymeric sensor membranes has been addressed by Lindner et al[65]: cellular responses during acute and chronic inflammation caused by ISE membranes that were cage-implanted on the backs of Sprague–Dawley rats were determined by counting leukocytes in their exudate. They found that there is a positive correlation between the amount of plasticizer in ISE membranes and the host inflammatory response. Among the membranes tested (e.g., PVC/DOS, PVC/NPOE, Tecoflex PU with and without plasticizer), plasticizer-free Tecoflex PU-based ISE membranes exhibited the least inflammatory response.

18.5. CONCLUDING REMARKS

In this chapter, we have summarized the recent advances in solvent polymeric ISE membranes. Particular emphasis was given to the utility of one-part room-temperature vulcanizing-type silicone rubber (RTV-SR)- and medical-grade poly-urethane (PU)-based membranes for the fabrication of all-solid-state electrodes and biomedical sensors.

RTV-SR, commercially available one-part adhesive or sealant, is readily dissolved or dispersed in common organic solvents (e.g., tetrahydrofuran) with most electroactive compounds (ionophores and lipophilic additives), and cured easily at room temperature by atmospheric moisture activation*: the resultant RTV-SR-based membranes are highly adhesive to most solid substrates, chemically stable, and mechanically sturdy. Thus, the RTV-SR membranes carefully formulated with appropriate electroactive compounds make excellent ISE membranes for all-solid-state electrodes. In this review, we have demonstrated that the all-solid-state-type Na^+-, K^+-, NH_4^+-, Ca^{2+}-, and CO_3^{2-}-selective electrodes can exhibit virtually the same potentiometric performance as that of corresponding conventional ISEs. We also discussed the use of low-impedance RTV-SR membranes doped with valinomycin for the fabrication of an ISFET-based differential pCO_2 sensor system.

The potentiometric performance of PU-based membranes is greatly dependent on their soft/hard-segment ratios; in general, the preferred ratio is about 60:40 in wt%. PU-based membranes also have stronger adhesion to most solid surfaces than PVC-based membranes and owing to the rubbery soft-segments, can be prepared without plasticizer. The PU membranes with a reduced amount or no plasticizer offer several advantages in biomedical applications: they exhibit less anionic interference from lipid-soluble anions, reduced sample-induced (typically, by protein adsorption) asymmetry potential, and enhanced biocompatibility (e.g., reduced

---

*Most RTV-SR contains a small amount of transition metal catalyst (e.g., titanium) to promote the curing process. It may be separated by centrifugation.[58] However, the potentiometric performance of RTV-SR-based ISE membranes is not significantly affected by the presence of metal catalyst.[59]

inflammatory response in the vicinity of the implanted sensor). To further enhance the blood compatibility of ISE membranes, a thin layer of photo-cross-linked PEO on the surface of Tecoflex PU-based membranes has been introduced. Meyerhoff et al. have also shown that the ISE membranes containing $DMHD/N_2O_2$ slowly emit NO and effectively reduce platelet adhesion.

Someday, all-solid-state ISEs based on various advanced methods may replace those based on solvent polymeric membranes. For the time being, however, solvent polymeric membranes are still the preferred method for fabricating all-solid-state ISEs and biomedical sensors. They are readily prepared from a simple mixture of polymer matrix and electroactive compounds, easily applied on any solid-state sensor device with standard industrial equipment (e.g., screen-printer and microdispenser) in batch scale, and provide high sensor-to-sensor reproducibility. Furthermore, their operational characteristics can be readily controlled by adding appropriate electroactive components (e.g., ionophores, lipophilic additives, plasticizers, adhesion promoters, and enzymes). However, all-solid-state ISEs based on solvent polymeric membranes developed to date still do not provide sufficient signal stability, lifetime, and biocompatibility. They also exhibit interference from dissolved gases (typically CO_2 and O_2). On theoretical side, the response mechanisms of all-solid-state ISEs are not fully understood. For these reasons, development of various solvent polymeric membranes with enhanced operational characteristics may remain one of the key themes in ion- and biosensor research.

ACKNOWLEDGMENTS: We gratefully acknowledge the financial support of the Korea Science and Engineering Foundation and the Ministry of Education. Dr. G. S. Cha expresses his sincere thanks to both Prof. Meyerhoff (Department of Chemistry) and Prof. Brown (Department of Electrical Engineering and Computer Science) of University of Michigan for providing research facilities. We are also indebted to our past and current graduate students for their commitment and hard work.

REFERENCES

1. (a) Anderson DJ, Guo B, Xu Y, et al. Clinical chemistry. *Anal Chem* 1997;69:165R–229R.; (b) Wang J. Electroanalysis and biosensors. *Anal Chem* 1995;67:487R–492R.
2. Collision ME, Meyerhoff ME. Chemical sensors for bedside monitoring of critically ill patients. *Anal Chem* 1990;62:425A–434A.
3. Yim HS, Kibbey CE, Ma SC, et al. Polymer membrane-based ion-, gas- and bio-selective potentiometric sensors. *Biosens Biolectron* 1993;8:1–38.
4. Oesch U, Amman D, Simon W. Ion-selective membrane electrodes for clinical use. *Clin Chem* 1986;32:1448–1459.
5. (a) Davies ML, Hamilton CJ, Murphy SM, et al. Polymer membranes in clinical sensor applications: I. An overview of membrane function. *Biomaterials* 1992;13:971–978; (b) Murphy SM, Davies ML, Hamilton CJ, et al. Polymer membranes in clinical sensor applications: II. The design and fabrication of permselective hydrogels for electrochemical devices. *Biomaterials* 1992;13:979–989; (c) Davies ML, Murphy SM, Hamilton CJ, et al. Polymer membranes in clinical sensor applications: III. Hydrogels as reactive matrix membranes in fiber optic sensors. *Biomaterials* 1992;13:991–998.
6. Freiser H. Coated wire ion-selective electrodes. In: Freiser H, ed. *Ion-Selective Electrodes in Analytical Chemistry*. New York: Plenum Press, 1980; vol 2, pp 85–105.

7. Cattrall RW, Drew DM, Hamilton IC. Some alkylphosphoric acid esters for use in coated-wire calcium-selective electrodes. *Anal Chim Acta* 1975;76:269–277.

8. Hulanicki A, Trojanowicz M. Calcium-selective electrodes with PVC membranes and solid internal contacts. *Anal Chim Acta* 1976;87:411–417.

9. Nikolskii BP, Materova EA. Solid contact in membrane ion-selective electrodes. *Ion-Selective Electrode Rev* 1985;7:3–39.

10. Fogt EJ, Untereker DF, Norenberg MS, et al. Response of ion-selective field effect transistors to carbon dioxide and organic acids. *Anal Chem* 1985;57:1998–2002.

11. Li X, Verpoorte EMJ, Harrison DJ. Elimination of neutral species interference at the ion-selective membrane/semiconductor device interference. *Anal Chem* 1988;60:493–498.

12. Dror M, Bergs EA, Rhodes RK. Potassium ion-selective electrodes based on valinomycin/PVC overlayered solid substrates. *Sens Act* 1987;11:23–36.

13. Miyahara Y, Simon W. Comparative studies between ion-selective field effect transistors and ion-selective electrodes with polymeric membranes. *Electroanalysis* 1991;3:287–292.

14. Yim HS, Meyerhoff ME. Reversible potentiometric oxygen sensors based on polymeric and metallic film electrodes. *Anal Chem* 1992;64:1777–1785.

15. Richard P, Nahir TM, Cosofret VV et al. Mechanism of transport in carrier-based ion-selective electrodes. *Anal Proc Anal Comm* 1994;31:301–312.

16. Pace SJ, Hamerslag JD. Thick film multilayer ion sensors for biomedical applications. *Polym Mater Sci Eng* 1991;64:368–369.

17. Lauks IR, Groves MR, Wieck HJ. Mass fabricated ion, gas and enzyme multispecies sensor using IC technology. *Proc Electrochem Soc (Electrochem Sens Biomed Appl)* 1986;116–128.

18. Knoll M, Cammann K, Dumschat C, et al. Micromachined ion-selective electrode with polymer matrix membranes. *Sens Act B* 1994;21:71–76.

19. Arquint Ph, Koudelka-Hep M, de Rooij NF. Organic membranes for miniaturized electrochemical sensors: fabrication of a combined pO_2 pCO_2 and pH sensor. *J Electroanal Chem* 1994;378:177–183.

20. (a) Lindner E, Cosofret VV, Ufer S, et al. Flexible (Kapton-based) microsensor arrays of high stability for cardiovascular applications. *J Chem Soc Faraday Trans* 1993;89:361–367; (b) Lindner E, Cosofret VV, Ufer S, et al. *In vivo* and *in vitro* testing of microelectronically fabricated planar sensors designed for applications in cardiology. *Fresenius J Anal Chem* 1993;342:584–588.

21. Mock T, Morrison D, Yatscoff R. Evaluation of the i-STAT system: a portable chemistry analyzer for the measurement of sodium, potassium, chloride, urea, glucose and hematocrit. *Clin Biochem* 1995;28:187–192.

22. Wahr JA, Lau W, Tremper KK, et al. Accuracy and precision of a new, portable, handheld blood gas analyzer, the IRMA. *J Clin Monit* 1996;12:317–324.

23. Cosofret VV, Buck RP, Kao WJ, et al. Electroanalytical and biocompatibility studies on carboxylated poly(vinyl chloride) membrane for microfabricated array sensors. *Analyst* 1994;119:2283–2292.

24. (a) Satchwill T, Harrison DJ. Synthesis and characterization of new poly(vinyl chloride) membranes for enhanced adhesion on electrode surfaces. *J Electroanal Chem* 1986;202:75–81; (b) Harrison JD, Cunningham LL, Xizhong L, et al. Enhanced lifetime and adhesion of potassium ion-, ammonium, ion- and calcium ion-sensitive membranes on solid surfaces using hydroxyl-modified poly(vinyl chloride) matrices. *J Electrochem Soc* 1988;135:2473–2478.

25. Moody GJ, Thomas JDR. Modified poly(vinyl chloride)matrix membranes for ion-selective field effect transistor sensors. *Analyst* 1988;113:1703–1706.

26. Blackburn G, Janata J. The suspended mesh ion selective field effect transistor. *J Electrochem Soc* 1982;129:2580–2584.

27. Bratov A, Abramova N, Alegret S, et al. Photocurable polymer matrices for potassium-sensitive ion-selective electrode membranes. *Anal Chem* 1995;67:3589–3595.

28. Dumschat C, Froemer R, Rautschek H, et al. Photolithographically patternable nitrate-sensitive acrylate-based membrane. *Anal Chim Acta* 1991;243:179–182.

29. Harrison DJ, Teclemariam A, Cunningham LL. Photopolymerization of plasticizer in ion-selective membranes on solid-state sensors. *Anal Chem* 1989;61:246–251.

30. (a) Cardell TJ, Cattrall RW, Iles PJ, et al. Photocured polymers in ion-selective electrode membranes: 4. An ultraviolet laser cured membrane for potassium. *Anal Chim Acta* 1989;219:135–140; (b) Cardell TJ, Cattrall RW, Iles PJ, et al. Photocured polymers ion-selective membranes: 3. A potassium

electrode for flow injection analysis. *Anal Chim Acta* 1988;204:329–332; (c) Cardell TJ, Cattrall RW, Iles PJ, et al. Photocured polymers ion-selective membranes: 2. A calcium electrode for flow injection analysis. *Anal Chim Acta* 1985;177:239–242; (d) Cattrall RW, Iles PJ, Hamilton IC. Photocured polymers in ion-selection electrode membranes. *Anal Chim Acta* 1985;169:403–406.

31. Moody GJ, Slater JM, Thomas JDR. Membrane design and photocuring encapsulation of flatback based ion-sensitive field-effect transistors. *Analyst* 1988;113:103–108.

32. (a) Ding L, Li J, Wang E, et al. K^+ sensors based on supported alkanethiol/phospholipid bilayers. *Thin Solid Films* 1997:293:153–158. (b) Li J, Ding L, Wang E, et al. The ion selectivity of monensin incorporated phospholipid/alkanethiol bilayers. *J Electroanal Chem* 1996;414:17–21.

33. Schoening MJ, Sauke M, Steffen A, et al. Ion-sensitive field-effect transistors with ultra thin Langmuir-Blodgett Membranes. *Sens Act B* 1995;27:325–328.

34. Kauffmann F, Hoffmann B, Erbach R, et al. Ca^{2+} sensor with amphiphilic Langmuir–Blodgett membranes. *Sens Act B* 1994;18/19:60–64.

35. Giladoni A, Margheri E, Gabrelli G. Building ion-selective membranes by insertion of ionophore in both lipid and polymer matrices. *Prog Colloid Polym Sci* 1993;93:233–236.

36. Osa T. Potentiometric response of lipid-modified ISFET. *Appl Biochem Biotechnol* 1993;41;35–40

37. Gilardoni A, Margheri E, Gabrelli G. Potassium ion-selective membranes built by the Langmuir–Blodgett technique. *Colloid Surf* 1992;68:235–242.

38. Arisawa S, Arise T, Yamamoto R. Concentration of enzymes adsorbed onto Langmuir films and characteristics of a urea sensor. *Thin Solid Films* 1992;209:259–263.

39. Erbach R, Vogel A, Hoffmann B. Ion-sensitive field-effect structures with Langmuir–Blodgett membranes. *GBF Monogr* 1992;17 (*Biosensors: Fundam Technol Appl*):353–357.

40. Howarth VA, Petty MC, Ancelin H, et al. Structural characterization of phospholipid Langmuir–Blodgett multilayers containing valinomycin. *Vib Spectrosc* 1990;1:29–33.

41. Yoshida S, Okawa Y, Watanabe T, et al. Ion-sensitive tin oxide electrodes carrying amphiphilic crown ether Langmuir–Blodgett films. *Chem Lett* 1989;243–246.

42. Anzai J, Furuya K, Chen C, et al. Enzyme sensors based on ion-sensitive field-effect transistor: use of Langmuir–Blodgett membrane as a support for immobilizing penicillinase. *Anal Sci* 1987;3:271–272.

43. Hauser PC, Chiang DWL, Wright GA. A potassium ion-selective electrode with valinomycin based poly(vinyl chloride) membrane and a poly(vinyl ferrocene) solid contact. *Anal Chim Acta* 1995;302:241–248.

44. Bobacka J, McCarrick M, Lewenstam A, et al. All solid-state poly(vinyl chloride) membrane ion-selective electrodes with poly(3-octylthiophene) solid internal contact. *Analyst* 1994;119:1985–1991.

45. Michalska A, Hulanicki A, Lewnstam A, et al. All solid-state hydrogen ion-selective electrode based on a conducting polypyrrole solid contact. *Analyst* 1994;119:2417–2420.

46. Pearson JF, Slater JM, Jovanovic V. Coated-wire and composite ion-selective electrodes based on doped polypyrrole. *Analyst* 1992;117:1885–1890.

47. Cadogan A, Gao Z, Lewnstam A, et al. All solid-state sodium-selective electrode based on calix[4]arene ionophore in a poly(vinyl chloride) membrane with a polypyrrole solid contact. *Anal Chem* 1992;64:2496–2501.

48. Sudhölster EJR, van der Wal PD, Skowronska-Ptasinska M, et al. Modification of ISFETs by covalent anchoring of poly(hydroxyethyl methacrylate)hydrogel: introduction of a thermodynamically defined semiconductor-sensing membrane interface. *Anal Chim Acta* 1990;230:59–65.

49. Bobacka J, Lindfors T, McCarrick M, et al. Single-piece all-solid-state ion-selective electrode. *Anal Chem* 1995;67:3819–3823.

50. van der Wal PD, Skowronska-Ptasisnska M, van den Berg A, et al. New membrane materials for potassium-selective ion-sensitive field-effect transistors. *Anal Chim Acta* 1990;231:41–52.

51. van der Wal PD, Sudhölter EJR, Boukamp BA, et al. Impedance spectroscopy and surface study of potassium-selective silicone rubber membranes. *J Electroanal Chem* 1991;317:153–168.

52. Kimura K, Matsuba T, Tsujimura Y, et al. Unsymmetrical calix[4]arene ionophore/silicone rubber composite membranes for high-performance sodium ion-sensitive field-effect transistors. *Anal Chem* 1992;64:2508–2511.

53. Brunink JAJ, Lugtenberg RJW, Brzøzka Z, et al. The design of durable Na^+-selective CHEMFETs based on polysiloxane membranes. *J. Electroanal Chem* 1994;378:185–200.

54. Malinowska E, Oklejas V, Hower RW, et al. Enhanced electrochemical performance of solid-state ion sensors based on silicone rubber membranes. *Transducers 95 — Eurosensors IX* 1995;851–854.

55. Oh BK, Kim CY, Lee HJ, et al. One-component room temperature vulcanizing-type silicone rubber-based calcium-selective electrodes. *Anal Chem* 1996;68:503–508.

56. Shin JH, Sakong DS, Nam H, et al. Enhanced serum carbon dioxide measurements with a silicone rubber-based carbonate ion-selective electrode and a high-pH dilution buffer. *Anal Chem* 1996;68:221–225.

57. Shin JH, Lee HJ, Kim CY, et al. ISFET-based differential pCO_2 sensors employing a low-resistance gas-permeable membrane. *Anal Chem* 1996;68:3166–3172.

58. Högg G, Lutze O, Camman K. Novel membrane material for ion-selective field-effect transistors with extended lifetime and improved selectivity. *Anal Chim Acta* 1996;335:103–108.

59. Poplawski ME, Brown RB, Rho KL, et al. One-component room temperature vulcanizing-type silicone rubber-based sodium-selective membrane electrodes. *Anal Chim Acta* 1997 [in press].

60. Lee HJ, Oh, HJ, Cui G, et al. All-solid-state sodium-selective electrodes based on room temperature vulcanizing-type silicone rubber matrix. *Anal Sci* 1997;13:289–294.

61. Meruva RK, Malinowska E, Hower RW, et al. Improved EMF stability of solid-state ion-selective sensors by incorporation of lipophilic silver-calix[4]arene complexes within polymeric films. *Transducers 95 — Eurosensors IX* 1995;855–858.

62. Hower RW, Shin JH, Cha GS, et al. New solvent system for the improved electrochemical performance of screen-printed polyurethane membrane-based solid-state sensors. *Transducers 95 — Eurosensors IX* 1995;859–862.

63. Cha GS, Liu D, Meyerhoff ME, et al. Electrochemical performance, biocompatibility, and adhesion of new polymer matrices for solid-state ion sensors. *Anal Chem* 1991;63:1666–1672.

64. Liu D, Meyerhoff ME. Potentiometric ion- and bioselective electrodes based on asymmetric polyurethane membranes. *Anal Chim Acta* 1993;274:37–46.

65. Lindner E, Cosofret VV, Ufer S, et al. Ion-selective membranes with low plasticizer content: electroanalytical characterization and biocompatibility studies. *J. Biomed Mater Res* 1994;28:591–601.

66. Park SB, Chung S, Cha GS, et al. Minimization of asymmetry potential in ETH 129-based calcium-selective membrane electrodes. *Bull Kor Chem Soc* 1995;16:1033–1037.

67. Espada-Torre C, Meyerhoff ME. Thrombogenic properties of untreated and poly(ethylene oxide)-modified polymeric matrices useful for preparing intraarterial ion-selective electrodes. *Anal Chem* 1995;67:3108–3114.

68. Espadas-Torre C, Oklejas V, Mowery K, et al. Thromboresistant chemical sensors using combined nitric oxide release/ion sensing polymer films. *J Am Chem Soc* 1997;119:2321–2322.

69. Shin JH, Lee JS, Lee HJ, et al. Potentiometric biosensors using immobilized enzyme layers mixed with hydrophilic polyurethane. *Sens Act B* 1998; 50:19–26.

70. Alvarez-Icaza M, Bilitewski U. Mass production of biosensors. *Anal Chem* 1993;65:525A–533A.

71. Lindner E, Niegreisz Z, Toth K, et al. Electrical and dynamic properties of non-plasticized potassium-selective membranes. *J Electroanal Chem* 1989;259:67–80.

72. Mostert IA, Anker P, Jenny HB, et al. Neutral carrier based silicone rubber membranes for H_3O^+, K^+, NH_4^+ and Ca_2^+ selective electrodes. *Mikrochim Acta* 1985;I:33–38.

73. Jenny HB, Riess C, Amman D, et al. Determination of K^+ in diluted and undiluted urine with ion-selective electrodes. *Mikrochim Acta* 1980;II:309–315.

74. Pick J, Toth K, Pungor E. A potassium-selective silicone-rubber membrane electrode based on a neutral carrier. *Anal Chim Acta* 1973;64:477–480.

75. Bakker E, Pretsch E. Lipophilicity of tetraphenylborate derivatives as anionic sites in neutral carrier-based solvent polymeric membranes and lifetime of corresponding ion-selective electrochemical and optical sensors. *Anal Chim Acta* 1995;309:7–17.

76. Amman D, Pretsch E, Simon W. Lipophilic salts as membrane additives and their influence on the properties of macro- and micro-electrodes based on neutral carriers. *Anal Chim Acta* 1985;171:119–129.

77. Herman HB, Rechnitz GA. Carbonate ion-selective membrane electrode. *Science* 1974;184:1074–1075.

78. Hong YK, Yoon WJ, Oh HJ, et al. Effects of varying quaternary ammonium salt concentration on the potentiometric properties of some trifluoroacetophenone derivative-based solvent-polymeric membranes. *Electroanalysis* 1997;9:865–868.

79. Lee KS, Shin JH, Han SH, et al. Asymmetric carbonate ion-selective cellulose acetate membrane electrodes with reduced salicylate interference. *Anal Chem* 1993;65:3151–3155.

80. Shin JH, Rho KL, Lee HJ, et al. unpublished results.

81. Oh HJ. Chloride selectivity of various quaternary ammonium salts incorporated in room-temperature vulcanizing-type silicone rubber matrix. BS Thesis, Kwangwoon University, 1995.

82. Coleman MM, Skrovanek DJ, Hu J, et al. Hydrogen bonding in polymer blends: 1. FTIR studies of urethane-ether blends. *Macromolecules* 1988;21:59–65.

83. Yun SY, Hong YK, Oh BK, et al. Potentiometric properties of ion-selective electrode membranes based on segmented polyurethane matrices. *Anal Chem* 1997;69:868–873.

84. Fiedler U, Ruzicka J. Selectrode—the universal ion-selective electrode: Part VII. A valinomycin-based potassium electrode with nonporous polymer membrane and solid-state inner reference system. *Anal Chim Acta* 1973;67:179–193.

85. Seki M, Sato K, Yosomiya R. Polyurethane elastomer-LiClO$_4$ complexes as a polymeric solid electrolyte. *Makromol Chem* 1992;193:2971–2978.

86. Cosofret VV, Erdosy M, Raleigh JS, et al. Aliphatic polyurethane as a matrix for pH sensors: effects of native sites and added proton carrier on electrical and potentiometric properties. *Talanta* 1996;43:143–151.

87 Covington AK, Zhou DM. Study of protein effect on calcium-ion-selective membranes by using a.c. impedance methods. *J Electroanal Chem* 1992;341:77–84.

88. Dürselen LFJ, Wegmann D, May K, et al. Elimination of the asymmetry in neutral-carrier-based solvent polymeric membranes induced by proteins. *Anal Chem* 1988;60:1455–1458.

89. Brooks KA, Allen JR, Pamela W, et al. Effect of surface-attached heparin on the response of potassium-selective electrodes. *Anal Chem* 1996;1439:1443.

19

Rapid Measurement of Biodegradable Substances in Water Using Novel Microbial Sensors

Reinhard Renneberg, Alex W. K. Kwong, Chiyui Chan, Gotthard Kunze, Maria L. Lung, and Klaus Riedel

19.1. INTRODUCTION

Microbial sensors based on microorganisms in direct contact with a transducer that converts the biochemical signal into a quantifiable electrical response signal are used for sensitive determinations of a broad range of substances, such as amino acids, saccharides, organic acids, aromatics, alcohols, vitamins, antibiotics, steroids, peptides, and inorganic molecules.[1-3] Each microorganism is able to recognize a specific set of substances, and this ability can help us determine many complex variables, such as the sum of biodegradable compounds in wastewater[4] and mutagenic compounds.[5] Moreover, the application of biosensor techniques also offers an elegant possibility for the physiological characterization of microorganisms, as has been demonstrated by the investigations of glucose uptake,[6] peptide metabolism,[7] and substrate adaptation[8] and by the determination of the alteration of the physiological state during fermentation.[9] Such physiological studies are based on determination of the respiration of microorganisms. In these cases a combination of a microorganism with an amperometric oxygen electrode is used in the biosensor.[10]

The function of such microbial sensors is as follows (Fig. 19.1): the oxygen diffuses from the air-saturated solution through the dialysis membrane, the mem-

---

Reinhard Renneberg, Alex W. K. Kwong and Chiyui Chan • Biosensor and Bioelectronics Laboratory, Department of Chemistry, Hong Kong University of Science and Technology, Clearwater Bay, Kowloon, Hong Kong.   Gotthard Kunze • Institute of Plant Genetics and Crop Plant Research, D-06466, Gatersleben, Germany.   Maria L. Lung • Department of Biology, Hong Kong University of Science and Technology, Clearwater Bay, Kowloon, Hong Kong.   Klaus Riedel • Dr. Bruno Lange GmbH Berlin, D-40549 Düsseldorf 11, Germany.

Biosensors and Their Applications, edited by Yang and Ngo, Kluwer Academic/Plenum Publishers, New York, 1999.

333

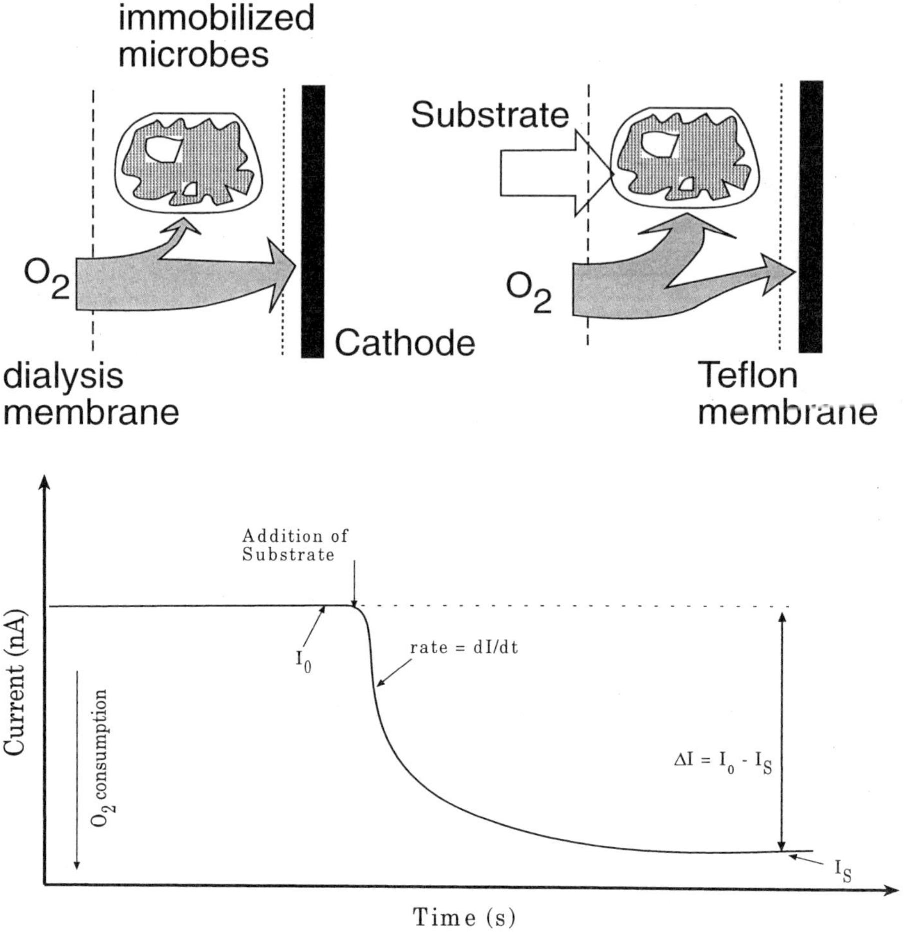

Figure 19.1.   Principle of measurement with microbial sensor. Oxygen is diffusing into the sensor membrane and the immobilized microorganisms at steady state (I_0). When the substrate is added, the immobilized microorganisms take up more oxygen by respiration and thus decrease the oxygen concentration in the layer. A decrease in current can be obtained by reaching a steady state (I_S). Measurement can be achieved by (i) the difference of currents (ΔI), or (ii) the first derivative of the current–time curve (dI/dt).

brane containing the microorganisms, and the Teflon membrane, and is then reduced at the platinum cathode at $-600\,\text{mV}$ vs. Ag/AgCl. Part of the oxygen is consumed by the microorganism. The steady-state current (I_0) represents the oxygen diffusion through the composite membrane and reflects the endogenous respiration of the microorganism. If assimilative substrate is added to the measuring solution, the substrate permeates through the dialysis membrane. Some of it is taken up by the microbial cells and is subsequently degraded. The respiration rate (dI/dt) increases, resulting in a decrease in the dissolved oxygen concentration and in the current until a new steady state (I_S) is reached. In principle, there are two possible types of respiration measurements:

1. End-point measurement where the difference in the currents (ΔI) reflects the respiration rate of the substrates.
2. Kinetic measurement (first derivative of the current–time curve corresponding to the acceleration of respiration (dI/dt)).

For physiological studies, kinetically controlled sensors, which have very low microbe loadings and suitable immobilization of microorganisms as well as thin membranes, have to be used.[11] The sensitivity of this type of sensor is determined mostly by cell activity, but not by diffusion limitation, which means that the substrate transport into the cell and substrate assimilation should be the rate-limiting process.[7] Monitoring pollution, however, requires a diffusion-controlled sensor with optimal cell loading.

19.2. APPLICATION OF MICROBIAL SENSORS IN POLLUTION CONTROL

Increasing attention is being paid to the use of microbial sensors in environmental monitoring, such as those for the estimation of biochemical oxygen demand (BOD). BOD as an indicator of the quantity of biodegradable organic compound is a widely used parameter for the control of wastewater. The conventional BOD method, BOD_5,[11] requires 5 days to determine BOD and is thus not suitable for process control, so a fast and precise sensor that is highly correlated to BOD_5 is needed.

Microbial sensors have been developed to determine BOD or a similar parameter (biodegradable substances) using cells with broad substrate ranges. Several microbial sensors were described by Karube et al.,[12] Hikuma et al.,[13] and Tan et al.[14] The most frequently used microorganism has been *Trichosporon cutaneum*. Measurements of total oxygen uptake required 15–20 min and the recovery time between measurements was as much as 3–4 hr.[15] In contrast, a fast determination of oxygen consumption was developed by Riedel et al.[16,17] The prerequisite for the use of microorganisms for BOD sensors is a broad substrate spectrum. A new BOD sensor has been developed using a combination of two microorganisms with different substrate spectra: the bacterium, *Rhodococcus erythropolis*, and the yeast, *Issatchenkia orientalis*.

The combisensor with *R. erythropolis* and *I. orientalis* associates the specificity of both strains and is able to handle a wide range of wastewater substrates in a range of minutes with high precision.[13] This combisensor was incorporated into a measuring tool to allow automatic measurement and marketed by Dr. Bruno Lange GmbH, Berlin (Fig. 19.2). This tool designated "ARAS" can be applied to the estimation of sensorBOD, an index to BOD_5, for various kinds of treated and untreated municipal and industrial wastewaters.

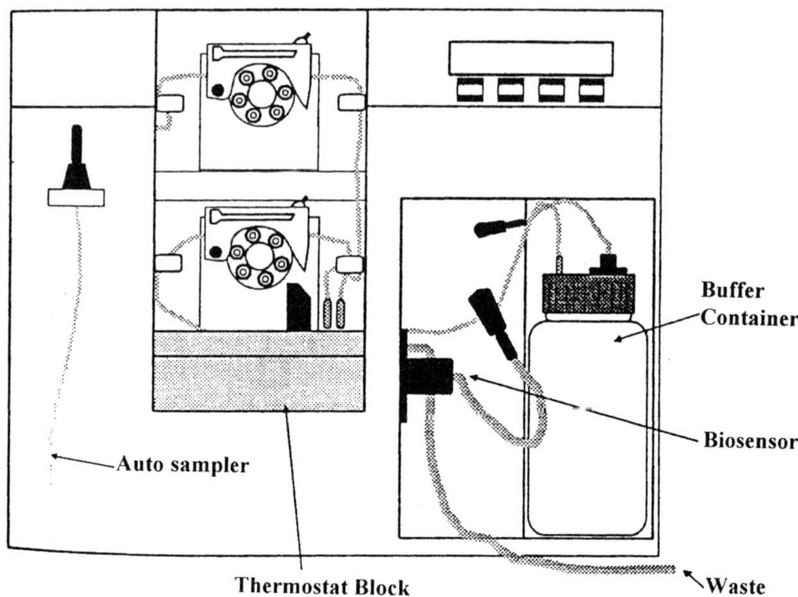

Figure 19.2.   BOD biosensor system ARAS developed by Dr. Bruno Lange GmbH Berlin, Düsseldorf. Automatic pumping of buffer and samples is controlled by the program provided.

The first successful application of the microbial combisensor was measuring the daily cycle of the organic load of a municipal sewage treatment plant in Germany in comparison to BOD_5 and chemical oxygen demand (COD). Similar results were obtained for sensorBOD values estimated by the microbial combi-sensor and determined by the conventional BOD_5 method. The range of sensorBOD values correlates quite well with COD and BOD_5.

The sensor was brought to the marshes of the Mai Po WWF Nature Reserve Area in Hong Kong to be tested under Asian conditions. The aim was to investigate the extent of water pollution caused by the rapid development of nearby new towns and the influence of the China Pearl River on an intact ecosystem.

Mai Po is located at the confluence of salt and fresh water and includes 40 fish ponds (see Fig. 19.3), which are the so-called "gei wai" for shrimp production along the seafront. In the first attempt, samples from all the gei wai were collected and analyzed. The data showed very low sensorBOD values, ranging from 1 to 33 mg/L, and did not indicate the sources of the pollution. The influence of salt and fresh water was also not clear. Therefore, in the second attempt, samples were collected at the bird-watching hide outside the gei wai at the sea front, and inside the gei wai in Mai Po. The sensorBOD values of the water samples were measured with the microbial sensor.

The preliminary results indicated that the water sample from outside the gei wai generally had a higher value of sensorBOD than that inside the gei wai. Nevertheless,

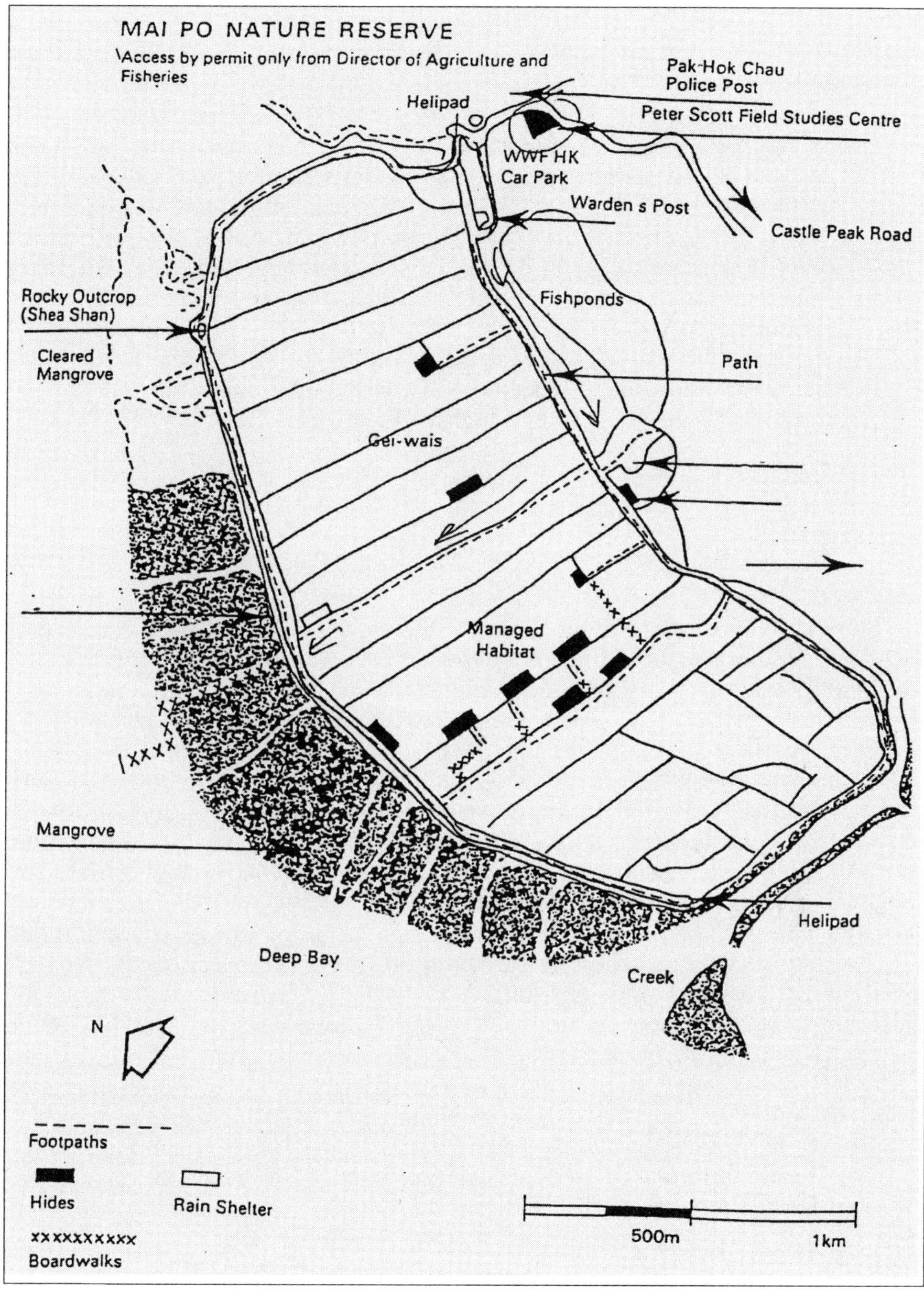

Figure 19.3.   Map of Mai Po Marshes Nature Reserve in Hong Kong. Pollution coming in from Shenzen and from Chinese rivers was investigated.

all the samples have lower sensorBOD values than those shown by the 5-day BOD test, which are 2 to 10 times higher. There are two possible interpretations of these observations:

1. Macromolecules (e.g., starch and cellulose from nearby pig farms) are present in the water samples that obviously cannot penetrate the outer protective membrane of the sensor. Therefore, only small-sized molecules can enter the pores of the membrane and be converted by the microbes. In contrast, in the 5-day BOD test, all biodegradable molecules, including macromolecules, can be digested by the delocalized seeding microbes in the bottle.
2. The microbes in the sensor are influenced by the salt content in the water samples, and owing to less dilution by river and groundwaters the sample from inside the gei wai has a higher salt content.

Thus there are at least two problems to be solved.

19.3. PREHYDROLYSIS OF MACROMOLECULES

Since large molecules cannot penetrate the outer semipermeable membrane of the microbial sensor, a pretreatment of water samples was suggested to break up the macromolecules (e.g., polysaccharides and proteins) into small molecules (e.g., monosaccharides and amino acids). These small molecules could then penetrate the membrane to be accessible to the microbes in the biosensor, which convert them under oxygen consumption to create a signal. Three substances were used to prepare a kind of "artificial wastewater": starch powder, milk powder, and cellulose, as they are the main constituents of domestic wastewater. The milk powder contains not only proteins, but also fat. We therefore have a "high-molecular-weight cocktail." A first attempt using hydrolytic enzymes (proteases, amylases, and cellulases) was not successful. Subsequently, acid hydrolysis was employed using various concentrations of hydrochloric acid and adjusting the duration of the process to create optimal conditions for the hydrolysis of artificial wastewater. First, the concentration of hydrochloric acid was kept constant and the time for hydrolysis was varied. In order to make the future hydrolysis method simpler, room temperature was chosen first. Figure 19.4 shows that the sensorBOD value became constant after 3 to 10 min of hydrolysis. After this, we determined the minimum concentration of acid required for hydrolysis treatment and hydrolysis was performed using different concentrations of acid for 10 min. From Fig. 19.5, we can see that, as a trend, the higher the concentration of acid used, the better the resulting signal of the sensor. By increasing the concentration of HCl from 3 to 6 M, the sensorBOD values increased nearly threefold for starch and cellulose. We can therefore conclude that it is better to use the highest concentration of 6 M hydrochloric acid.

The next step was to optimize the temperature of hydrolysis. We hydrolyzed a starch, milk powder, and cellulose "cocktail" at different temperatures. Fig. 19.6 shows that the higher the temperature, the higher the sensorBOD value obtained.

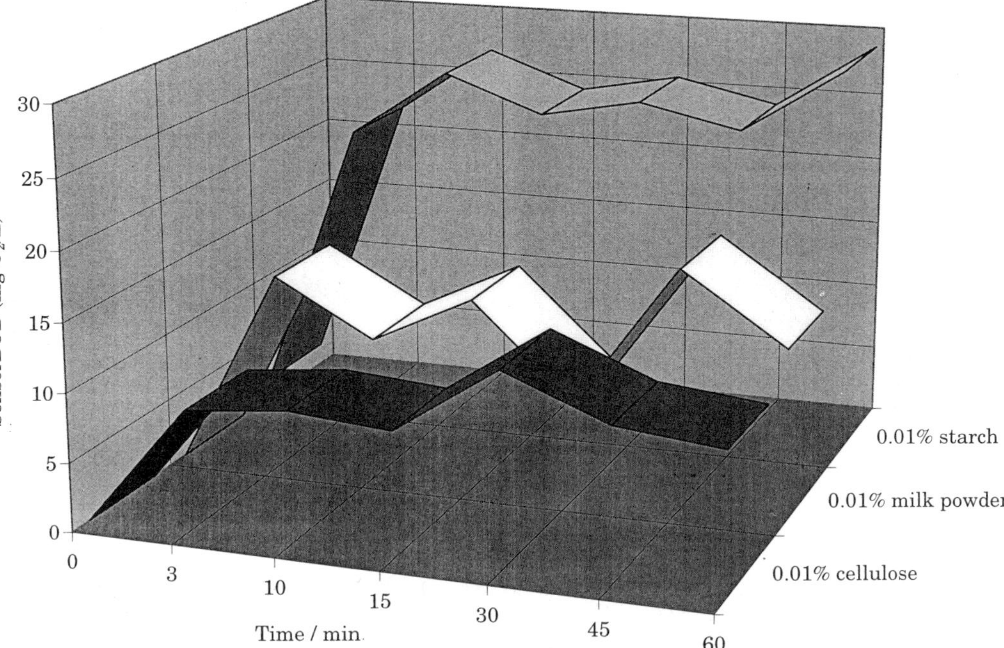

Figure 19.4.   Hydrolysis of artificial wastewater by 6 M HCl at room temperature with different time frames. Three common substances in household wastewater (starch, milk powder, and cellulose) were hydrolyzed. With longer time of hydrolysis, the sensorBOD value increases owing to more macromolecules being broken down and taken up by the immobilized microbes. All macromolecules were broken down dramatically within 3 min. Starch is said to be the easiest to hydrolyze into small molecules.

Since we could not increase the water temperature beyond the boiling point at normal pressure, 100°C was finally chosen as the optimal temperature for acid hydrolysis.

Knowing that 100°C was the optimal temperature for hydrolysis at normal pressure, we returned to the incubation time. Originally it was assumed that 10 min would be sufficient for acid hydrolysis. However, a longer duration was found to be optimal for complete acid hydrolysis (Fig. 19.7). Thus, a one-hour hydrolysis with 6 M HCl at 100°C was finally chosen as the optimal way to hydrolyze most of the compounds.

With this procedure, the sensor can also detect high-molecular-weight substrates with pretreated samples. The water samples from Mai Po marshes, after prehydrolysis, gave final results similar to the BOD$_5$ values.

19.4.   EFFECT OF SALT ON THE VALUE OF SENSORBOD

Different concentrations of sodium chloride have been combined with the same amount of biodegradable substance (standard solution) and tested. Sodium chloride was chosen because it is the major content in seawater. Salt content in the seawater,

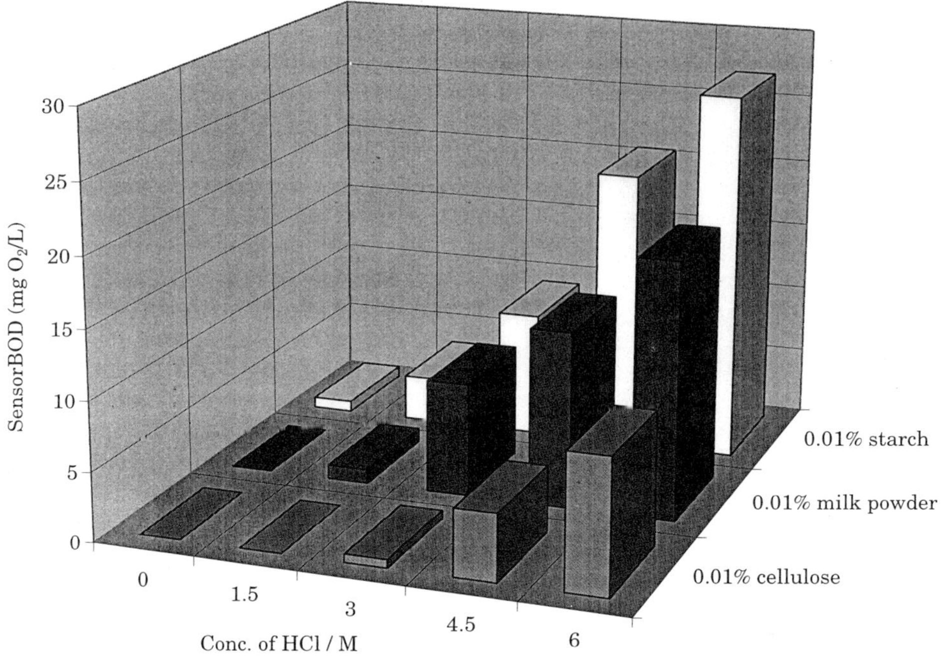

Figure 19.5.   Hydrolysis of artificial wastewater at room temperature in 10 min with different concentrations of acid. With increasing concentration of hydrochloric acid, the sensorBOD value increases owing to more macromolecules being broken down by the stronger acid and taken up by the immobilized microbes.

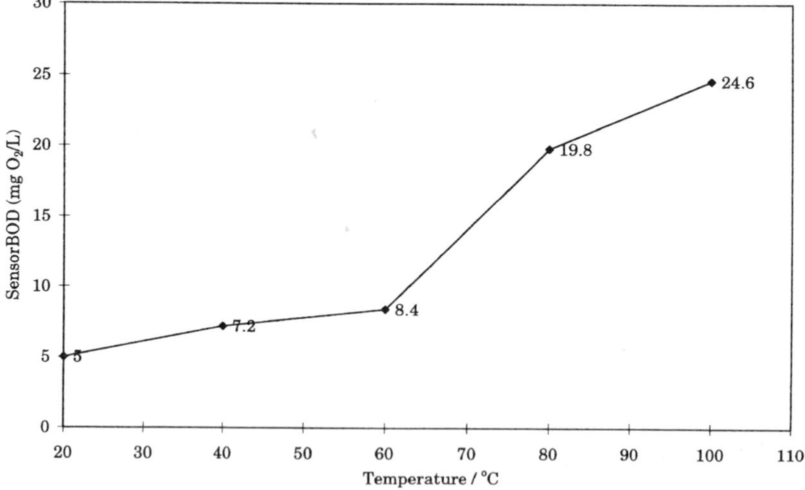

Figure 19.6.   Hydrolysis of a substrate "cocktail" by 6 M HCl in 10 min at different temperatures. The mixture of starch, milk powder, and cellulose was hydrolyzed.

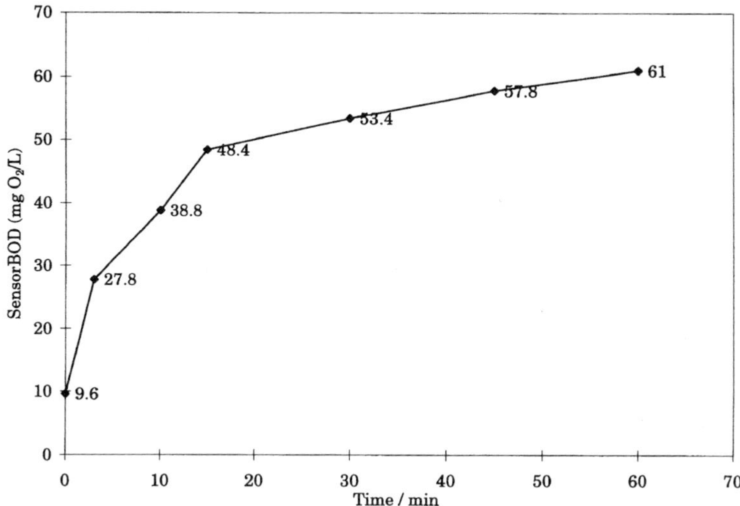

Figure 19.7. Hydrolysis of substrate "cocktail" by 6 M HCl at 100°C with different times. With longer duration of hydrolysis, the sensorBOD value increases owing to more macromolecules being broken down and taken up by the immobilized microbes. The sensorBOD increases by sixfold from no pretreatment to 1-h hydrolysis.

even in small concentrations, was thought to influence the microbes in the BOD sensor as both microorganisms, *I. orientalis* and *R. erythropolis*, were extracted from fresh water mediums and neither is salt-tolerant. When salt diffuses into the immobilized microbes, the uptake rate of the biodegradable substances increases and, consequently, the metabolic rate of the microbes is raised above normal, which enhances the rate of oxygen uptake and accordingly leads to a higher sensorBOD value. On the other hand, salt in higher concentrations may inhibit microbial metabolism. The investigation of the effect of salt on the biosensor is essential for further development of the sensor for particular areas of the world. For instance, seawater is used for flushing water in Hong Kong, while elsewhere fresh water is used. In Hong Kong all wastewater is a mixture of saline and fresh water. At 50 mM NaCl the combisensor recorded a 10% weaker signal compared to a solution containing no salt. Obviously, this problem can be solved only by using salt-tolerant microorganisms, and will be discussed below.

19.5. ANALYSIS OF WASTEWATER FROM A UNIVERSITY IN HONG KONG

In order to test the capability of the microbial sensor with sewage water samples, we decided to measure the sensorBOD of sewage water of the Hong Kong University of Science and Technology (HKUST) in Clear Water Bay in the East of

Hong Kong. For elementary analyses, sewage water samples were collected monthly from the wastewater pipe by the Safety and Environmental Protection Office (SEPO) of HKUST. The sensorBOD test with and without pretreatment and the BOD_5 test were performed.

After a 3-month trial, we found good correlation between sensorBOD and BOD_5 values (Fig. 19.8). As can be clearly seen, a better correlation was obtained using prehydrolyzed samples. In addition, such a correlation can be used to predict the result of the 5-day BOD test from the 5-min sensorBOD data.

For a direct analysis of the polluted water, samples were freshly collected from the main wastewater pipe of HKUST. Both untreated and hydrolyzed samples were tested (Fig. 19.9). The findings can be summarized as follows:

1. A first peak value of sensorBOD was found at 8 a.m., when people get up, have breakfast and bathe.
2. In the afternoon, the peak is at 4 p.m.; and in the evening at 9 p.m., the time when dishwashing is usually done in the HKUST restaurants.
3. After 10 p.m., the pollution levels go down.
4. The trends for hydrolyzed wastewater are quite similar to those for untreated wastewater except that there is a very high peak at 1 p.m. when there are large molecules (e.g., starch) in the water. The starch may come mainly from cooked rice.
5. The sensor values of the pretreated wastewater sample are always higher because the samples include large molecules.

We compared the sensorBOD data with data from the traditional 5-day BOD

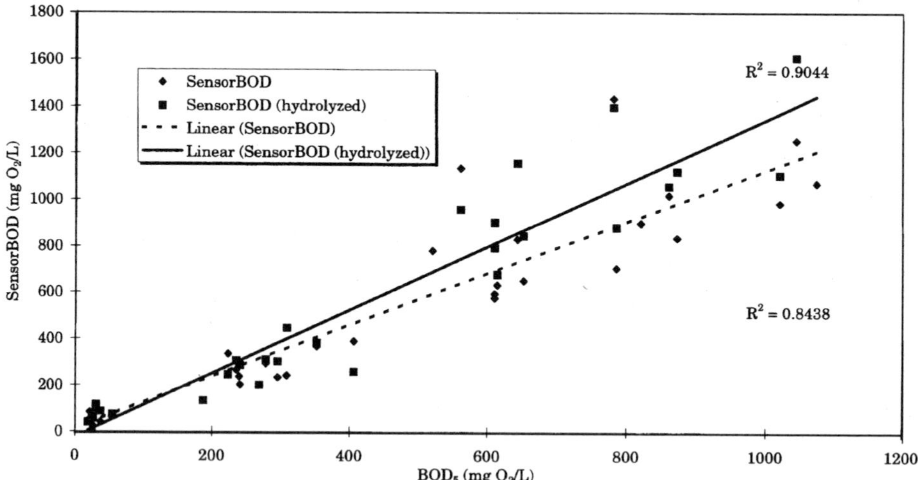

Figure 19.8. BOD_5 vs. sensorBOD. Hydrolyzed and nonhydrolyzed samples are compared with the best correlation being obtained by the pretreated samples ($R^2 = 0.9044$) compared to untreated samples ($R^2 = 0.8438$).

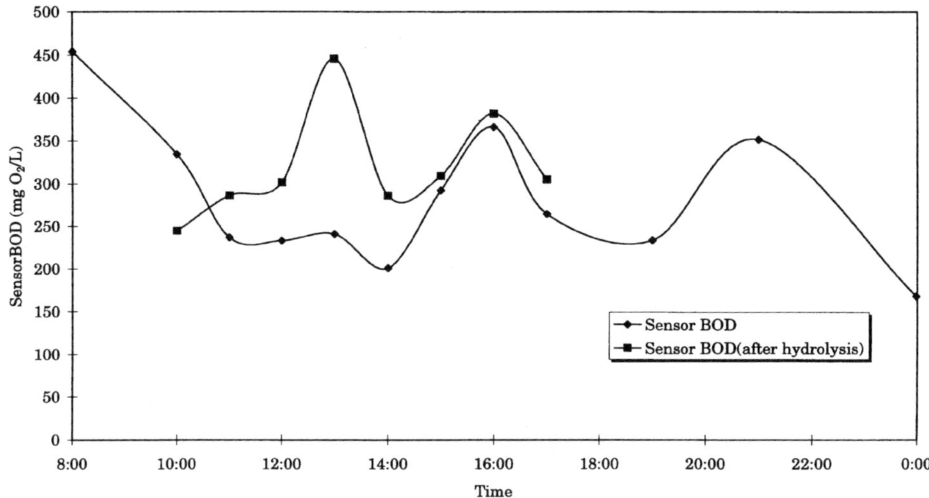

Figure 19.9. Day profile of wastewater at HKUST measured by ARAS SensorBOD during the summer vacation. This preliminary test shows that macromolecules in sewage water samples were detected by sensorBOD after being pretreated with acid hydrolysis.

test. We also measured the sensorBOD of all samples after 5 days of storage at 4°C to investigate any possible degradation of sample during storage. As shown in Fig. 19.10, the result is quite satisfactory. The peaks with all three methods are reached at the same time. A recalculation factor between sensorBOD and BOD_5 can be found from the correlation. The remeasured sensorBOD after 5 days of storage was lower than that of the fresh directly measured samples. Since the remeasured

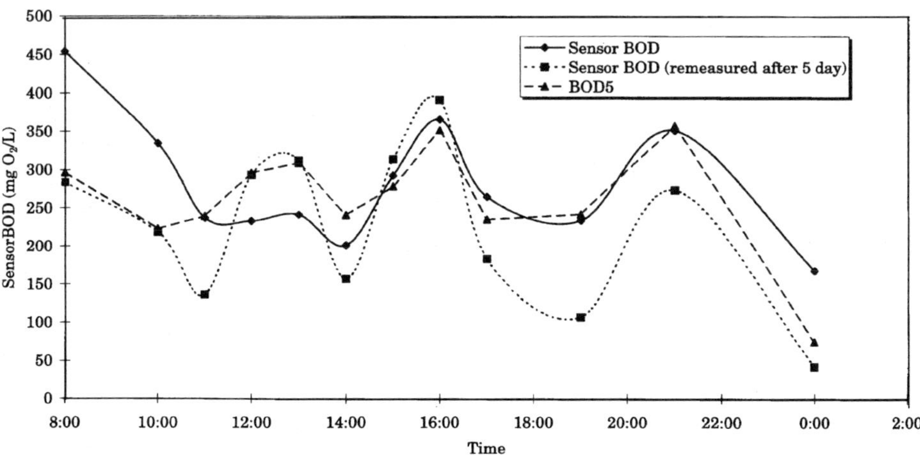

Figure 19.10. Day profile of wastewater at HKUST measured by ARAS SensorBOD and BOD_5.

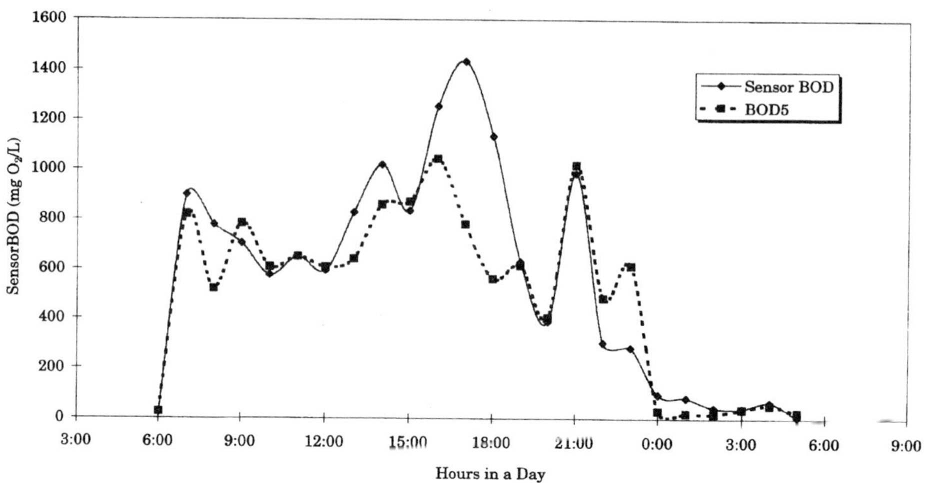

Figure 19.11.   Pollution level measurement at HKUST during the semester.

sensorBOD and BOD$_5$ tests were done on the same day, we used these two for a direct comparison. Obviously, wastewater microbes in samples (even stored at 4°C) can digest the organic substances in the samples. The largest difference is 140 mg/L. Most of them have differences within 50 mg/L and the margin of error is less than 20% (Fig. 19.10). These results are quite remarkable. It is also interesting to note that the sensorBOD value at lunchtime (12 p.m.) was increased by storage, indicating a hydrolysis of macromolecules. At 9 p.m., however, the fresh sensor value was higher than the stored one. One might speculate that there were fewer macromolecules in the wastewater at night.

The data shown were obtained at the end of a semester when most of the students had already left the campus. However, we collected samples during a more active period, over 24 h on October 2 and 3, 1996. It was a normal university day, and the sensorBOD values were expected to be much higher than for the previous tests when students were on holiday. Indeed, from our profile of wastewater (Fig. 19.11), the sensorBOD values were on average four times higher than in the previous trial. However, the trends were the same as in the earlier experiment. In the new test, moreover, wastewater discharged after midnight was also checked and its sensorBOD value was found to be very low (lower than 50 mg/L). At 6:00 on the following morning, the sensorBOD value increased again.

19.6.   A NOVEL APPROACH USING THE SALT-TOLERANT YEAST ARXULA

The microorganism used in commercial sensors (Nisshin Denki, Japan) until now has been mainly *Trichosporon cutaneum*, but it has been shown to be pathogenic. In the commercial ARAS BOD sensor produced by Dr. Bruno Lange (GmbH

Berlin) the bacteria *R. erythropolis* and the yeast *I. orientalis* are coimmobilized to broaden the substrate specificity. However, the combisensor also has disadvantages.

Each microorganism prefers specific substrates. In order to have a broad substrate range, one approach would be to mix many different microorganisms, as in activated sludge. However, we found that with populations of mixed microorganisms, the sensor signal changed after several days of measurements and gave nonreproducible results. Therefore, we tried using a salt-tolerant microorganism with broad substrate capability. The salt-tolerant yeast *Arxula adeninivorans* was found to have an amazingly broad substrate range: it assimilates all sugars, polyalcohols, and organic acids except for ribose, lactose, lactic acid, and methanol.[18] As a result, an *Arxula*-based sensor should correlate with BOD_5 for many substrates and real samples.[19]

A. adeninivorans isolated from wood hydrolysates in Siberia was supplied from the yeast collection of the Institut fuer Pflanzengenetik und Kulturpflanzenforschung (IPK) in Gatersleben, Germany.[20] Poly(carbamoyl)sulfonate (PCS) was used to immobilize the microorganism. PCS, which was introduced and tested by Vorlop et al.[21] and Muscat et al.,[22] is prepared by isocyanate terminated polyurethane (PUR) prepolymer with bisulfite and is a better gel for entrapment than PUR and calcium alginate.[21] It combines the advantageous properties of calcium alginate (low toxicity) and the PUR matrix (high elasticity).

The *Arxula* sensor has very high stability and can be used for more than 40 days (Fig. 19.12). A gradually increasing signal obtained during the first 5 days was assumed to be due to the growth of the cells. The gradually decreasing background current proves this assumption. The sensor showed a very stable signal up to 40 days. For each set of data, five measurements were made. The sensor was stable for

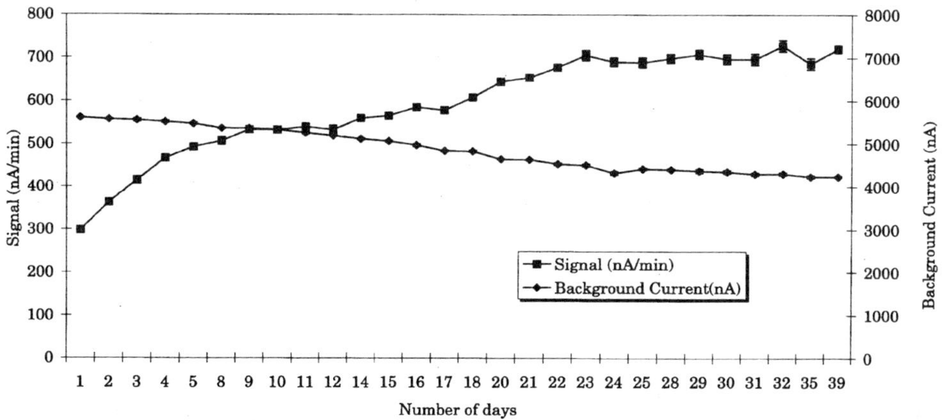

Figure 19.12. The changes of signal and background current of an *Arxula* sensor. The signal is determined at 70 s after sample injection. The signal increases from 300 nA/min up to a steady range of 680 nA/min after 40 days. The substrate for each measurement is 214.4 mg/L glucose and glutamic acid (GGA) and all data were based on five measurements. The background current indicates the saturated oxygen concentration in the gel layer in front of the oxygen sensor in the absence of the substrate.

over 150 measurements. The percentage of error for each set of data was within 10%. This is very important because we need a stable, accurate, and precise microbial sensor for long-term wastewater monitoring. Apart from measurement stability, *Arxula* is easily stored for long period of time. A 2-month storage at 4°C did not cause any problems.

We use 214.4 mg/L of glucose and glutamic acid as a calibration standard for a 275 mg O_2/L BOD value. The sensor has a linear response up to 550 mg/L (Fig. 19.13). If we want to measure the BOD of samples with concentrations higher than 550 mg/L, the sample must be diluted first. The linearity of this range is satisfactory ($r^2 = 0.9911$). The limit of determination was calculated to 13 mg/L and the limit of detection was 8 mg/L BOD.

In order to have a strong signal, an optimal number of cells should be immobilized. However, a thicker cell layer hinders the diffusion of oxygen because the diffusion is directly proportional to the distance to be traversed, and the sensor then produces a weaker signal instead. Since we want the substrate concentration to be the only parameter causing a signal, the external oxygen concentration should not affect the signal and oxygen diffusion and should be at the maximum level. The background current at the maximum range is over 4000 nA. For a gel layer without any cells, the background current is 5000 nA. For a 180 μg/mm^2 or higher concentration of *Arxula*, however, the background current drops to 2000 nA.

From our findings, the optimal cell loading for the *Arxula* sensor is between 75 and 100 μg/mm^2. (Fig. 19.14), which actually corresponds to 0.5–0.75 g/mL wet weight of *Arxula* in a PCS gel mixture. We took measurements on the first day and the fifth day after activation, and the signal on the fifth day was much stronger than that on the first, which is compatible with our previous results for stability measurements. We finally chose 100 μg/mm^2 for further measurements.

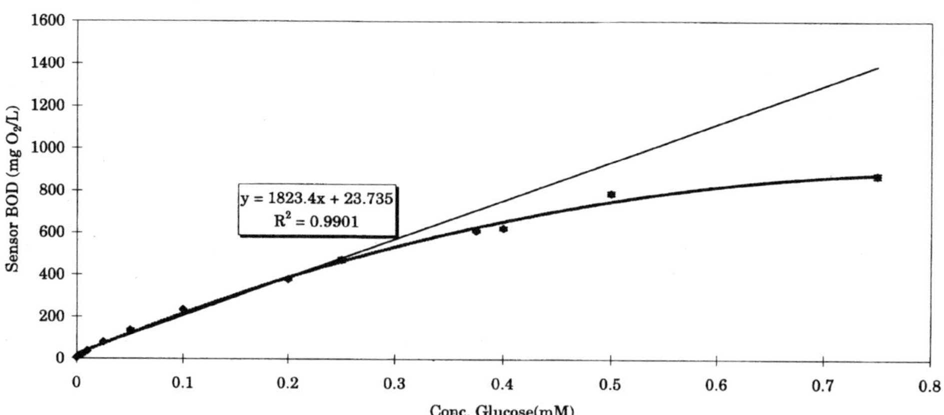

Figure 19.13.   Calibration curve for the *Arxula* sensor. The sensor is calibrated by the GGA standard. The concentration of glucose is the final concentration in the measuring cell. The range from 0 to 0.25 mM of glucose is chosen for linear regression and the r^2 value is 0.9911.

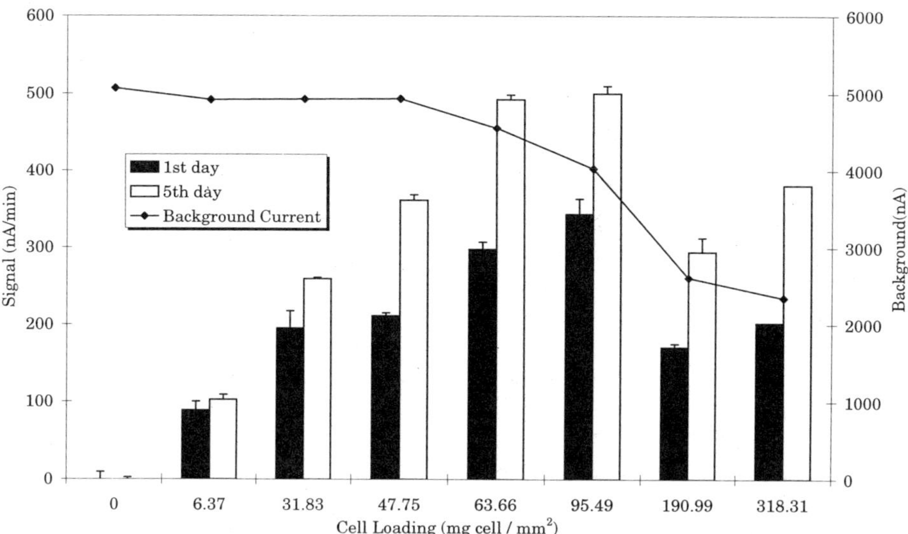

Figure 19.14. Profile of cell loading of the *Arxula* sensor. The dotted line represents data from the first day of sensing. The solid line represents data from the fifth day of sensing. The substrate for each measurement is 214.4 mg/L GGA and each set of data was based on five measurements. The peak region is at 75–100 μg cell/mm². The diameter of the electrode is 5 mm. The cell loading was calculated from the volume of the cell and the gel mixture (20 μL), the cell concentration, and the electrode surface area.

Figure 19.15 shows a comparison between the BOD_5 and sensorBOD value. The *Arxula* sensorBOD value correlates well with the BOD_5 below 550 mg/L. After reaching the linear range, the *Arxula* sensor is saturated. This problem can be overcome by diluting the concentrated sample back to the linear range. We also compared the result with the commercial BOD sensor that utilizes *R. erythropolis* and *I. orientalis*. We found that the commercial sensor has a lower saturation level of glucose concentration. When we compared the error in 550 mg/L, the *Arxula* sensor had a 10% error but the commercial sensor had an error of up to 40% compared to BOD_5, showing that the *Arxula* sensor has a better correlation with BOD_5 than the commercial sensor. *Arxula* also has broader substrate specificity compared with the strains combined in the commercial sensor with various organic substances.[23–25]

One of the main reasons for using *Arxula* was to overcome the salt-dependence of the commercial sensor ARAS. In many coastal or island places, seawater is used in toilet flushing. In wastewater measurement with microbial sensors, the salt concentration affects the BOD measurement. The commercial sensor ARAS is very sensitive to salt and is activated by low concentrations of salt. At high salt concentrations, it is deactivated. A salt-independent microbial sensor for BOD is needed. *Arxula* is salt-tolerant and stable for very high concentrations of salt, up to 120 mM of sodium chloride (Fig. 19.16), while the commercial sensor is 18% deactivated at this concentration. For sea salt concentration ($\sim$40 mM NaCl), the

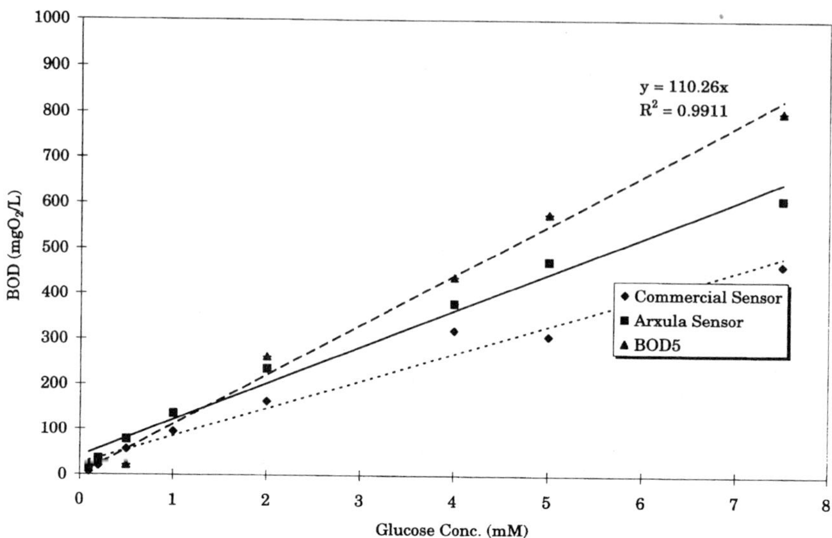

Figure 19.15. Comparison of accuracy between the novel *Arxula* sensor and the commercial ARAS Sensor BOD. The sensor is calibrated by the GGA standard. The concentration of glucose is the final concentration in the measuring cell.

Arxula sensor was not affected at all by salt, but the commercial sensor was 10% deactivated.

Arxula has an amazingly high stability. The *Arxula* sensor is stable for 40 days under working conditions without any decrease in activity. It reveals a linear range up to 550 mg/L and shows a better correlation with BOD_5 than the commercial

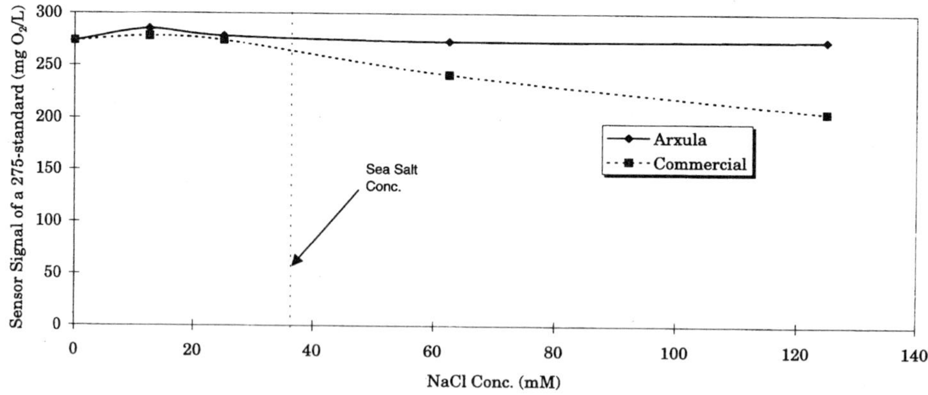

Figure 19.16. Comparison of the effect of salt on the *Arxula* sensor and the commercial sensor using *R. erythropolis* and *I. orientalis* cells. The sodium chloride concentration is the concentration in the measuring cell. The salt concentration of seawater is normally 3.8%, which corresponds roughly to 50 mM of NaCl.

sensor with *R erythropolis* and *I. orientalis.* Unlike the commercial sensor, it is stable for certain salt concentrations, which is very important for wastewater measurement in coastal areas and islands.

From our results, *Arxula* is a suitable strain for microbial sensor development. After optimization, it will certainly perform better for wastewater measurements. From the analysis and experiments being done, the microbial wastewater sensor can be used as a rapid screening and direct determination method for water pollution.

The fast response time of the microbial sensor, in minutes, can be of great help to wastewater treatment operations. When there is a large influx of wastewater into the waste treatment plant, the operator can now monitor the extent of waste in real time. Action can be taken instantly, such as stepping up the process of biodegradation by activated sludge, or by increasing the air-pumping rate of the treatment plant. On the other hand, energy-consuming air pumps can be switched off if clean water flows into the treatment plant.

It is clear that microbial wastewater sensors can contribute a great deal to environmental protection and energy conservation.

REFERENCES

1. Riedel K, Renneberg R, Wollenberger U, et al. Microbial sensors: fundamentals and application for process control. *J Chem Biotechnol* 1989;31:502–504.
2. Riedel K., (1994) Whole-cell and tissue sensors. In: Wagner G, Guilbault GG. (eds.) *Food Biosensor Analysis.* New York: Marcel Dekker, Inc. 1994, pp 123–150.
3. Riedel K. Application of biosensors to environmental samples. In: Ramsay G, ed. *Commercial Biosensors: Application to Clinical, Bioprocess and Environmental Samples.* New York: John Wiley & Sons, 1996.
4. Riedel K. Microbial sensors and their applications in environment. *Exp Techn Physics* 1994;40:63–76.
5. Karube I, Nakahara T, Matsunaga T, et al. *Salmonella* electrode for screening mutagens. *Anal Chem* 1982;54:1725–1727.
6. Riedel K. Biochemical fundamentals and improvement of selectivity of microbial sensors: mini-review. *Bioelectrochem Bioenerg* 1991;25:19–30.
7. Riedel K., Renneberg R, Scheller F. Studies in peptide utilization by microorganisms using biosensor techniques. *Bioelectrochem Bioenerg* 1989;22:113–125.
8. Riedel K, Lange KP, Stein HJ, et al. A microbial sensor for BOD. *Water Res* 1990;24:883–887.
9. Riedel K, Liebs P, Renneberg R, et al. Characterization of the physiological state of microorganisms using the respiration electrode. *Anal Lett* 1988;21:1305–1322.
10. Riedel K, Renneberg R, Liebs P. An electrochemical method for determination of cell respiration. *J Basic Microbial Biotech* 1985;25:51–56.
11. *Standard Methods for the Examination of Waters and Wastewater.* Washington, DC: American Public Health Association, 1986, 16th Ed, pp. 523–531.
12. Karube I, Matsuaga T, Suzuki S. A new microbial electrode for BOD estimation, *J Solid Phase Biochem* 1977;2:97–104.
13. Hikuma M, Suzuki H, Yasuda T, et al. Amperometric estimation of BOD by using living immobilized yeasts. *Eur J Appl Microbiol* 1979;8:289–297.
14. Tan TC, Li F, Neoh KG. Measurement of BOD by initial rate of response of a microbial sensor, *Sens Act B* 1993;10:137–142.
15. Tan TC, Li F, Neoh KG, et al. Microbial membrane-modified dissolved oxygen probe for rapid biochemical oxygen demand measurement, *Sens Act B* 1992;8:167–172.
16. Riedel K, Renneberg R, Kuehn M, et al. A fast estimation of biochemical oxygen demand using microbial sensors. *Appl Microbiol Biotech* 1988;28:316–318.
17. Riedel K. Microbial sensors and their applications in environment, *Exp Tech Phys* 1994;40(1):63–76.

18. Middelhoven WJ, de Jong IM, de Winter. *Arxula adeninivorans,* a yeast assimilating many nitrogenous and aromatic compounds. *Antonie Leeuwenhoek J Microbiol* 1991;59:129–137.

19. Riedel K, Lehmann M, Renneberg R. *Arxula adeninivorans* based sensor for the estimation of BOD. *Appl Microbiol Biotech* (in press).

20. Gienov U, Kunze G, Schauer F, et al. The yeast genus *Trichosporon spec. LS3*: molecular characterization of genomic complexity. *Zbl Microbiol* 145:3–12.

21. Vorlop K-D, Muscat A, Beyersdorf J, Entrapment of microbial cells within polyurethane hydrogel beads with the advantage of low toxicity. *Biotechnol Tech* 1992;6:483.

22. Muscat A, Beyersdorf J, Vorlop K-D, Poly(carbamoylsulphonate) hydrogel, a new polymer material for cell entrapment. *Biosens Bioelectron* 1995;10:11–14.

23. Riedel K, Lehmann M, Renneberg R, et al. *Arxula adeninivorans* based sensor for the estimation of BOD. *Appl Microbiol Biotech* (in press).

24. Renneberg R, Chan Ch, Wai K, et al. Novel microbial sensors for waste water control. In: Turner, APF and Renneberg, R, Eds. *Advances in Biosensors, vol 4: A Chinese Perspective.* JAI Press Inc. 1999, 195–213.

25. Lehmann M, Chan Ch, Lo A, et al. Measurement of biodegradable substances using the salt-tolerant yeast *Arxula adeninivorans* for a microbial sensor immobilized with poly(carbamoyl) sulfonate (PCS). Part II: Application of the novel biosensor to real samples from coastal and island regions. *Biosens Bioelectron* 1999; 295–302.

Index